Springer Proceedings in Mathematics & Statistics

Volume 369

This book series features volumes composed of selected contributions from workshops and conferences in all areas of current research in mathematics and statistics, including operation research and optimization. In addition to an overall evaluation of the interest, scientific quality, and timeliness of each proposal at the hands of the publisher, individual contributions are all refereed to the high quality standards of leading journals in the field. Thus, this series provides the research community with well-edited, authoritative reports on developments in the most exciting areas of mathematical and statistical research today.

More information about this series at https://link.springer.com/bookseries/10533

Toshiyasu Arai · Makoto Kikuchi · Satoru Kuroda ·
Mitsuhiro Okada · Teruyuki Yorioka
Editors

Advances in Mathematical Logic

Dedicated to the Memory of Professor Gaisi Takeuti, SAML 2018, Kobe, Japan, September 2018, Selected, Revised Contributions

Springer

Editors
Toshiyasu Arai
The University of Tokyo
Tokyo, Japan

Makoto Kikuchi
Kobe University
Kobe, Japan

Satoru Kuroda
Gunma Prefectural Women's University
Gunma, Japan

Mitsuhiro Okada
Keio University
Tokyo, Japan

Teruyuki Yorioka
Shizuoka University
Shizuoka, Japan

ISSN 2194-1009 ISSN 2194-1017 (electronic)
Springer Proceedings in Mathematics & Statistics
ISBN 978-981-16-4172-5 ISBN 978-981-16-4173-2 (eBook)
https://doi.org/10.1007/978-981-16-4173-2

Mathematics Subject Classification: 03Axx, 03Bxx, 03Cxx, 03Dxx, 03Exx, 03Fxx, 03Gxx

This Springer imprint is published by the registered company Springer Nature Singapore Pte Ltd.
The registered company address is: 152 Beach Road, #21-01/04 Gateway East, Singapore 189721, Singapore

Preface

Gaisi Takeuti was one of the most brilliant, genius, and influential logicians of the twentieth century. He was born on January 25, 1926, in Ishikawa Prefecture in Japan, graduated from the University of Tokyo in 1947, and received his Ph.D. in mathematical logic from the University of Tokyo in 1956. He was an assistant professor at Tokyo University of Education from 1950, promoted to a professor in 1962, moved to United States in 1966, and had been a professor in mathematics at the University of Illinois at Urbana-Champaign, USA, where he became a professor emeritus after his retirement. He was a member of the Institute for Advanced Study in Princeton during 1959–1960, 1966–1968, and 1971–1972, and the president of the Kurt Gödel Society from 2003 to 2009. He received the Asahi Prize in 1982, and the Bolzano Medal from the Czech Academy of Science in 1996. He passed away on May 10, 2017, at the age of 91.

Takeuti was one of the founders of proof theory, a branch of mathematical logic that originated from Hilbert's program about the consistency of mathematics. Based on Gentzen's pioneering works of proof theory in the 1930s, he proposed a conjecture in 1953 concerning the essential nature of formal proofs of higher-order logic, which is now known as Takeuti's fundamental conjecture. The conjecture is especially important for the foundations of mathematics since a positive solution of the conjecture brings about a finitary proof of the consistency of formalized mathematics, and he gave a partial positive solution of the conjecture by introducing the concept of ordinal diagrams. His arguments on the conjecture and proof theory in general have had great influence on the later developments of mathematical logic, philosophy of mathematics, and applications of mathematical logic to theoretical computer science.

Takeuti's work ranged over the whole spectrum of mathematical logic. In particular, he made significant contributions to set theory, computability theory, Boolean-valued analysis, fuzzy logic, bounded arithmetic, and theoretical computer science. He wrote many monographs and textbooks both in English and in Japanese, and his monumental monograph *Proof Theory*, published in 1975, has long been a standard reference of proof theory. *Introduction to Axiomatic Set Theory*, written with Zaring in 1971, was published as the first volume of Springer Graduate Texts in Mathematics. He had a wide range of interests covering virtually all areas of mathematics and extending to physics. His publications include many Japanese books for students and general readers about mathematical logic, mathematics in general, and connections between mathematics and physics, as well as many essays for Japanese science magazines.

This volume is a collection of papers based on the Symposium on Advances in Mathematical Logic 2018. The symposium had been held from September 18 to 20, 2018, at Kobe University, Japan, and was dedicated to the memory of Professor Gaisi Takeuti. The program and organizing committees are Toshiyasu Arai, Makoto Kikuchi, Satoru Kuroda, Mitsuhiro Okada, and Teruyuki Yorioka. The invited speakers are Samuel R. Buss, Jean-Yves Girard, Kanji Namba, Masanao Ozawa, Wilfried Sieg, and Mariko Yasugi. The contributed speakers are Sakaé Fuchino, Daichi Hayashi, Daisuke Ikegami, Sohei Iwata, Mamoru Kaneko, Takayuki Kihara, Taishi Kurahashi, Hidenori Kurokawa, Yo Matsubara, Tadatoshi Miyamoto, NingNing Peng, Norbert Preining, Toshimichi Usuba, Mariko Yasugi, Keita Yokoyama, and Yasuo Yoshinobu. All the papers contained in this volume are original and refereed.

Tokyo, Japan — Toshiyasu Arai
Kobe, Japan — Makoto Kikuchi
Gunma, Japan — Satoru Kuroda
Tokyo, Japan — Mitsuhiro Okada
Shizuoka, Japan — Teruyuki Yorioka
March 2021 — SAML2018

Organization

SAML 2018 (Symposium on Advances in Mathematical Logic 2018)

Organized by Section of Foundation of Mathematics and History of Mathematical Sciety of Japan

Commitee

Organizer:

Makoto Kikuchi, Kobe University, Japan
Hiroshi Sakai, Kobe University, Japan
Teruyuki Yorioka, Shizuoka University, Japan

Program Committee:

Toshiyasu Arai, Chiba University, Japan
Makoto Kikuchi, Kobe University, Japan
Satoru Kuroda, Gunma Prefectural Women's University, Japan
Mitsuhiro Okada, Keio University, Japan
Teruyuki Yorioka, Shizuoka University, Japan

Invited Speakers:

Samuel R. Buss, University of California San Diego, USA
Jean-Yves Girard, Université d'Aix-Marseille, France
Kanji Namba, Prof. Emeritus, University of Tokyo
Masanao Ozawa, Prof. Emeritus, Nagoya University
Wilfried Sieg, Carnegie Mellon University, USA
Mariko Yasugi, Prof. Emeritus, Kyoto Sangyo University

Sponsoring Institutions

KAKENHI: JSPS Grant-in-Aid for Scientific Research
Japan Association for Philosophy of Science

Symposium on Advances in Mathematical Logic 2018 Dedicated to the Memory of Professor Gaisi Takeuti Program

September 18, 2018

11:20–12:20	Jean-Yves Girard, Université d'Aix-Marseille, On Second Order Logic
13:20–14:20	Kanji Namba, Prof. Emeritus, University of Tokyo, η function, Gauss-AGM and computational complexity
14:35–15:05	Yasuo Yoshinobu, Nagoya University, Indestructible and absolute properness
15:10–15:40	Yo Matsubara, Nagoya University, Precipitousness of Countable Stationary Towers
15:55–16:25	Daisuke Ikegami, Shibaura Institute of Technology, On large cardinal properties of ω_1
16:30–17:00	Toshimichi Usuba, Waseda University, Multiverse and large cardinals
17:05–17:35	Hidenori Kurokawa, Kanazawa University, On Takeuti's view of the concept of set

September 19, 2018

10:00–10:30	Daichi Hayashi, Hokkaido University, On upper bounds for some complete and consistent theories of truth
10:35–11:05	Mamoru Kaneko, Waseda University, Small Infinitary Epistemic Logics
11:20–12:20	Wilfried Sieg, Carnegie Mellon University, Proofs and objects: Hilbert's pivotal thought
13:20–14:20	Masanao Ozawa, Prof. Emeritus, Nagoya University, From Boolean-Valued Analysis to Quantum Set Theory
14:35–15:05	Norbert Preining, Accelia Inc. and Technische Universität Wien, First Order Gödel Logics with Propositional Quantifiers

15:10–15:40	Keita Yokoyama, Japan Advanced Institute of Science and Technology, Ekeland's variational principle in reverse mathematics
15:55–16:25	NingNing Peng, Tohoku University, The eigen-distribution for weighted trees on various kinds of distributions
16:30–17:00	Takayuki Kihara, Nagoya University, Computability-theoretic methods in descriptive set theory
17:05–17:35	Mariko Yasugi, Prof. Emeritus, Kyoto Sangyo University, Irrational based computability of functions

September 20, 2018

10:00–10:30	Sohei Iwata, Kobe University, Interpolation properties for fragments of Gödel-Löb logic GL
10:35–11:05	Taishi Kurahashi, Kisarazu College, Rosser provability and the second incompleteness theorem
11:20–12:20	Samuel R. Buss, University of California San Diego, Bounded Arithmetic, Expanders, and Monotone Propositional Logic
13:20–14:20	Mariko Yasugi, Prof. Emeritus, Kyoto Sangyo University, A reflection on Takeuti's finitist standpoint
14:35–15:05	Tadatoshi Miyamoto, Nanzan University, Partitioning the Tripes of the Countable Ordinals and Morasses
15:10–15:40	Sakae Fuchino, Kobe Univerisity, Reflection principles and continuum hypothesis

Contents

Reflection Principles, Generic Large Cardinals, and the Continuum Problem

Sakaé Fuchino and André Ottenbreit Maschio Rodrigues

Abstract Strong reflection principles with the reflection cardinal $\leq \aleph_1$ or $< 2^{\aleph_0}$ imply that the size of the continuum is either $\aleph_1$ or $\aleph_2$ or very large. Thus, the stipulation, that a strong reflection principle should hold, seems to support the trichotomy on the possible size of the continuum. In this article, we examine the situation with the reflection principles and related notions of generic large cardinals.

Keywords Continuum problem · Laver-generically large cardinals · Forcing axioms · Reflection principles

1 Gödel's Program and Large Cardinals

The Continuum Problem has been considered to be one of the central problems in set theory. Georg Cantor tried till the end of his mathematical carrier to prove his "theorem" which claims, formulated in present terminology, the continuum, the cardinality $2^{\aleph_0}$ of the set of all real numbers, is the first uncountable cardinal $\aleph_1$. This statement is now called the Continuum Hypothesis (CH). By Gödel [27–29], and Cohen [4–6], it is proven hat CH is independent from the axiom system ZFC of Zermelo-Fraenkel set theory with the Axiom of Choice.[1]

Although the majority of the non-set theorists apparently believes that the results by Gödel and Cohen were the final solutions of the Continuum Problem, Gödel maintained in [30] that the conclusive solution to the problem is yet to be obtained in that a "right" extension of ZFC will be found which will decide the size of the continuum. Today the research program of searching for possible legitimate extensions of ZFC to settle the Continuum Problem is called Gödel's Program. Now

[1]Due to the Incompleteness Theorems, if we would like to formulate this statement[†] precisely, we have to put it under[‡] the assumption that ZFC is consistent (which we not only assume but do believe).

S. Fuchino (✉) · A. O. M. Rodrigues
Kobe University, Kobe, Japan
e-mail: fuchino@diamond.kobe-u.ac.jp
URL: https://fuchino.ddo.jp/

T. Arai et al. (eds.), *Advances in Mathematical Logic*, Springer Proceedings in Mathematics & Statistics 369, https://doi.org/10.1007/978-981-16-4173-2_1

that, besides CH, a multitude of mathematically significant statements is known to be independent from ZFC, the program should aim to decide not only the size of the continuum but also many of these independent mathematical statements. For modern views on Gödel's Program, the reader may consult e.g. Bagaria [2], Steel [39].

Gödel suggested in [30] that the large cardinal axioms are good candidates of axioms to be added to the axiom system ZFC. Unfortunately large cardinals do not decide the size of the continuum which Gödel also admits in the postscript to [30] added in 1966. Nevertheless, it is known today that some notable structural aspects of the continuum like the Projective Determinacy are decided under the existence of certain large large cardinals.

In this paper, we discuss about a new notion of generic large cardinals introduced in Fuchino, Ottenbreit and Sakai [20] and called there Laver-generic large cardinals (see Sect. 6). Reasonable instances of (the existential statement of a) Laver-generic large cardinal decide the size of the continuum to be either $\aleph_1$ or $\aleph_2$ or fairly large. We show that these three possible scenarios of Laver-generic large cardinal are in accordance with respective strong reflection properties with reflection cardinal $< \aleph_2$ or $< 2^{\aleph_0}$.

In connection with the view-point of set-theoretical multiverse (see Fuchino [16]), our trichotomy theorems, or some further development of them, have certain possibility to become the final answer to the Continuum Problem. As is well-known, Hugh Woodin is creating a theory which should support CH from the point of view of what should hold in a canonical model of the set theory. It should be emphasized that our trichotomy is not directly in contradiction with the possible outcome of his research program. In any case, it should be mathematical results in the future which should decide the matter definitively (if ever?).

2 Reflection Principles

The following type of mathematical reflection properties are considered in many different mathematical contexts.

(2.1) If a structure $\mathfrak{A}$ in the class $\mathcal{C}$ has the property $\mathcal{P}$, then there is a structure $\mathfrak{B}$ in relation $\mathcal{Q}$ to $\mathfrak{A}$ such that $\mathfrak{B}$ has the cardinality $< \kappa$ and $\mathfrak{B}$ also has the property $\mathcal{P}$.

We shall call "$< \kappa$" in (2.1) above the *reflection cardinal* of the reflection property. If κ is a successor cardinal μ^+ we shall also say that the reflection cardinal is $\leq \mu$.

An example of an instance of (2.1) is, when $\mathcal{C}$ = "compact Hausdorff topological spaces", $\mathcal{P}$ = "non-metrizable", $\mathcal{Q}$ = "subspace" and $\kappa = \aleph_2$, that is, with the reflection cardinal $\leq \aleph_1$. In this case, we obtain the statement:

(2.2) For any compact Hausdorff topological space, if X is non-metrizable, then there is a subspace Y of X of cardinality $< \aleph_2$ such that Y is also non-metrizable.

This assertion is known to be a theorem in ZFC (see Dow [11]).

If we extend the class $\mathcal{C}$ in (2.2) to $\mathcal{C}$ = "locally compact Hausdorff space", the statement thus obtained

(2.3) For any locally compact Hausdorff topological space, if X is non-metrizable, then there is a subspace Y of X of cardinality $< \aleph_2$ such that Y is also non-metrizable

is no more a theorem in ZFC: we can construct a counterexample to (2.3), using a non-reflecting stationary subset S of $E^{\kappa}_{\omega} = \{\alpha < \kappa : cf(\alpha) = \omega\}$ for some regular $\kappa > \omega_1$ (Fuchino, Juhász, Soukup, Szentmiklóssy and Usuba [18]). Note that $\Box_\lambda$ for any uncountable λ implies that there is such S for $\kappa = \lambda^+$. In particular, (2.3) implies the total failure of the square principles and thus we need very large large cardinals to obtain the consistency of this reflection principle. Actually, a known consistency proof of this principle requires the existence of a strongly compact cardinal.[2]

(2.3) is equivalent to the stationarity reflection principle called Fodor-type Reflection Principle (FRP) introduced in [18].[3] This principle can be formulated as follows (see [24]).

For a regular uncountable cardinal λ and $E \subseteq E^{\lambda}_{\omega} = \{\gamma \in \lambda : cf(\gamma) = \omega\}$, a mapping $g : E \to [\lambda]^{\aleph_0}$ is said to be a *ladder system on* E if, for all $\alpha \in E$, $g(\alpha)$ is a cofinal subset of α and $otp(g(\alpha)) = \omega$.

(FRP): *For any regular* $\lambda > \aleph_1$, *stationary* $E \subseteq E^{\lambda}_{\omega}$, *and a ladder system* $g : E \to \lambda^{\aleph_0}$ *on* E, *there is an* $\alpha^* \in E^{\lambda}_{\omega_1}$ *such that*

$$\{x \in [\alpha^*]^{\aleph_0} : \sup(x) \in E,\ g(\sup(x)) \subseteq x\}$$

is stationary in $[\alpha^*]^{\aleph_0}$.

Besides (2.3), there are many mathematical reflection principles in the literature which have been previously studied rather separately but which are now all shown to be equivalent to FRP and hence also equivalent to each other (see [14, 15, 22, 24]). The equivalence of (2.3) to FRP is established in [24] via a further characterization of FRP by non existence of a ladder system with a strong property of disjointness from which a counterexample to (2.3) (and other reflection properties proved to be equivalent to FRP) can be constructed. Here we want to mention only a couple of other reflection statements equivalent to FRP:

For a graph $G = \langle G, \mathcal{E}\rangle$, where $\mathcal{E} \subseteq G^2$ is the adjacency relation of the graph, is said to be *of countable coloring number* if there is a well-ordering $\sqsubset$ on G such that, for each $g \in G$, $\{h \in G : h\ \mathcal{E}$ and $h \sqsubset g\}$ is finite.

The following assertion is also equivalent to FRP ([18], Fuchino, Sakai Soukup and Usuba [24]):

(2.4) For any graph G, if G is not of countable coloring number, then there is a subgraph H of cardinality $< \aleph_2$ such that H is neither of countable coloring number.

---

[2] The existence of a strongly compact cardinal is enough to force Rado's Conjecture discussed below and Rado's Conjecture implies the reflection statement (2.3).

[3] Here, we are not only talking about equiconsistency but really about equivalence over ZFC.

In particular, it follows that the assertions (2.3) and (2.4) are equivalent to each other over ZFC.

(Strong) Downward Löwenheim Skolem Theorems of extended logics can be seen also as instances of the scheme (2.1). The following is a theorem in ZFC:

SDLS($\mathcal{L}(Q)$, $< \aleph_2$): *For any uncountable first-order structure $\mathfrak{A}$ in a countable signature, there is an elementary submodel $\mathfrak{B}$ of $\mathfrak{A}$ with respect to the logic $\mathcal{L}(Q)$ of cardinality*[4] $< \aleph_2$ *where the quantifier Q in a formula "$Qx\,\varphi$" is to be interpreted as "there are uncountably many x* such that φ".

Adopting the notation of Fuchino, Ottenbreit and Sakai [19], let $\mathcal{L}^{\aleph_0}_{stat}$ be the logic with monadic (weak) second order variable where the second order variables are to be interpreted as they are running over countable subsets of the structure in consideration. The logic has the built-in predicate ε where atomic formulas of the form $x\varepsilon X$ is allowed for first and second order variables x and X respectively. The logic also has the unique second order quantifier $stat$ which is interpreted by

(2.5) for a structure $\mathfrak{A} = \langle A, ...\rangle$, $\mathfrak{A} \models stat\; X\,\varphi[X, ...]$ holds if and only if $\{U \in [A]^{\aleph_0} : \mathfrak{A} \models \varphi[U, ...]\}$ is stationary in $[A]^{\aleph_0}$.

Note that $\mathcal{L}^{\aleph_0}_{stat}$ extends $\mathcal{L}(Q)$ above, since $Qx\,\varphi$ can be expressed by $stat X \exists x\, (x \not\varepsilon X \wedge \varphi)$.

In $\mathcal{L}^{\aleph_0}_{stat}$ we have two natural generalizations of the notion of elementary substructure. For (first order) structures $\mathfrak{A} = \langle A, ...\rangle$ and $\mathfrak{B} = \langle B, ...\rangle$ with $\mathfrak{B} \subseteq \mathfrak{A}$, let

(2.6) $\mathfrak{B} \prec_{\mathcal{L}^{\aleph_0}_{stat}} \mathfrak{A}$ if and only if, for all $\mathcal{L}^{\aleph_0}_{stat}$-formula $\varphi = \varphi(x_0, ..., X_0, ...)$ in the signature of $\mathfrak{A}$, $b_0, ... \in B$, and $U_0, ... \in [B]^{\aleph_0}$, we have $\mathfrak{B} \models \varphi[b_0, ..., U_0, ...] \Leftrightarrow \mathfrak{A} \models \varphi[b_0, ..., U_0, ...]$.

(2.7) $\mathfrak{B} \prec^{-}_{\mathcal{L}^{\aleph_0}_{stat}} \mathfrak{A}$ if and only if, for all $\mathcal{L}^{\aleph_0}_{stat}$-formula $\varphi = \varphi(x_0, ...)$ in the signature of $\mathfrak{A}$ without any free second order variables, and $b_0, ... \in B$, we have $\mathfrak{B} \models \varphi[b_0, ...] \Leftrightarrow \mathfrak{A} \models \varphi[b_0, ...]$.

By the remark after (2), the following principles are generalizations of SDLS($\mathcal{L}(Q)$, $< \aleph_2$):

SDLS($\mathcal{L}^{\aleph_0}_{stat}$, $< \aleph_2$): *For any uncountable first-order structure $\mathfrak{A}$ in a countable signature, there is a submodel $\mathfrak{B}$ of $\mathfrak{A}$ of cardinality $< \aleph_2$* such that $\mathfrak{B} \prec_{\mathcal{L}^{\aleph_0}_{stat}} \mathfrak{A}$.

SDLS$^-$($\mathcal{L}^{\aleph_0}_{stat}$, $< \aleph_2$): *For any uncountable first-order structure $\mathfrak{A}$ in a countable signature, there is a submodel $\mathfrak{B}$* of $\mathfrak{A}$ *of cardinality $< \aleph_2$* such that $\mathfrak{B} \prec^{-}_{\mathcal{L}^{\aleph_0}_{stat}} \mathfrak{A}$.

M. Magidor noticed that SDLS$^-$($\mathcal{L}^{\aleph_0}_{stat}$, $< \aleph_2$) implies (2.4) (see Magidor [37]). By the equivalence of (2.4) to FRP, we obtain

[4] The cardinality of a structure is defined to be the cardinality of the underlying set.

Theorem 1 SDLS$^-(\mathcal{L}^{\aleph_0}_{stat}, <\aleph_2)$ *implies* FRP. □

Actually, it is also easy to see that the stationarity reflection principle RP (which is a strengthening of RP in Jech [31]) follows from SDLS$^-(\mathcal{L}^{\aleph_0}_{stat}, <\aleph_2)$.

FRP follows from our RP ([18]) which is defined as follows:

RP: *For every regular $\lambda \geq \aleph_2$, stationary $S \subseteq [\lambda]^{\aleph_0}$, and $X \in [\lambda]^{\aleph_1}$, there is $Y \in [\lambda]^{\aleph_1}$ such that $cf(Y) = \omega_1$, $X \subseteq Y$ and $S \cap [Y]^{\aleph_0}$ is stationary in $[Y]^{\aleph_0}$.*

Jech's RP is just as our RP as defined above but without demanding the property "$cf(Y) = \omega_1$" for the reflection point Y.

Theorem 2 SDLS$^-(\mathcal{L}^{\aleph_0}_{stat}, <\aleph_2)$ *implies* RP.

Sketch of the proof. Let λ, S, X be as in the definition of RP. Let $\mu > \lambda^{\aleph_0}$ be regular and $\mathfrak{A} = \langle \mathcal{H}(\mu), \lambda, S, X, \in \rangle$ where λ, S and X are thought to be interpretations of unary predicate symbols. Let $\mathfrak{B} = \langle B, ... \rangle$ be such that B is of cardinality $\aleph_1$ and $\mathfrak{B} \prec^-_{\mathcal{L}^{\aleph_0}_{stat}} \mathfrak{A}$. Then $Y = \lambda \cap B$ is as desired. For example, $cf(Y) = \omega_1$ follows from the fact that $\mathfrak{B} \models \psi$ by elementarity where ψ is the $\mathcal{L}^{\aleph_0}_{stat}$-sentence: $stat\ X\, \exists y (y \mathrel{\underline{\in}} \underline{\lambda} \wedge \forall z\, ((z \,\varepsilon\, X \wedge z \mathrel{\underline{\in}} \underline{\lambda}) \rightarrow z \mathrel{\underline{\in}} y))$ where $\underline{\lambda}$ and $\underline{\in}$ are constant and binary relation symbols corresponding to λ and $\in$ in the structure $\mathfrak{A}$. □ (Theorem 2)

By a theorem of Todorčević, RP in the sense of Jech implies $2^{\aleph_0} \leq \aleph_2$ (see Theorem 37.18 in [31]). Thus

Corollary 3 SDLS$^-(\mathcal{L}^{\aleph_0}_{stat}, <\aleph_2)$ *implies* $2^{\aleph_0} \leq \aleph_2$. □

In contrast to Corollary 3, FRP does not put almost any restriction on the cardinality of the continuum since FRP is preserved by ccc forcing (see [18]).

A proof similar to that of Theorem 2 shows that SDLS$^-(\mathcal{L}^{\aleph_0}_{stat}, <\aleph_2)$ implies the Diagonal Reflection Principle down to an internally club reflection point of cardinality $<\aleph_2$ of S. Cox [8]. Conversely, we can also easily prove that the Diagonal Reflection Principle down to an internally club reflection point of cardinality $<\aleph_2$ implies SDLS$^-(\mathcal{L}^{\aleph_0}_{stat}, <\aleph_2)$. The internally clubness of the reflection point is used to guarantee that the internal interpretation of the stationary logic coincides with the external correct interpretation of the logic in the small substructure to make it an elementary substructure (in the sense of $\prec^-_{\mathcal{L}^{\aleph_0}_{stat}}$) of the original structure. Thus we obtain (1) of the following theorem.

Theorem 4 (Theorem 1.1, (3) and (4) in [19])

(1) SDLS$^-(\mathcal{L}^{\aleph_0}_{stat}, <\aleph_2)$ *is equivalent to the Diagonal Reflection Principle down to an internally club reflection point of cardinality $<\aleph_2$.*

(2) SDLS$(\mathcal{L}^{\aleph_0}_{stat}, <\aleph_2)$ *is equivalent to* SDLS$^-(\mathcal{L}^{\aleph_0}_{stat}, <\aleph_2)$ *plus* CH. □

S. Cox proved in [8] that the Diagonal Reflection Principle down to an internally club reflection point of cardinality $\aleph_1$ follows from MA$^{+\omega_1}$(σ-closed). Thus,

Corollary 5 (1) MA$^{+\omega_1}$(*σ-closed*) *implies* SDLS$^-(\mathcal{L}^{\aleph_0}_{stat}, <\aleph_2)$.

(2) MA$^{+\omega_1}$(*σ-closed*) + CH *implies* SDLS$(\mathcal{L}^{\aleph_0}_{stat}, <\aleph_2)$. □

The reflection cardinal $< \aleph_2$ (or equivalently $\leq \aleph_1$) in the reflection principles above can be considered to be significant and even natural since, with this reflection cardinal, the reflection principles can be seen as statements claiming that the cardinality $\aleph_1$ is archetypical among uncountable cardinals, and hence that $\aleph_1$ already captures various phenomenon in uncountability in the sense that a certain type of properties of an uncountable structure can be reflected down to a substructure of the cardinality $\aleph_1$. From that point of view, it is interesting that one of the strongest reflection principles, namely the Strong Downward Löwenheim-Skolem Theorem for stationary logic with this reflection cardinal implies CH.

In a similar way, we can also argue that the reflection with the reflection cardinal $< 2^{\aleph_0}$ or $\leq 2^{\aleph_0}$ should be regarded as significant and even natural since we can interpret the reflection with these reflection cardinals as a pronouncement of the richness of the continuum.

Let $\mathsf{SDLS}(\mathcal{L}^{\aleph_0}_{stat}, < 2^{\aleph_0})$ and $\mathsf{SDLS}^-(\mathcal{L}^{\aleph_0}_{stat}, < 2^{\aleph_0})$ be the principles obtained from $\mathsf{SDLS}(\mathcal{L}^{\aleph_0}_{stat}, < \aleph_2)$ and $\mathsf{SDLS}^-(\mathcal{L}^{\aleph_0}_{stat}, < \aleph_2)$ by replacing "$< \aleph_2$" with "$< 2^{\aleph_0}$".

Theorem 6 (Proposition 2.1, Corollaries 2.3, 2.4 in [20])

(1) $\mathsf{SDLS}^-(\mathcal{L}^{\aleph_0}_{stat}, < 2^{\aleph_0})$ *implies* $2^{\aleph_0} = \aleph_2$. *In particular, if* $2^{\aleph_0} > \aleph_2$, *then* $\mathsf{SDLS}^-(\mathcal{L}^{\aleph_0}_{stat}, < 2^{\aleph_0})$ *does not hold.*

(2) $\mathsf{SDLS}(\mathcal{L}^{\aleph_0}_{stat}, < 2^{\aleph_0})$ *is inconsistent.* □

Note that $\mathsf{SDLS}^-(\mathcal{L}^{\aleph_0}_{stat}, < 2^{\aleph_0})$ follows from $\mathsf{MA}^{+\omega_1}(\sigma\text{-closed}) + \neg\mathsf{CH}$ which is e.g. a consequence of $\mathsf{PFA}^{+\omega_1}$.

Note that Lemma 9 implies that $\mathsf{GRP}^{\omega,\omega_1}(< 2^{\aleph_0})$ is also inconsistent.

In contrast to the reflection down to $< 2^{\aleph_0}$ whose strong version implies that the continuum is $\aleph_2$ (see Theorem 6, (2) above), the reflection down to $\leq 2^{\aleph_0}$ does not exert any such restriction on the size of the continuum as we will see this in the next section.

A slightly different type of reflection principle with reflection cardinal $< 2^{\aleph_0}$ implies that the continuum is very large. We will see this in Sect. 5.

3 Game Reflection Principles and Generically Large Cardinals

There is a further strengthening of $\mathsf{SDLS}^-(\mathcal{L}^{\aleph_0}_{stat}, < \aleph_2)$ which is called (Strong) Game Reflection Principle[5] (GRP) introduced in B. König [34]. The following is a generalization of the principle:

For a regular uncountable cardinal μ, a set A, and $\mathcal{A} \subseteq {}^{\mu>}A$, $\mathcal{G}^{{}^{\mu>}A}(\mathcal{A})$ is the following game of length μ for players I and II. A match in $\mathcal{G}^{{}^{\mu>}A}(\mathcal{A})$ looks like:

[5] In [34], B. König originally called the principle introduced here the Strong Game Reflection Principle and the local version of the principle the Game Reflection Principle.

$$\begin{array}{c|cccccccc} I & a_0 & & a_1 & & a_2 & \cdots a_\xi & & \cdots \\ \hline II & & b_0 & & b_1 & & b_2 \cdots & b_\xi & \cdots \end{array} \qquad (\xi < \mu)$$

where $a_\xi, b_\xi \in A$ for $\xi < \mu$.

II wins this match if

(3.1) $\langle a_\xi, b_\xi : \xi < \eta \rangle \in \mathcal{A}$ and $\langle a_\xi, b_\xi : \xi < \eta \rangle ^\frown \langle a_\eta \rangle \notin \mathcal{A}$ for some $\eta < \mu$; or $\langle a_\xi, b_\xi : \xi < \mu \rangle \in [\mathcal{A}]$

where $\langle a_\xi, b_\xi : \xi < \eta \rangle$ denotes the sequence $f \in {}^{2\cdot\eta}A$ such that $f(2 \cdot \xi) = a_\xi$ and $f(2 \cdot \xi + 1) = b_\xi$ for all $\xi < \eta$ and $[\mathcal{A}] = \{f \in {}^{\mu}A : f \restriction \alpha \in \mathcal{A}\}$ for all $\alpha < \mu$.

For regular cardinals μ, κ with $\mu < \kappa$ $\mathcal{C} \subseteq [A]^{<\kappa}$ is said to be μ-club if $\mathcal{C}$ is cofinal in $[A]^{<\kappa}$ with respect to $\subseteq$ and closed with respect to the union of increasing $\subseteq$-chain of length ν for any regular $\mu \leq \nu < \kappa$.

$\mathsf{GRP}^{<\mu}(<\kappa)$: *For any set A of regular cardinality $\geq \kappa$ and μ-club $\mathcal{C} \subseteq [A]^{<\kappa}$, if the player II has no winning strategy in $\mathcal{G}^{{}^{\mu>}A}(\mathcal{A})$ for some $\mathcal{A} \subseteq {}^{\mu>}A$, there is $B \in \mathcal{C}$ such that the player II has no winning strategy in $\mathcal{G}^{{}^{\mu>}B}(\mathcal{A} \cap {}^{\mu>}B)$.*

B. König's *Game Reflection Principle* (GRP) is $\mathsf{GRP}^{<\omega_1}(<\aleph_2)$.

Sometimes, the following variation of the games and the principles is useful: For a limit ordinal δ, a set A, and $\mathcal{A} \subseteq {}^{\delta\geq}A$, $\mathcal{G}^{{}^{\delta\geq}A}(\mathcal{A})$ is the following game of length δ for players I and II. A match in $\mathcal{G}^{{}^{\delta\geq}A}(\mathcal{A})$ looks like:

$$\begin{array}{c|cccccccc} I & a_0 & & a_1 & & a_2 & \cdots a_\xi & & \cdots \\ \hline II & & b_0 & & b_1 & & b_2 \cdots & b_\xi & \cdots \end{array} \qquad (\xi < \delta)$$

where $a_\xi, b_\xi \in A$ for $\xi < \delta$.

II wins this match if

(3.2) $\langle a_\xi, b_\xi : \xi < \eta \rangle \in \mathcal{A}$ and $\langle a_\xi, b_\xi : \xi < \eta \rangle ^\frown \langle a_\eta \rangle \notin \mathcal{A}$ for some $\eta < \delta$; or $\langle a_\xi, b_\xi : \xi < \eta \rangle \in \mathcal{A}$ for all $\eta \leq \delta$.

where $\langle a_\xi, b_\xi : \xi < \eta \rangle$ is defined as above.

For a limit ordinal δ, and uncountable regular cardinals μ, κ with $\delta \leq \mu < \kappa$,

$\mathsf{GRP}^{\delta,\mu}(<\kappa)$: *For any A of regular cardinality $\geq \kappa$ and μ-club $\mathcal{C} \subseteq [A]^{<\kappa}$, if the player II has no winning strategy in $\mathcal{G}^{{}^{\delta\geq}A}(\mathcal{A})$ for some $\mathcal{A} \subseteq {}^{\delta\geq}A$, there is $B \in \mathcal{C}$ such that the player II has no winning strategy in $\mathcal{G}^{{}^{\delta\geq}B}(\mathcal{A} \cap {}^{\delta\geq}B)$.*

The next Lemma follows immediately from the definitions:

Lemma 7 *Suppose that δ and δ' are limit ordinals and μ, μ', κ, κ' are regular cardinals such that $\delta \leq \delta' < \mu \leq \mu' < \kappa$. Then we have*

(3.3) $\mathsf{GRP}^{<\mu'}(<\kappa) \Rightarrow \mathsf{GRP}^{<\mu}(<\kappa) \Rightarrow \mathsf{GRP}^{\delta',\mu}(<\kappa) \Rightarrow \mathsf{GRP}^{\delta,\mu}(<\kappa)$ □

GRP is indeed a strengthening of $\mathsf{SDLS}(\mathcal{L}^{\aleph_0}_{stat}, <\aleph_2)$. The following Theorem 8, Lemma 9 and Corollary 10 are slight generalizations of results in B. König [34].

Theorem 8 (Theorem 4.7 in [19]) *Suppose that κ is a regular uncountable cardinal such that*

(3.4) $\mu^{\aleph_0} < \kappa$ *for all* $\mu < \kappa$, *and*
(3.5) $\mathsf{GRP}^{\omega,\omega_1}(<\kappa)$ *holds.*

Then $\mathsf{SDLS}(\mathcal{L}^{\aleph_0}_{stat}, <\kappa)$ *holds.*[6] □

Lemma 9 (Lemma 4.2 in [19]) *For a regular cardinal κ,* $\mathsf{GRP}^{\omega,\omega_1}(<\kappa)$ *implies* $2^{\aleph_0} < \kappa$. □

Remember that GRP is the principle $\mathsf{GRP}^{<\omega_1}(<\aleph_2)$. For a regular cardinal $\kappa > \aleph_1$ we shall write $\mathsf{GRP}(<\kappa)$ for $\mathsf{GRP}^{<\omega_1}(<\kappa)$. Thus GRP is $\mathsf{GRP}(<\aleph_2)$.

Corollary 10 (1) GRP *implies* $\mathsf{SDLS}(\mathcal{L}^{\aleph_0}_{stat}, <\aleph_2)$.
(2) $\mathsf{GRP}(<(2^{\aleph_0})^+)$ *implies* $\mathsf{SDLS}(\mathcal{L}^{\aleph_0}_{stat}, \leq 2^{\aleph_0})$.

Proof (1) : By Lemma 9, GRP implies CH. Thus, under GRP, (8) holds for $\kappa = \aleph_2$. By Lemma 7, GRP implies $\mathsf{GRP}^{\omega,\omega_1}(<\aleph_2)$. By Theorem 8, it follows that $\mathsf{SDLS}(\mathcal{L}^{\aleph_0}_{stat}, <\aleph_2)$.

(2) : Note that, for $\mu < (2^{\aleph_0})^+$, $\mu^{\aleph_0} \leq 2^{\aleph_0} < (2^{\aleph_0})^+$ holds. By Lemma 7, $\mathsf{GRP}(<(2^{\aleph_0})^+)$ implies $\mathsf{GRP}^{\omega,\omega_1}(<(2^{\aleph_0})^+)$. Thus, by Theorem 8, it follows that $\mathsf{SDLS}(\mathcal{L}^{\aleph_0}_{stat}, <(2^{\aleph_0})^+)$, or $\mathsf{SDLS}(\mathcal{L}^{\aleph_0}_{stat}, \leq 2^{\aleph_0})$ in the other notation, holds.

□ (Corollary 10)

GRP also implies another prominent reflection principle which is called Rado's Conjecture.

We call a partial ordering $T = \langle T, \leq_T \rangle$ a *tree* if the initial segment below any element is a well-ordering. A tree $T = \langle T, \leq_T \rangle$ is said to be *special* if it can be partitioned into countably many antichains (i.e. pairwise incomparable sets). Note that every special tree has height $\leq \omega_1$.

For a regular cardinal $\kappa > \aleph_1$, we define Rado's Conjecture with reflection cardinal $<\kappa$ as

$\mathsf{RC}(<\kappa)$: *For any tree T, if T is not special then there is* $B \in [T]^{<\kappa}$ such that *B (as the tree $\langle B, \leq_T \cap B^2 \rangle$) is not special.*

The original *Rado's Conjecture* (RC) is $\mathsf{RC}(<\aleph_2)$.

Theorem 11 (B. König [34], see also Theorem 4.3 in [19]) *For a regular cardinal* $\kappa > \aleph_1$, $\mathsf{GRP}^{<\omega_1}(<\kappa)$ *implies* $\mathsf{RC}(<\kappa)$. □

[6] Actually we can prove a slight strengthening of $\mathsf{SDLS}(\mathcal{L}^{\aleph_0}_{stat}, <\kappa)$ (see [19]).

FRP is also a consequence of GRP. This is simply because FRP follows from RC (see [25]).

Game Reflection Principles are characterizations of certain instances of the existence of generically supercompact cardinals.

Let $\mathcal{P}$ be a class of posets. A cardinal κ is said to be a *generically supercompact cardinal by* $\mathcal{P}$, if, for any regular λ, there is a poset $\mathbb{P} \in \mathcal{P}$ such that, for any $(\mathsf{V}, \mathbb{P})$-generic filter $\mathbb{G}$, there are classes M, $j \subseteq \mathsf{V}[\mathbb{G}]$ such that M is an inner model of $\mathsf{V}[\mathbb{G}]$, $j : \mathsf{V} \xrightarrow{\preceq} M$, $crit(j) = \kappa$, $j(\kappa) > \lambda$ and $j''\lambda \in M$.

Theorem 12 ([19]) *For a regular uncountable* κ*, the following are equivalent:*

(a) $2^{<\kappa} = \kappa$ *and* $\mathsf{GRP}^{<\kappa}(<\kappa^+)$ *holds.*

(b) κ^+ *is generically supercompact by* $<\kappa$*-closed posets.* □

Corollary 13 (B. König [34]) *The following are equivalent:*

(a) GRP *holds.*

(b) $\aleph_2$ *is generically supercompact by* σ*-closed posets.*

Proof Assume that GRP holds (remember that GRP denotes $\mathsf{GRP}^{<\omega_1}(<\aleph_2)$). Then, by Corollary 10, (1), $2^{<\aleph_1} = 2^{\aleph_0} = \aleph_1$. Thus, by Theorem 12, "(a) $\Rightarrow$ (b)" for $\kappa = \aleph_1$, it follows that $\aleph_2 = (\aleph_1)^+$ is generically supercompact by σ-closed forcing. The implication "(b) $\Rightarrow$ (a)" follows from "(b) $\Rightarrow$ (a)" of Theorem 12 for $\kappa = \aleph_1$.
□ (Corollary 13)

4 Simultaneous Reflection down to $< 2^{\aleph_0}$ and $\leq 2^{\aleph_0}$

As we discussed in Sect. 2, the reflection down to $< 2^{\aleph_0}$ as well as the reflection down to $\leq 2^{\aleph_0}$ can be regarded as significant being principles which claim certain richness of the continuum.

One of the strong form of reflection principles with reflection cardinal $< 2^{\aleph_0}$ implies that the continuum is equal to $\aleph_2$ (Theorem 6, (2)) while there is a limitation on the possible types of reflection (Theorem 6, (3)).

In contrast, as we see below, the reflection down to $\leq 2^{\aleph_0}$ can be established in one of its strongest forms without almost any restriction on the size of the continuum: (a) of Theorem 12 can be easily realized starting from a supercompact cardinal.

The following is well-known.

Lemma 14 (Lemma 4.10 in [19]) *If* κ *is a supercompact and* $\mu < \kappa$ *is an uncountable regular cardinal then for* $\mathbb{P} = \mathrm{Col}(\mu, \kappa)$ *and* $(\mathsf{V}, \mathbb{P})$*-generic filter* $\mathbb{G}$*, we have* $\mathsf{V}[\mathbb{G}] \models$ "$\kappa = \mu^+$ *and* κ *is generically supercompact by* $<\mu$*-closed posets*". □

Suppose now that κ_1 is a supercompact cardinal and $2^{\aleph_0}$ is a regular cardinal. Let $\mathbb{Q} = \mathrm{Col}(2^{\aleph_0}, \kappa_1)$ and let $\mathbb{H}$ be a $(\mathsf{V}, \mathbb{Q})$-generic filter. By $< 2^{\aleph_0}$-closedness of $\mathbb{Q}$, we have $(2^{\aleph_0})^{\mathsf{V}} = (2^{\aleph_0})^{\mathsf{V}[\mathbb{H}]}$ and $\mathsf{V}[\mathbb{H}] \models \kappa_1 = (2^{\aleph_0})^+$. By Lemma 14, $\mathsf{V}[\mathbb{H}] \models$

"$(2^{\aleph_0})^+$is a generically supercompact cardinal by $< 2^{\aleph_0}$-closed posets". By Theorem 12, it follows that $\mathsf{V}[\mathbb{H}] \models$"$\mathsf{GRP}^{<2^{\aleph_0}}(<(2^{\aleph_0})^+)$".

By Corollary 10, (2), Lemma 7 and Theorem 11, we have, in particular,

(4.1) $\mathsf{V}[\mathbb{H}] \models$"$\mathsf{SDLS}(\mathcal{L}^{\aleph_0}_{stat}, \leq 2^{\aleph_0}) \wedge \mathsf{RC}(\leq 2^{\aleph_0})$".

Note that the continuum can be forced to be practically anything of uncountable cofinality below κ_1 prior to the generic extension by $\mathbb{Q}$.

The following Proposition 15 should also belong to the folklore (for similar statements, see Theorem 4.1 in König and Yoshinobu [35] or Theorem 4.3 in Larson [36]).

Recall that, for a regular cardinal μ, a poset $\mathbb{P}$ is $<\mu$ *-directed closed* if any downward directed subset of $\mathbb{P}$ of cardinality $<\mu$ has a lower bound (in $\mathbb{P}$).

Proposition 15 *Suppose that* $\mathsf{MA}^{+\omega_1}$*(σ-closed) (or* $\mathsf{PFA}^{+\omega_1}$*, or* $\mathsf{MM}^{+\omega_1}$*, resp.) holds. If* $\mathbb{P}$ *is* $<\aleph_2$*-directed closed, then we have*

(4.2) $\Vdash_{\mathbb{P}}$ "$\mathsf{MA}^{+\omega_1}$*(σ-closed) (or* $\mathsf{PFA}^{+\omega_1}$*, or* $\mathsf{MM}^{+\omega_1}$*, resp.)*".

Proof We prove the case of $\mathsf{MA}^{+\omega_1}$(σ-closed). Other cases can be proved by the same argument.

Suppose that $\mathbb{P}$ is a $<\aleph_2$-directed closed poset and let $\underset{\sim}{\mathbb{Q}}$, $\langle \underset{\sim}{D}_\alpha : \alpha < \omega_1 \rangle$, $\langle \underset{\sim}{S}_\beta : \beta < \omega_1 \rangle$ be $\mathbb{P}$-names such that

(4.3) $\Vdash_{\mathbb{P}}$ "$\underset{\sim}{\mathbb{Q}}$ is a σ-closed poset,
each $\underset{\sim}{D}_\alpha$ $(\alpha < \omega_1)$ is a dense subset of $\underset{\sim}{\mathbb{Q}}$, and
each $\underset{\sim}{S}_\beta$ $(\beta < \omega_1)$ is a stationary subset of ω_1"

Let $\mathbb{P}^* = \mathbb{P} * \underset{\sim}{\mathbb{Q}}$. For $\alpha < \omega_1$, let

(4.4) $D^*_\alpha = \{\langle \mathbb{p}, \underset{\sim}{\mathbb{q}} \rangle \in \mathbb{P}^* : \mathbb{p} \Vdash_{\mathbb{P}}$ "$\underset{\sim}{\mathbb{q}} \in \underset{\sim}{D}_\alpha$"$\}$.

For $\beta < \omega_1$, let

(4.5) $\underset{\sim}{S}^*_\beta = \{\langle \langle \mathbb{p}, \underset{\sim}{\mathbb{q}} \rangle, \check{\alpha} \rangle : \langle \mathbb{p}, \underset{\sim}{\mathbb{q}} \rangle \in \mathbb{P}^*, \mathbb{p} \Vdash_{\mathbb{P}}$ "$\underset{\sim}{\mathbb{q}} \Vdash_{\underset{\sim}{\mathbb{Q}}}$ "$\check{\alpha} \in \underset{\sim}{S}_\beta$""$\}$.

By the definition of $\mathbb{P}^*$, $\langle D^*_\alpha : \alpha < \omega_1 \rangle$, and $\langle \underset{\sim}{S}^*_\beta : \beta < \omega_1 \rangle$, the following is easy to show:

Claim 16 $\mathbb{P}^*$ is a σ-closed poset, D^*_α is a dense subset of $\mathbb{P}^*$ for all $\alpha < \omega_1$, and $\underset{\sim}{S}^*_\beta$ is a $\mathbb{P}^*$-name with $\Vdash_{\mathbb{P}^*}$ "$\underset{\sim}{S}^*_\beta$ is a stationary subset of ω_1" for all $\beta < \omega_1$. □

Let $\mathcal{D}^* = \{D^*_\alpha : \alpha < \omega_1\}$. By $\mathsf{MA}^{+\omega_1}$(σ-closed), there is a $\mathcal{D}^*$-generic filter $\mathbb{G}^*$ on $\mathbb{P}^*$ such that $\underset{\sim}{S}^*_\beta[\mathbb{G}^*]$ is a stationary subset of ω_1 for all $\beta < \omega_1$.

Let θ be a sufficiently large regular cardinal and let $M \prec \mathcal{H}(\theta)$ be of cardinality $\aleph_1$ such that $\omega_1 \subseteq M$ and M contains everything relevant (in particular, $\mathbb{G}^* \in M$).

Let $\mathbb{G}_0 = \mathbb{G}^* \cap M$ and let $\mathbb{G}$ be the filter on $\mathbb{P}^*$ generated by $\mathbb{G}_0$. By the choice of M, we have $\underset{\sim}{S}^*_\beta[\mathbb{G}^*] = \underset{\sim}{S}^*_\beta[\mathbb{G}_0] = \underset{\sim}{S}^*_\beta[\mathbb{G}]$.

Let $G = \{\mathbb{p} \in \mathbb{P} : \langle \mathbb{p}, \underset{\sim}{\mathbb{q}} \rangle \in \mathbb{G}$ for some $\underset{\sim}{\mathbb{q}}\}$. Since $|G| \leq |M| < \aleph_2$ and G is downward directed, there is a lower bound $\mathbb{p}_0 \in \mathbb{P}$ of G.

Let

(4.6) $\underset{\sim}{\mathbb{H}} = \{\langle \underset{\sim}{\mathbb{q}}, \mathbb{1}_{\mathbb{P}}\rangle : \langle \mathbb{p}, \underset{\sim}{\mathbb{q}}\rangle \in \mathbb{G}$ for some $\mathbb{p} \in \mathbb{P}\}$.

Then we have

(4.7) $\mathbb{p}_0 \Vdash_{\mathbb{P}}$ "$\underset{\sim}{\mathbb{H}}$ is a $\{\underset{\sim}{D}_\alpha : \alpha < \omega_1\}$-generic filter on $\underset{\sim}{\mathbb{Q}}$ such that $\underset{\sim}{S}_\beta[\underset{\sim}{\mathbb{H}}]$ is a stationary subset of ω_1 for all $\beta < \omega_1$ ".

Since the argument above can be also performed in $\mathbb{P} \restriction \mathbb{r}$ instead of $\mathbb{P}$ for any $\mathbb{r} \in \mathbb{P}$. It follows that

(4.8) $\Vdash_{\mathbb{P}}$ "there is a $\{\underset{\sim}{D}_\alpha : \alpha < \omega_1\}$-generic filter H on $\underset{\sim}{\mathbb{Q}}$ such that $\underset{\sim}{S}_\beta[H]$ is a stationary subset of ω_1 for all $\beta < \omega_1$ ".

□ (Proposition 15)

Theorem 16 *Suppose that κ and κ_1 with $\kappa < \kappa_1$ are two supercompact cardinals. Then there is a generic extension* $\mathsf{V}[\mathbb{G} * \mathbb{H}]$ *such that*

$$\mathsf{V}[\mathbb{G} * \mathbb{H}] \models \mathsf{MM}^{+\omega_1} + \mathsf{GRP}^{<2^{\aleph_0}}(\leq 2^{\aleph_0}).$$

Note that, by Corollary 5, (1), we have

$$\mathsf{V}[\mathbb{G} * \mathbb{H}] \models \mathsf{SDLS}^-(\mathcal{L}^{\aleph_0}_{stat}, < 2^{\aleph_0}) + \mathsf{GRP}^{<2^{\aleph_0}}(\leq 2^{\aleph_0}).$$

Proof of Theorem 16 Let $\mathsf{V}[\mathbb{G}]$ be a standard model of MM obtained by a reverse countable iteration of length κ along with a fixed Laver-function $\kappa \to V_\kappa$. It is easy to see that $\mathsf{V}[\mathbb{G}]$ also satisfies $\mathsf{MM}^{+\omega_1}$. Note that we have $\mathsf{V}[\mathbb{G}] \models \kappa = \aleph_2 = 2^{\aleph_0}$. In $\mathsf{V}[\mathbb{G}]$, κ_1 is still supercompact. Thus, working in $\mathsf{V}[\mathbb{G}]$, let $\mathbb{Q} = \mathrm{Col}(2^{\aleph_0}, \kappa_1)$. Let $\mathbb{H}$ be a $(\mathsf{V}[\mathbb{G}], \mathbb{Q})$-generic filter. By Proposition 15, we have $\mathsf{V}[\mathbb{G} * \mathbb{H}] \models \mathsf{MM}^{+\omega_1}$. By Lemma 14 and Theorem 12, we have $\mathsf{V}[\mathbb{G} * \mathbb{H}] = (\mathsf{V}[\mathbb{G}])[\mathbb{H}] \models \mathsf{GRP}^{<2^{\aleph_0}}(\leq 2^{\aleph_0})$.

□ (Theorem 16)

5 Reflection Principles Under Large Continuum

The continuum can be "very large" as a cardinal number. For example, this is the case in the model $\mathsf{V}[\mathbb{G}]$ obtained by starting from a supercompact κ and then adding κ many Cohen reals. In this model, we have $2^{\aleph_0} = \kappa$ and there is a countably saturated normal fine filter over $\mathcal{P}_\kappa(\lambda)$ for all regular $\lambda \geq \kappa$. The last property of $\mathsf{V}[\mathbb{G}]$ implies that κ there is still fairly large (e.g. κ-weakly Mahlo and more, see e.g. Proposition 16.8 in Kanamori [32]).

If the ground model satisfies FRP then $\mathsf{V}[\mathbb{G}]$ also satisfies FRP since FRP is preserved by ccc extensions (see [18]). On the other hand, as we already have seen, $\mathsf{SDLS}^-(\mathcal{L}^{\aleph_0}_{stat}, < \aleph_2)$ or even $\mathsf{SDLS}^-(\mathcal{L}^{\aleph_0}_{stat}, < 2^{\aleph_0})$ is incompatible with large continuum. In particular, these reflection principles do not hold in our model $\mathsf{V}[\mathbb{G}]$.

A weakening of $\mathsf{SDLS}^-(\mathcal{L}^{\aleph_0}_{stat}, < 2^{\aleph_0})$ is compatible with large continuum. Let us begin with the diagonal reflection principle which characterizes the version of the strong downward Löwenheim-Skolem theorem with reflection points of cardinality $<$ large continuum. The following is a weakening of Cox's Diagonal Reflection Principle down to an internally club reflection point.

For regular cardinals κ, λ with $\kappa \leq \lambda$, let

$(*)^{int+}_{<\kappa,\lambda}$: *For any countable expansion* $\tilde{\mathfrak{A}}$ of $\langle\mathcal{H}(\lambda), \in\rangle$ *and sequence* $\langle S_a : a \in \mathcal{H}(\lambda)\rangle$ such that S_a *is a stationary subset of* $[\mathcal{H}(\lambda)]^{\aleph_0}$ for all $a \in \mathcal{H}(\lambda)$, *there are stationarily many* $M \in [\mathcal{H}(\lambda)]^{<\kappa}$ such that

(1) $\tilde{\mathfrak{A}} \restriction M \prec \tilde{\mathfrak{A}}$; *and*
(2) $S_a \cap M$ *is stationary in* $[M]^{\aleph_0}$ *for all* $a \in M$.

Note that (1) implies that $c \subseteq M$ holds for all $c \in [M]^{\aleph_0} \cap M$.

In the notation above, "int" (internal) refers to the condition (2) in which not $S_a \cap [M]^{\aleph_0}$ but $S_a \cap M$ is declared to be stationary in $[M]^{\aleph_0}$; "+" refers to the condition that $M \in [\mathcal{H}(\lambda)]^{<\kappa}$ with (1) and (2) not only exists but there are stationarily many such M.

That $(*)^{int+}_{<\kappa,\lambda}$ is compatible with $\kappa = 2^{\aleph_0}$ and it is arbitrarily large is seen in the following Theorem 17 together with Lemma 18 below:

Theorem 17 (Theorem 2.10 in [20]) *Suppose that* κ *is a generically supercompact cardinal by proper posets. Then* $(*)^{int+}_{<\kappa,\lambda}$ *holds for all regular* $\lambda \geq \kappa$. □

Similarly to Lemma 14, starting from a supercompact cardinal, it is easy to force that the continuum is generically supercompact cardinal by ccc-posets. Let us call a poset $\mathbb{P}$ appropriate for κ, if we have $j''\mathbb{P} \lessdot j(\mathbb{P})$ for all supercompact embedding j for κ.

Lemma 18 *If* κ *is a supercompact and* $\mu < \kappa$ *is an uncountable regular cardinal then for any* $< \mu$*-cc poset* $\mathbb{P}$ *appropriate for* κ*, adding* $\geq \kappa$ *many reals, we have* $\mathsf{V}[\mathbb{G}] \models$ "$\kappa \leq 2^{\aleph_0}$ *and* κ *is generically supercompact by* $< \mu$*-cc posets*". □

"$(*)^{int+}_{<\kappa,\lambda}$ holds for all regular $\lambda \geq \kappa$" characterizes the strong downward Löwenheim-Skolem theorem for internal interpretation of stationary logic defined in the following.

For a structure $\mathfrak{A} = \langle A, ...\rangle$ of a countable signature, an $\mathcal{L}^{\aleph_0}_{stat}$-formula $\varphi = \varphi(x_0, ..., X_0, ...)$[7] and $a_0, ... \in A$, $U_0, ... \in [A]^{\aleph_0} \cap A$, we define the internal interpretation of $\varphi(a_0, ..., U_0, ...)$ in $\mathfrak{A}$ (notation: $\mathfrak{A} \models^{int} \varphi(a_0, ..., U_0, ...)$ for "$\varphi(a_0, ..., U_0, ...)$ holds internally in $\mathfrak{A}$") by induction on the construction of φ as follows:

If φ is "$x_i \varepsilon X_j$" then

(5.1) $\mathfrak{A} \models^{int} \varphi(a_0, ..., U_0, ...) \Leftrightarrow a_i \in U_j$

for a structure $\mathfrak{A} = \langle A, ...\rangle$, $a_0, ... \in A$ and $U_0, ... \in [A]^{\aleph_0} \cap A$.

For first-order connectives and quantifiers in $\mathcal{L}^{\aleph_0}_{stat}$, the semantics "$\models^{int}$" is defined exactly as for the first order "$\models$".

---

[7] As before, when we write $\varphi = \varphi(x_0, ..., X_0, ...)$, we always assume that the list $x_0, ...$ contains all the free first order variables of φ and $X_0, ...$ all the free weak second order variables of φ.

For an $\mathcal{L}^{\aleph_0}_{stat}$ formula φ with $\varphi = \varphi(x_0, ..., X_0, ..., X)$, assuming that the notion of $\mathfrak{A} \models^{int} \varphi(a_0, ..., U_0, ..., U)$ has been defined for all $a_0, ... \in A$, $U_0, ..., U \in [A]^{\aleph_0} \cap A$, we stipulate

(5.2) $\mathfrak{A} \models^{int} stat\, X\, \varphi(a_0, ..., U_0, ..., X) \Leftrightarrow$
$\{U \in [A]^{\aleph_0} \cap A \,:\, \mathfrak{A} \models^{int} \varphi(a_0, ..., U_0, ..., U)\}$ is stationary in $[A]^{\aleph_0}$

for a structure $\mathfrak{A} = \langle A, ...\rangle$ of a relevant signature, $a_0, ... \in A$ and $U_0, ... \in [A]^{\aleph_0} \cap A$.

For structures $\mathfrak{A}$, $\mathfrak{B}$ of the same signature with $\mathfrak{B} = \langle B, ...\rangle$ and $\mathfrak{B} \subseteq \mathfrak{A}$, we define

(5.3) $\mathfrak{B} \prec^{int}_{\mathcal{L}^{\aleph_0}_{stat}} \mathfrak{A} \Leftrightarrow$
$\mathfrak{B} \models^{int} \varphi(b_0, ..., U_0, ...)$ *if and only if* $\mathfrak{A} \models^{int} \varphi(b_0, ..., U_0, ...)$
for all $\mathcal{L}^{\aleph_0}_{stat}$-formulas φ *in the signature of the structures with* $\varphi = \varphi(x_0, ..., X_0, ...)$, $b_0, ... \in B$ *and* $U_0, ... \in [B]^{\aleph_0} \cap B$.

Finally, for a regular $\kappa > \aleph_1$, the internal strong downward Löwenheim-Skolem Theorem $\mathsf{SDLS}^{int}_{+}(\mathcal{L}^{\aleph_0}_{stat}, <\kappa)$ is defined by

$\mathsf{SDLS}^{int}_{+}(\mathcal{L}^{\aleph_0}_{stat}, <\kappa)$: *For any structure* $\mathfrak{A} = \langle A, ...\rangle$ *of countable signature with* $|A| \geq \kappa$, *there are stationarily many* $M \in [A]^{<\kappa}$ such that $\mathfrak{A} \restriction M \prec^{int}_{\mathcal{L}^{\aleph_0}_{stat}} \mathfrak{A}$.

Similarly to the $+$ in "$(*)^{int+}_{<\kappa,\lambda}$", '$+$' in "$\mathsf{SDLS}^{int}_{+}(\mathcal{L}^{\aleph_0}_{stat}, <\kappa)$" refers to the existence of "stationarily many" reflection points M. This additional condition can be dropped if $\kappa = \aleph_2$. This is because the quantifier $Qx\,\varphi$ defined by $stat\, X \exists x\, (x \notin X \wedge \varphi$, $\mathfrak{A} \models^{int} Qx\,\varphi(x, ...))$ still implies that "there are uncountably many $a \in A$ with $\varphi(a, ...)$". Note that, if $\mathfrak{A} \models^{int} \neg stat\, X\, (x \equiv x)$, for a structure $\mathfrak{A} = \langle A, ...\rangle$, we can easily find even club many $X \in [A]^{<\kappa}$ for any regular $\aleph_1 \leq \kappa \leq |A|$ such that $\mathfrak{A} \restriction X \prec^{int}_{\mathcal{L}^{\aleph_0}_{stat}} \mathfrak{A}$.

Proposition 19 (Proposition 3.1 in [20]) *For a regular cardinal* $\kappa > \aleph_1$, *the following are equivalent:*

(a) $(*)^{int+}_{<\kappa,\lambda}$ *holds for all regular* $\lambda \geq \kappa$.

(b) $\mathsf{SDLS}^{int}_{+}(\mathcal{L}^{\aleph_0}_{stat}, <\kappa)$ *holds.* □

Although $\mathsf{SDLS}^{int}_{+}(\mathcal{L}^{\aleph_0}_{stat}, <2^{\aleph_0})$ is compatible with large continuum, as a weakening of $\mathsf{SDLS}^{-}(\mathcal{L}^{\aleph_0}_{stat}, <2^{\aleph_0})$, this principle does not imply the largeness of the continuum. The strong Löwenheim-Skolem theorem for the following variation of stationary logic does.

For sets s and t we denote with $\mathcal{P}_s(t)$ the set $[t]^{<|s|} = \{a \in \mathcal{P}(t) \,:\, |a| < |s|\}$. We say $S \subseteq \mathcal{P}_s(t)$ is stationary if it is stationary in the sense of Jech [31].

The logic $\mathcal{L}^{\mathsf{PKL}}_{stat}$ has a built-in unary predicate symbol $\underset{\sqcup}{K}(\cdot)$.[8] For a structure $\mathfrak{A} = \langle A, \underset{\sqcup}{K}^{\mathfrak{A}}, ...\rangle$, the weak second-order variables X, Y,... run over elements of $\mathcal{P}_{\underset{\sqcup}{K}^{\mathfrak{A}}}(A)$.

[8] PKL stands here for "pi-kappa-lambda" in the sense of "$\mathcal{P}_\kappa(\lambda)$".

We shall call a structure $\mathfrak{A}$ with $\underline{K}$ in its signature as a unary predicate symbol such that $|\underline{K}^{\mathfrak{A}}|$ is a regular uncountable cardinal, a PKL-*structure*.

$\mathcal{L}^{\mathsf{PKL}}_{stat}$ has the unique second-order quantifier "$stat$" and the internal interpretation $\models^{int}$ of formulas in this logic is defined similarly to $\mathcal{L}^{\aleph_0}_{stat}$ with the crucial step in the inductive definition being

(5.4) $\mathfrak{A} \models^{int} stat\, X\, \varphi(a_0, ..., U_0, ..., X) \Leftrightarrow$
$\{U \in \mathcal{P}_{\underline{K}^{\mathfrak{A}}}(A) \cap A : \mathfrak{A} \models^{int} \varphi(a_0, ..., U_0, ..., U)\}$ is stationary in $\mathcal{P}_{\underline{K}^{\mathfrak{A}}}(A)$

for an $\mathcal{L}^{\mathsf{PKL}}_{stat}$-formula $\varphi = \varphi(x_0, ..., X_0, ..., X)$ (for which the relation $\models^{int}$ has been defined), a PKL-structure $\mathfrak{A} = \langle A, \underline{K}^{\mathfrak{A}}, ...\rangle$ of a relevant signature, $a_0, ... \in A$ and $U_0, ... \in \mathcal{P}_{\underline{K}^{\mathfrak{A}}}(A) \cap A$.

For PKL-structures $\mathfrak{A}$, $\mathfrak{B}$ of the same signature with $\mathfrak{B} = \langle B, \underline{K}^{\mathfrak{B}}, ...\rangle$ and $\mathfrak{B} \subseteq \mathfrak{A}$, we define:

(5.5) $\mathfrak{B} \prec^{int}_{\mathcal{L}^{\mathsf{PKL}}_{stat}} \mathfrak{A} \Leftrightarrow$
$\mathfrak{B} \models^{int} \varphi(b_0, ..., U_0, ...)$ *if and only if* $\mathfrak{A} \models^{int} \varphi(b_0, ..., U_0, ...)$
for all $\mathcal{L}^{\mathsf{PKL}}_{stat}$*-formulas* φ *in the signature of the structures with*
$\varphi = \varphi(x_0, ..., X_0, ...)$, $b_0, ... \in B$ *and* $U_0, ... \in \mathcal{P}_{\underline{K}^{\mathfrak{B}}}(B) \cap B$.

Finally, we define the internal strong downward Löwenheim-Skolem theorem for this logic as follows:

Suppose that κ is a regular cardinal $> \aleph_1$.

$\mathsf{SDLS}^{int}_{+}(\mathcal{L}^{\mathsf{PKL}}_{stat}, < \kappa)$: *For any* PKL*-structure* $\mathfrak{A} = \langle A, \underline{K}^{\mathfrak{A}}, ...\rangle$ *of countable signature with* $|A| \geq \kappa$ *and* $|\underline{K}^{\mathfrak{A}}| = \kappa$*,there are stationarily many* $M \in [A]^{<\kappa}$ *such that* $\mathfrak{A} \restriction M$ *is a* PKL*-structure and* $\mathfrak{A} \restriction M \prec^{int}_{\mathcal{L}^{\mathsf{PKL}}_{stat}} \mathfrak{A}$.

The following diagonal reflection characterizes $\mathsf{SDLS}^{int}_{+}(\mathcal{L}^{\mathsf{PKL}}_{stat}, < \kappa)$. For regular cardinals κ, λ with $\kappa \leq \lambda$, let

$(*)^{int+\mathsf{PKL}}_{<\kappa,\lambda}$: *For any countable expansion* $\mathfrak{A}$ *of the structure* $\langle \mathcal{H}(\lambda), \kappa, \in\rangle$ *and any family* $\langle S_a : a \in \mathcal{H}(\lambda)\rangle$ *such that* S_a *is a stationary subset of* $\mathcal{P}_\kappa(\mathcal{H}(\lambda))$ *for all* $a \in \mathcal{H}(\lambda)$*, there are stationarily many* $M \in \mathcal{P}_\kappa(\mathcal{H}(\lambda))$ *such that* $|\kappa \cap M|$ *is regular,* $\mathfrak{A} \restriction M \prec \mathfrak{A}$ *and* $S_a \cap \mathcal{P}_{\kappa\cap M}(M) \cap M$ *is stationary in* $\mathcal{P}_{\kappa\cap M}(M)$.

Proposition 20 (Proposition 4.1 in [20]) *For a regular cardinal* $\kappa > \aleph_1$*, the following are equivalent:*

(a) $(*)^{int+\mathsf{PKL}}_{<\kappa,\lambda}$ *holds for all regular* $\lambda \geq \kappa$.

(b) $\mathsf{SDLS}^{int}_{+}(\mathcal{L}^{\mathsf{PKL}}_{stat}, < \kappa)$ *holds.* □

For a regular cardinal κ and a cardinal $\lambda \geq \kappa$, $\mathcal{S} \subseteq \mathcal{P}_\kappa(\lambda)$ is said to be 2-*stationary* if, for any stationary $\mathcal{T} \subseteq \mathcal{P}_\kappa(\lambda)$, there is an $a \in \mathcal{S}$ such that $|\kappa \cap a|$ is a regular uncountable cardinal and $\mathcal{T} \cap \mathcal{P}_{\kappa \cap a}(a)$ is stationary in $\mathcal{P}_{\kappa \cap a}(a)$. A regular cardinal κ has the 2-*stationarity property* if $\mathcal{P}_\kappa(\lambda)$ is 2-stationary (as a subset of itself) for all $\lambda \geq \kappa$.

Since the property (a) in Proposition 20 is an extension of the 2-stationarity of κ, we obtain:

Lemma 21 *For a regular uncountable* κ, $\mathsf{SDLS}^{int}_{+}(\mathcal{L}^{\mathsf{PKL}}_{stat}, < \kappa)$ *implies that* κ *is 2-stationary.* □

This implies that a regular uncountable κ with $\mathsf{SDLS}^{int}_{+}(\mathcal{L}^{\mathsf{PKL}}_{stat}, < \kappa)$ must be a fairly large cardinal:

Lemma 22 (Lemma 4.3 in [20]) *Suppose that* κ *is a regular uncountable cardinal. If* κ *is 2-stationary then* κ *is a weakly Mahlo cardinal.* □

Actually the proof of Lemma 22 (in [20]) shows that κ is weakly hyper Mahlo, weakly hyper hyper Mahlo, etc.

Corollary 23 $\mathsf{SDLS}^{int}_{+}(\mathcal{L}^{\mathsf{PKL}}_{stat}, < 2^{\aleph_0})$ *implies that* $2^{\aleph_0}$ *is weakly Mahlo, weakly hyper Mahlo, weakly hyper hyper Mahlo, etc.* □

Using the characterization Proposition 20 of $\mathsf{SDLS}^{int}_{+}(\mathcal{L}^{\mathsf{PKL}}_{stat}, < \kappa)$, a proof similar to that of Theorem 12 shows the following:

Theorem 24 *Suppose that* κ *is a generically supercompact cardinal by* $< \mu$*-cc posets for some* $\mu < \kappa$. *Then* $\mathsf{SDLS}^{int}_{+}(\mathcal{L}^{\mathsf{PKL}}_{stat}, < \kappa)$ *holds.* □

Since ccc posets are proper, we obtain the following by Theorem 17, Proposition 19 and Theorem 24:

Corollary 25 *Suppose that* κ *is a generically supercompact cardinal by ccc posets. Then* $\mathsf{SDLS}^{int}_{+}(\mathcal{L}^{\aleph_0}_{stat}, < \kappa)$ *and* $\mathsf{SDLS}^{int}_{+}(\mathcal{L}^{\mathsf{PKL}}_{stat}, < \kappa)$ *hold.* □

By Lemma 18, it follows that

Corollary 26 *If* ZFC + *"there is a supercompact cardinal" is consistent then so is* $\mathsf{ZFC} + \mathsf{SDLS}^{int}_{+}(\mathcal{L}^{\aleph_0}_{stat}, < 2^{\aleph_0})$ *and* $\mathsf{SDLS}^{int}_{+}(\mathcal{L}^{\mathsf{PKL}}_{stat}, < 2^{\aleph_0})$. *Note that the continuum is fairly large in the latter axiom system by Corollary 23.* □

6 Laver-Generic Large Cardinals

The reflection properties we presented so far in connection with the size of the continuum can be summarized in three possible scenarios:

(A) GRP .... This implies CH (Lemma 9).

(B) $\mathsf{SDLS}^{-}(\mathcal{L}^{\aleph_0}_{stat}, < 2^{\aleph_0})$ .... This implies $2^{\aleph_0} = \aleph_2$ (Theorem 6, (2)).

(C) $\mathsf{SDLS}^{int}_{+}(\mathcal{L}^{\mathsf{PKL}}_{stat}, < 2^{\aleph_0})$ This implies that $2^{\aleph_0}$ is fairly large (Corollary 23).

These three possible scenarios can be treated in a uniform way from the point of view of the Laver-generic large cardinals defined below.

We shall call a class $\mathcal{P}$ of posets *iterable* if

(6.1) $\mathcal{P}$ is closed with respect to forcing equivalence. That is, if $\mathbb{P} \in \mathcal{P}$ and $\mathbb{P}'$ is forcing equivalent to $\mathbb{P}$ then $\mathbb{P}' \in \mathcal{P}$;
(6.2) For any $\mathbb{P} \in \mathcal{P}$ and $\mathbb{p} \in \mathbb{P}$, $\mathbb{P} \restriction \mathbb{p} \in \mathcal{P}$;
(6.3) If $\mathbb{P} \in \mathcal{P}$ and $\Vdash_{\mathbb{P}}$ "$\underset{\sim}{\mathbb{Q}} \varepsilon \mathcal{P}$" then $\mathbb{P} * \underset{\sim}{\mathbb{Q}} \in \mathcal{P}$.

Note that most of natural classes of posets like σ-closed posets, ccc posets, proper posets, stationary preserving posets etc. are iterable.

For a cardinal κ and an iterable class $\mathcal{P}$ of posets, we call κ a *Laver-generically supercompact for* $\mathcal{P}$ if, for any $\lambda \geq \kappa$ and any $\mathbb{P} \in \mathcal{P}$, there is a poset $\mathbb{Q} \in \mathcal{P}$ with $\mathbb{P} \lessdot \mathbb{Q}$ such that, for any $(\mathsf{V}, \mathbb{Q})$-generic filter $\mathbb{H}$, there are $M, j \subseteq \mathsf{V}[\mathbb{H}]$ such that

(6.4) M is an inner model of $\mathsf{V}[\mathbb{H}]$,
(6.5) $j : \mathsf{V} \overset{\prec}{\rightarrow} M$,
(6.6) $crit(j) = \kappa$, $j(\kappa) > \lambda$,
(6.7) $\mathbb{P}, \mathbb{H} \in M$ and
(6.8) $j''\lambda \in M$.[9]

κ is *Laver-generically superhuge* (*Laver-generically super almost-huge* resp.) *for* $\mathcal{P}$ if κ satisfies the definition of Laver-generic supercompactness for $\mathcal{P}$ with (6) replaced by

(6.8)' $j''j(\kappa) \in M$ ($j''\mu \in M$ for all $\mu < j(\kappa)$ resp.).

κ is *tightly Laver-generically supercompact* (*tightly Laver-generically superhuge, tightly Laver-generically super almost-huge*, resp.) if the definition of Laver-generically supercompact (Laver-generically superhuge, Laver-generically super almost-huge, resp.) holds with (6) replaced by

(6.6)' $crit(j) = \kappa$, $j(\kappa) = |\mathbb{Q}| > \lambda$.

All consistency proofs of the existence of Laver-generic large cardinals we know actually show the existence of tightly Laver-generic large cardinals (see the proof of Theorem 5.2 in [20]).

The consistency of the existence of a Laver-generic large cardinal can be proved from the assumption of the existence of the corresponding genuine large cardinals except the case of the Laver-generic large cardinals by proper posets. This case will be further discussed in [21].

Theorem 27 ([20]) (1) *Suppose that* ZFC + *"there exists a supercompact cardinal" is consistent. Then* ZFC + *"there exists a Laver-generically supercompact cardinal for* σ*-closed posets" is consistent as well.*

---

[9] This definition of Laver-generic supercompactness for $\mathcal{P}$ is different from the one given in [20]. However, it is easy to show that the present definition is equivalent to the one in [20] for an iterable $\mathcal{P}$. Note that, strictly speaking, this equivalence is used at the end of the proof of Theorem 34 below.

(2) *Suppose that* ZFC + *"there exists a superhuge cardinal" is consistent. Then* ZFC + *"there exists a Laver-generically super almost-huge cardinal for proper posets" is consistent as well.*

(3) *Suppose that* ZFC + *"there exists a supercompact cardinal" is consistent. Then* ZFC + *"there exists a Laver-generically supercompact cardinal κ for c.c.c. posets" with $\kappa = 2^{\aleph_0}$ is consistent as well.*

Sketch of the proof. Let us consider (2). The other assertions are similarly and easier to prove.

Starting from a model of ZFC with a superhuge cardinal κ, we can obtain models of respective assertions by iterating in countable support with proper posets κ times along a Laver function for super almost-hugeness which exists by a result in Corazza [7].

In the resulting model, we obtain Laver-generically super almost-hugeness in terms of proper poset $\mathbb{Q}$ in each respective inner model $M[\mathbb{G}]$ of $\mathsf{V}[\mathbb{G}]$. The closedness of M in V in terms of super almost-hugeness implies that $\mathbb{Q}$ is also proper in $\mathsf{V}[\mathbb{G}]$ (this is the place where we need the super almost-hugeness: for (1) and (3) we do not need this much closedness of M).

This shows that κ is Laver-generically super almost-huge of proper posets.
□ (Theorem 27)

In contrast to simple generic supercompactness, a Laver-generically supercompact cardinal for a natural class $\mathcal{P}$ of posets is determined uniquely if it exists:

Proposition 28 ([20]) (0) *If κ is generically measurable for some poset $\mathbb{P}$, then κ is regular.*

(1) *Suppose that κ is generically measurable by a ω_1 preserving $\mathbb{P}$. Then $\kappa > \omega_1$.*

(2) *Suppose that κ is Laver-generically supercompact for a class $\mathcal{P}$ of posets with* $\mathrm{Col}(\omega_1, \{\omega_2\}) \in \mathcal{P}$. *Then $\kappa \leq \omega_2$. If all elements of $\mathcal{P}$ are ω_1-preserving, then we have $\kappa = \omega_2$.*

(3) *Suppose that $\mathcal{P}$ is a class of posets containing a poset $\mathbb{P}$ such that any* $(\mathsf{V}, \mathbb{P})$-*generic filter $\mathbb{G}$ codes a new real. If κ is a Laver-generically supercompact for $\mathcal{P}$, then $\kappa \leq 2^{\aleph_0}$.*

(4) *Suppose that $\mathcal{P}$ is a class of posets such that elements of $\mathcal{P}$ do not add any reals. If κ is Laver-generically supercompact by $\mathcal{P}$, then we have $2^{\aleph_0} < \kappa$.* □

For a class $\mathcal{P}$ of posets and cardinals μ, κ,

$\mathsf{MA}^{+\mu}(\mathcal{P}, <\kappa)$: *For any $\mathbb{P} \in \mathcal{P}$, any family $\mathcal{D}$ of dense subsets of $\mathbb{P}$ with $|\mathcal{D}| < \kappa$ and any family $\mathcal{S}$ of* $\mathbb{P}$-names such that $|\mathcal{S}| \leq \mu$ *and* $\Vdash_{\mathbb{P}}$ " $\underset{\sim}{S}$ is a stationary subset of ω_1 " *for all $\underset{\sim}{S} \in \mathcal{S}$, there is a $\mathcal{D}$-generic filter $\mathbb{G}$ over $\mathbb{P}$* such that $\underset{\sim}{S}[\mathbb{G}]$ *is a stationary subset of ω_1 for all $\underset{\sim}{S} \in \mathcal{S}$.*

The following strengthening of the Laver-genericity is needed to obtain "++" versions of forcing axioms.

For a cardinal κ and an iterable class $\mathcal{P}$ of posets, we call κ a *strongly Laver-generically supercompact for* $\mathcal{P}$ *if, for any* $\lambda \geq \kappa$ *and any* $\mathbb{P} \in \mathcal{P}$, *there is a* $\mathbb{P}$*-name of a poset* $\underset{\sim}{\mathbb{Q}}$ *with* $\Vdash_{\mathbb{P}}$ "$\underset{\sim}{\mathbb{Q}} \varepsilon \mathcal{P}$" such that, *for any* $(\mathsf{V}, \mathbb{P} * \underset{\sim}{\mathbb{Q}})$*-generic filter* $\mathbb{H}$, *there are* $M, j \subseteq \mathsf{V}[\mathbb{H}]$ *with*

(6.9) M is an inner model of $\mathsf{V}[\mathbb{H}]$,
(6.10) $j : \mathsf{V} \overset{\preccurlyeq}{\to} M$,
(6.11) $crit(j) = \kappa$, $j(\kappa) > \lambda$,
(6.12) $\mathbb{P}, \mathbb{H} \in M$ and
(6.13) $({}^{\lambda}M)^{\mathsf{V}[\mathbb{H}]} \subseteq M$.

The notions of *strongly Laver-generically superhuge* and *strongly Laver-generically super almost-huge* are defined correspondingly. For example, κ is strongly Laver-generically super almost-huge if the definition of strongly Laver-generically supercompact cardinal holds with (6) replaced with

(6.14) $({}^{\mu}M)^{\mathsf{V}[\mathbb{H}]} \subseteq M$ holds for all $\mu < j(\kappa)$.

Note that, if $\mathcal{P}$ is the class of ccc posets, then the "strongly" version of the Laver-generically large cardinal is equivalent to the original version of the corresponding Laver-generic largeness. Note also that the construction in the proof of Theorem 27 actually provides models of strongly Laver-genericity.

Theorem 29 (Theorem 5.7 in [20]) *For a class* $\mathcal{P}$ *of proper posets, if* $\kappa > \aleph_1$ *is a strongly Laver-generically supercompact for* $\mathcal{P}$, *then* $\mathsf{MA}^{+\mu}(\mathcal{P}, < \kappa)$ *holds for all* $\mu < \kappa$. □

Lemma 30 *Suppose that* κ *is generically supercompact by a class* $\mathcal{P}$ *of posets such that all* $\mathbb{P} \in \mathcal{P}$ *has the* $< \mu$*-cc for some* $\mu < \kappa$. *Then for any regular* $\lambda \geq \kappa$, $\mathcal{P}_\kappa(\lambda)$ *has a* μ*-saturated normal fine filter* F *over* $\mathcal{P}_\kappa(\lambda)$.

Proof For a regular $\lambda \geq \kappa$, let $\mathbb{P}$ be such that there are $(\mathsf{V}, \mathbb{P})$-generic $\mathbb{G}$, and M, $j \subseteq \mathsf{V}[\mathbb{G}]$ such that $j : \mathsf{V} \overset{\preccurlyeq}{\to} M$, $crit(j) = \kappa$, $j(\kappa) > \lambda$ and $j''\lambda \in M$.

In V, let $F = \{A \subseteq \mathcal{P}_\kappa(\lambda) : \Vdash_{\mathbb{P}} \text{“} j''\lambda \,\varepsilon\, j(\check{A}) \text{”}\}$. By the $< \mu$-cc of $\mathbb{P}$, this F is as desired. □ (Lemma 30)

Combining Proposition 28, Theorem 29 and Lemma 30, we obtain:

Theorem 31 *Suppose that* κ *is strongly Laver-generically supercompact cardinal for an iterable class* $\mathcal{P}$ *of posets.*

(A′) *If all elements of* $\mathcal{P}$ *are* ω_1*-preserving and do not add any reals, and* $\mathrm{Col}(\omega_1, \{\omega_2\}) \in \mathcal{P}$, *then* $\kappa = \aleph_2$ *and* CH *holds.*

(B′) *If all elements of* $\mathcal{P}$ *are* ω_1*-preserving and* $\mathcal{P}$ *contains all proper posets then* $\mathsf{PFA}^{+\omega_1}$ *holds and* $\kappa = 2^{\aleph_0} = \aleph_2$.

(C′) *If all elements of* $\mathcal{P}$ *are* $< \mu$*-cc for some fixed* $\mu < \kappa$ *and* $\mathbb{P}$ *contains a poset which adds a new real then* κ *is fairly large (in the sense of Lemma 30) and* $\kappa \leq 2^{\aleph_0}$.

Proof (A$'$): By Proposition 28, (2) and (4) .

(B$'$) : $\kappa = \aleph_2$ by Proposition 28, (2) . $\mathsf{PFA}^{+\omega_1}$ holds by Theorem 29. PFA implies $2^{\aleph_0} = \aleph_2$.

(C$'$) : $\kappa \leq 2^{\aleph_0}$ by Proposition 28, (3) . κ is large by Lemma 30. □ (Theorem 31)

The three cases in Theorem 31 can be further modified to fit to the reflection principles discussed in earlier sections.

Theorem 32 (A$''$) *Suppose that κ is Laver-generically supercompact for σ-closed posets. Then $2^{\aleph_0} = \aleph_1$, $\kappa = \aleph_2$, $\mathsf{MA}^{+\omega_1}$(σ-closed) and* GRP *holds. It follows that* RC *and* $\mathsf{SDLS}(\mathcal{L}^{\aleph_0}_{stat}, <\aleph_2)$ *hold.*

(B$''$) *Suppose that elements of $\mathcal{P}$ are ω_1-preserving and $\mathcal{P}$ contains all proper posets. If κ is strongly Laver-generically supercompact for $\mathcal{P}$, then $2^{\aleph_0} = \kappa = \aleph_2$,* $\mathsf{PFA}^{+\omega_1}$ *and hence also* $\mathsf{SDLS}^-(\mathcal{L}^{\aleph_0}_{stat}, < 2^{\aleph_0})$ *holds.*

(C$''$) *Suppose that κ is Laver-generically supercompact for ccc posets. Then $\kappa \leq 2^{\aleph_0}$ and $\mathcal{P}_\kappa(\lambda)$ for any regular $\lambda \geq \kappa$ carries an $\aleph_1$-saturated normal ideal. In particular, κ is κ-weakly Mahlo.* $\mathsf{MA}^{+\mu}(ccc, <\kappa)$ *for all $\mu < \kappa$,* $\mathsf{SDLS}^{int}(\mathcal{L}^{\aleph_0}_{stat}, <\kappa)$ *and* $\mathsf{SDLS}^{int}_+(\mathcal{L}^{\mathsf{PKL}}_{stat}, <\kappa)$ *also hold.*

Proof(A$''$): $2^{\aleph_0} = \aleph_1$ and $\kappa = \aleph_2$ follows from Theorem 31, (A). $\mathsf{MA}^{+\omega_1}$(σ-closed) holds by Theorem 29. GRP holds by Corollary 13. RC and $\mathsf{SDLS}(\mathcal{L}^{\aleph_0}_{stat}, <\aleph_2)$ follow from GRP by Theorem 11 and Corollary 10, (1) .

(B$''$) : This is just as (B) in Theorem 31. $\mathsf{SDLS}^-(\mathcal{L}^{\aleph_0}_{stat}, < 2^{\aleph_0})$ holds by Corollary 5, (1) .

(C$''$) : The first half of the assertion follows from Theorem 31, (C). $\mathsf{MA}^{+\mu}(ccc, <\kappa)$ for all $\mu < \kappa$ holds by Theorem 29. $\mathsf{SDLS}^{int}(\mathcal{L}^{\aleph_0}_{stat}, <\kappa)$ and $\mathsf{SDLS}^{int}_+(\mathcal{L}^{\mathsf{PKL}}_{stat}, <\kappa)$ hold by Corollary 25. □ (Theorem 32)

At the moment we do not know whether the assumption in (C) in Theorem 32 implies $\kappa = 2^{\aleph_0}$. The following partial answer is obtained in [20]:

Theorem 33 (Theorem 5.3 in [20]) *If κ is tightly Laver-generically superhuge for ccc posets, then $\kappa = 2^{\aleph_0}$.* □

The following Theorem 34 is a Laver-generic version of Theorem 16. The proof of the theorem is a typical application of the master condition argument (see e.g. Cummings [9]).

Theorem 34 *Suppose that κ is strongly Laver-generically almost super-huge for an iterable $\mathcal{P}$ which provably contains all posets of the form* $\mathrm{Col}(\mu, \lambda)$ *for all regular μ, λ with $\kappa \leq \mu < \lambda$. For a regular λ_0, let $\mathbb{P}_0 = \mathrm{Col}(\kappa, \lambda_0)$ and let $\mathbb{G}_0$ be a* $(\mathsf{V}, \mathbb{P}_0)$*-generic filter.*

Then we have

(6.15) $\mathsf{V}[\mathbb{G}_0] \models$ "*κ is strongly Laver-generically super almost-huge for $\mathcal{P}$*".

Proof Suppose that $\mathsf{V}[\mathbb{G}_0] \models$"$\mathbb{P} \in \mathcal{P}$" and let $\underset{\sim}{\mathbb{P}}$ be a $\mathbb{P}_0$-name of $\mathbb{P}$ such that $\Vdash_{\mathbb{P}_0}$"$\underset{\sim}{\mathbb{P}} \mathrel{\varepsilon} \mathcal{P}$". Let $\lambda \geq \kappa$ be regular. Without loss of generality, we may assume that $\lambda \geq |\mathbb{P}| \geq \lambda_0$. $\mathbb{P}_0 * \underset{\sim}{\mathbb{P}} \in \mathcal{P}$ by iterability of $\mathcal{P}$. Since κ is strongly Laver-generically super almost-huge for $\mathcal{P}$, there are $\mathbb{P}_0 * \underset{\sim}{\mathbb{P}}$-name $\underset{\sim}{\mathbb{Q}}$ with $\Vdash_{\mathbb{P}_0 * \underset{\sim}{\mathbb{P}}}$"$\underset{\sim}{\mathbb{Q}} \mathrel{\varepsilon} \mathcal{P}$" and $(\mathsf{V}, \mathbb{P}_0 * \underset{\sim}{\mathbb{P}} * \underset{\sim}{\mathbb{Q}})$-generic filter $\mathbb{H}$ such that $\mathbb{G}_0 \subseteq \mathbb{H}$ and such that there are M, $j \subseteq \mathsf{V}[\mathbb{H}]$ with (6.9), (6.10), (6.11), (6.14) and

(6.16) $\mathbb{P}_0 * \underset{\sim}{\mathbb{P}}, \mathbb{H} \in M$.

We have $j''\mathbb{G}_0 \in M$. Let $\mathbb{P}_1 = j(\mathbb{P}_0)$ by (6.14).

By elementarity, we have $M \models$"$\mathbb{P}_1 = \mathrm{Col}(j(\kappa), j(\lambda_0))$". Note that we also have $\mathsf{V}[\mathbb{H}] \models$"$\mathbb{P}_1 = \mathrm{Col}(j(\kappa), j(\lambda_0))$" by (6.14). Since $M \models$"$j''\mathbb{G}_0$ has the fip" by elementarity and $M \models$"$|j''\mathbb{G}_0| \leq |\mathbb{P}| \leq \lambda < j(\kappa)$", there is $\mathbb{q}^* \in \mathbb{P}_1$ in M such that $\mathbb{q}^* \leq_{\mathbb{P}_1} j(\mathbb{p})$ for all $\mathbb{p} \in \mathbb{G}_0$. Let $\mathbb{G}_1$ be $(\mathsf{V}[\mathbb{H}], \mathbb{P}_1)$-generic filter with $\mathbb{q}^* \in \mathbb{G}_1$. In $\mathsf{V}[\mathbb{H} * \mathbb{G}_1]$, let

(6.17) $\tilde{j} : \mathsf{V}[\mathbb{G}_0] \overset{\preccurlyeq}{\to} M[\mathbb{G}_1]; \underset{\sim}{a}[\mathbb{G}_0] \mapsto j(\underset{\sim}{a})[\mathbb{G}_1]$.

Since $\mathbb{P} \in \mathcal{P}^{\mathsf{V}[\mathbb{G}_0]}$ and λ were arbitrary, the elementary embedding $\tilde{j}$ above witnesses the Laver-generic super almost-hugeness of κ for $\mathcal{P}$ in $\mathsf{V}[\mathbb{G}_0]$. □ (Theorem 34)

Corollary 35 *Suppose that $\mathcal{P}$ is an iterable class of posets which provably contains all posets of the form* $\mathrm{Col}(\mu, \lambda)$ *for all regular uncountable μ, λ with $\kappa \leq \mu < \lambda$. If the theory* ZFC + *"there is a strongly Laver-generically super almost-huge κ for $\mathcal{P}$" + "there is a supercompact $\kappa_1 > \kappa$" is consistent, then so is the theory* ZFC + *"there is a strongly Laver-generically super almost-huge κ for $\mathcal{P}$" + "κ^+ is generically supercompact by $< \kappa$-closed posets". In particular,* $\mathsf{GRP}^{<\kappa}(<\kappa^+)$ *follows from this theory.*

Proof Suppose that κ is strongly Laver-generically supercompact for $\mathcal{P}$ and κ_1 is a supercompact cardinal. Let $\mathbb{P}_0 = \mathrm{Col}(\kappa, \kappa_1)$ and let $\mathbb{G}_0$ be a $(\mathsf{V}, \mathbb{P}_0)$-generic filter. By Theorem 34, $\mathsf{V}[\mathbb{G}_0] \models$ "κ is strongly Laver generically supercompact for $\mathcal{P}$". $\mathsf{V}[\mathbb{G}_0] \models \kappa_1 = \kappa^+$ and $\mathsf{V}[\mathbb{G}_0] \models$"$\kappa^+$ is generically supercompact by $< \kappa$-closed posets" by Lemma 14. By Theorem 12, $\mathsf{GRP}^{<\kappa}(<\kappa^+)$ follows. □ (Corollary 35)

Theorem 34 and Corollary 35 have many variants with similar proofs. For example:

Theorem 36 *Suppose that κ is strongly Laver-generically supercompact for an iterable $\mathcal{P}$ which provably contains all σ-closed posets. For a regular λ_0, let $\mathbb{P}_0 = \mathrm{Col}(\kappa, \lambda_0)$ and let $\mathbb{G}_0$ be a $(\mathsf{V}, \mathbb{P}_0)$-generic filter.*

Then we have

(6.18) $\mathsf{V}[\mathbb{G}_0] \models$ "*κ is strongly Laver-generically supercompact for $\mathcal{P}$*". □

Corollary 37 *Suppose that $\mathcal{P}$ is an iterable class of posets which provably contains all σ-closed posets. If the theory* ZFC + *"there is a strongly Laver-generically supercompact κ for $\mathcal{P}$" + "there is a supercompact $\kappa_1 > \kappa$" is consistent, then so is the* *theory* ZFC + *"there is a strongly Laver-generically supercompact* κ *for* $\mathcal{P}$*"* + *"*κ^+ *is generically supercompact by* $<\kappa$*-closed posets". In particular,* $\mathsf{GRP}^{<\kappa}(<\kappa^+)$ *follows from this theory.* □

Note that, by Proposition 28, (2), we have $\kappa \leq \aleph_2$ in Theorem 36 and Corollary 37 above.

In [21], we show that the combination of the principles $\mathsf{SDLS}^{int}(\mathcal{L}^{\aleph_0}_{stat}, < 2^{\aleph_0})$ and $\mathsf{GRP}^{<2^{\aleph_0}}(\leq 2^{\aleph_0})$ is also consistent under large continuum assuming the consistency of two supercompact cardinals.

7 Some Open Problems and Takeuti's Account on Gödel's Contribution to the Continuum Problem

Let us mention some open problems. Some of them will be addressed in [21].

The following problem is already mentioned in the previous section:

Problem 1 If κ is Laver-generically supercompact for ccc posets, does this imply $\kappa = 2^{\aleph_0}$?

Forcing axioms have some characterizations which may be interpreted as suggestions of the correctness of the axioms. See e.g. Bagaria [1], Fuchino [13].

Problem 2 Is there any nice characterizations of "$+\mu$" versions of forcing axioms?

Any meaningful answer to this problem would enhance the relevance of the trichotomy in Theorem 32.

The trichotomy (A), (B), (C) of reflection principles mentioned at the beginning of Sect. 6 has an alternative trichotomy (A), (B*), (C*), where

(B*) $\mathsf{SDLS}^{-}(\mathcal{L}^{\aleph_0}_{stat}, < 2^{\aleph_0})$ *and* RC *hold.*

(C*) $\mathsf{SDLS}^{int}_{+}(\mathcal{L}^{\mathsf{PKL}}_{stat}, < 2^{\aleph_0})$ and $\mathsf{RC}(< 2^{\aleph_0})$ hold.

Note that RC in (B*) is equivalent to $\mathsf{RC}(< 2^{\aleph_0})$ since $\mathsf{SDLS}^{-}(\mathcal{L}^{\aleph_0}_{stat}, < 2^{\aleph_0})$ implies $2^{\aleph_0} = \aleph_2$.

(B*) and (C*) are not compatible with (B″) and (C″) respectively, since $\mathsf{MA}(\kappa)$—i.e. MA (for ccc posets) for $\leq \kappa$ many dense sets—implies the negation of $\mathsf{RC}(\leq \kappa^+)$ (see Sect. 5 in Fuchino [17]).

(B*) and (C*) can be realized by starting from a supercompact cardinal and then forcing with a Mitchell type mixed support iteration (for (B*), this is mentioned in [43], for (C*), see [21]). Actually, (C*) can be realized much easier by simply adding supercompact many Cohen reals, but we do need mixed support iteration to obtain a model of

(C†) $\mathsf{SDLS}^{int}_{+}(\mathcal{L}^{\mathsf{PKL}}_{stat}, < 2^{\aleph_0})$, $\mathsf{RC}(< 2^{\aleph_0})$ *and* $\mathsf{GRP}^{<2^{\aleph_0}}(\leq 2^{\aleph_0})$ *hold.*

Models obtained in this way seem to be much more artificial than the models for (B″) and (C″) as in the proof of Theorem 27. Even so, we have the feeling that we do not understand yet much about the models constructed by the mixed support iteration and its variations. Thus

Problem 3 What is possible with variations of mixed support iteration?

The result of König in [34] cited here as Corollary 13 can be also regarded as a characterization of $\aleph_2$ being generically supercompact by σ-closed posets. Thus we may further ask:

Problem 4 Is there any characterizations of Laver-generically large cardinals in terms of some strong reflection principles?

The following essential problem might be much harder than the other problems:

Problem 5 Are Laver-generically large cardinals equiconsistent with corresponding genuine large cardinals?

The consistency of the reflection of non-metrizability of a first countable topological spaces down to $< \aleph_2$ is an open problem known as Hamburger's problem. The consistency of the reflection of the property of partial orderings that they are not represented as countable union of chains down to $< \aleph_2$ is also an open problem known as Galvin's conjecture. In case of Hamburger's problem, it is known that the reflection of non-metrizability of a first countable topological spaces down to $< 2^{\aleph_0}$ is realized in the model obtained by adding supercompact many Cohen reals (Dow, Tall and Weiss [12]). The reflection cardinal of uncountable chromatic number of graphs is known to be $\geq \beth_\omega$ (Erdős and Hajnal, see [23] for a detailed proof in ZFC). There are many open problems in connection with reflection of these and some other mathematical properties. Some of them seem to be extremely difficult. Let us mention here merely one problem which may have some connection to Problem 3:

Problem 6 Is the reflection of non-metrizability of first countable topological spaces down to $< 2^{\aleph_0}$ consistent with $\mathsf{RC}(< 2^{\aleph_0})$?

The first author of this article belongs to the generation of Japanese logicians who were strongly inspired by the writings of late Professor Gaishi Takeuti who published many expository articles and books in Japanese from 1960s to the end of 1990s. Although set theory was not his main field, Professor Takeuti wrote many expositions and told his views on the subject. Gödel's program was one of the issues he discussed repeatedly there.

Gödel, who usually refused to publish papers which he thought was not yet perfect, tried once to publish a quite unfinished note in 1970 with the title "Some considerations leading to the probable conclusion that the true power of the continuum is $\aleph_2$." It is said that, being seriously ill, he did so under the fear that he would soon die. After this crisis, Gödel withdrew the note finding out some inaccuracy in it but he continued the study on the problem. Oskar Morgenstern noticed in his diary on 20. September, 1975 that Gödel told him in a telephone call that he was finally convinced that (the newest version of) his axiom implies that the continuum is "different from $\aleph_1$" and that he will write it up ([38]). Dawson [10] contains some accounts about this development.

Takeuti wrote about the details of what he understood from the 1970 note on pp. 99–124 in his book [40] published in 1972. According to Takeuti [42], he was then invited by Gödel in 1975 or 1976 shortly before Gödel's retirement from IAS and discussed with him about the results on the Continuum Problem. Takeuti [41] in 1978 must be closely related to this discussion.

Twenty years later, in May 1998, the first author of the present article obtained a letter from Professor Takeuti with a copy of his handwritten manuscript in Japanese, which contained a further development of the material in [41] among other things. Unfortunately, the first author could not give any reasonable comments to the manuscript at that time. The part of the manuscript on "Gödel's Continuum Hypothesis" was then published in the new edition of [42] as an appendix in September 1998.

Modern treatment of Gödel's axioms is to be found in Brendle, Larson and Todorčević [3]. Takeuti [41] is cited in [3] but neither [40] nor [42] is mentioned there.

Though the technical details of the present article are rather orthogonal to the Gödel–Takeuti line of the support of $2^{\aleph_0} = \aleph_2$, the first author considers the results presented in this article as his belated reply to the letter in 1998 and would like to dedicate this article to the memory of Professor Gaishi Takeuti.

References

1. Bagaria, J.: A characterization of Martin's axiom in terms of absoluteness. J. Symb. Logic **62**(2), 366–372 (1997)
2. Bagaria, J.: Natural axioms of set theory and the continuum problem. In: Proceedings of the 12th International Congress of Logic, Methodology, and Philosophy of Science, pp. 43–64. King's College, London (2005)
3. Brendle, J., Larson, P., Todorčević, S.: Rectangular axioms, perfect set properties and decomposition, Bulletin (Académie serbe des sciences et des arts. Classe des sciences mathématiques et naturelles. Sciences mathématiques) of Serbian Academy of Sciences and Arts, vol. 33, pp. 91–130 (2008)
4. Cohen, P.: The independence of the continuum hypothesis I. Proc. Natl. Acad Sci. U.S.A. **50**, 1143–1148 (1963)
5. Cohen, P.: The independence of the continuum hypothesis I. Proc. Natl. Acad Sci. U.S.A. **51**, 105–110 (1964)
6. Cohen, P.: Set Theory and the Continuum Hypothesis. Benjamin, New York (1966)
7. Corazza, Paul: Laver sequences for extendible and super-almost-huge cardinals. J. Symb. Logic **64**(3), 963–983 (1999)
8. Cox, Sean: The diagonal reflection principle. Proc. Am. Math. Soc. **140**(8), 2893–2902 (2012)
9. Cummings, J.: Large cardinals, forcing and reflection, 京都大学数理解析研究所講究録 (RIMS Kôkyûroku) **1895**, 26–36 (2014)
10. Dawson, J.W.: Logical Dilemmas, the Life and Work of Kurt Gödel. AK Peters (1997)
11. Dow, Alan: An introduction to applications of elementary submodels to topology. Topol. Proce. **13**(1), 17–72 (1988)
12. Dow, A., Tall, F.D., Weiss, W.A.R.: New proofs of the consistency of the normal Moore space conjecture I. II, Topol. Appl. **37**(33–51), 115–129 (1990)

13. Fuchino, S.: On potential embedding and versions of Martin's axiom. Notre Dame J. Logic **33**(4), 481–492 (1992)
14. Fuchino, S.: Left-separated topological spaces under Fodor-type Reflection Principle, 京都大学数理解析研究所講究録 (RIMS Kôkyûroku), No.1619, 32–42 (2008). https://fuchino.ddo.jp/papers/RIMS08-fleissner.pdf
15. Fuchino, S.: Fodor-type Reflection Principle and Balogh's reflection theorems, 京都大学数理解析研究所講究録 (RIMS Kôkyûroku), No. 1686, 41–58 (2010). Extended version of the paper: https://fuchino.ddo.jp/papers/balogh-x.pdf
16. Fuchino, Sakaé: The set-theoretic multiverse as a mathematical plenitudinous Platonism viewpoint. Ann. Jpn Assoc. Philos. Sci. **20**, 49–54 (2012)
17. Fuchino, S.: On reflection numbers under large continuum, 京都大学数理解析研究所講究録 (RIMS Kôkyûroku), No. 1988, 1–16 (2016). Extended version of the paper: https://fuchino.ddo.jp/papers/RIMS15-large-continuum-x.pdf
18. Fuchino, Sakaé: Juhász, Istvan, Soukup, Lajos, Szentmiklóssy, Zoltán, and Usuba, Toshimichi: Fodor-type reflection principle and reflection of metrizability and meta-Lindelöfness. Topol. Appl. **157**(8), 1415–1429 (2010)
19. Fuchino, S.: Ottenbereit Maschio Rodriques, André, and Sakai, Hiroshi: Strong downward Löwenheim-Skolem theorems for stationary logics, I. Arch. Math. Logic **60**, 1–2 (2021), 17–47. Extended version of the paper: https://fuchino.ddo.jp/papers/SDLS-x.pdf
20. Fuchino, S., et al.: Strong downward Löwenheim-Skolem theorems for stationary logics, II — reflection down to the continuum. Arch. Math. Log. **60**(3-4), 495–523 (2021). Extended version of the paper: https://fuchino.ddo.jp/papers/SDLS-II-x.pdf
21. Fuchino, S., et al.: Strong downward Löwenheim-Skolem theorems for stationary logics, III, to appear in the Proceedings of the Asian Logic Conference 2019
22. Fuchino, Sakaé, Rinot, Assaf: Openly generated Boolean algebras and the Fodor-type Reflection Principle. Fund. Math. **212**, 261–283 (2011)
23. Fuchino, S., Sakai, H.: On reflection and non-reflection of countable list-chromatic number of graphs, 京都大学数理解析研究所講究録 (RIMS Kôkyûroku), No. 1790, 31–44 (2012). Extended version of the paper: https://fuchino.ddo.jp/papers/RIMS11-list-chromatic-x.pdf
24. Fuchino, S., Sakai, H., Soukup, L., Usuba, T.: More about the Fodor-type Reflection Principle, submitted. Extended version of the paper: https://fuchino.ddo.jp/papers/moreFRP-x.pdf
25. Fuchino, S., Sakai, T.P., Victor, Usuba, T.: Rado's Conjecture and the Fodor-type Reflection Principle, in preparation. A note on the implication of FRP from RC: https://fuchino.ddo.jp/notes/RCimpliesFRP2.pdf
26. Gödel, K.: Über formal unentscheidbare Sätze der Principia Mathematica und verwandter Systeme, I. Monatshefte für Mathematik und Physik, **38**(1), 173–198 (1931)
27. Gödel, K.: The consistency of the axiom of choice and of the generalized continuum-hypothesis. Proc. Natl. Acad. Sci. U.S.A. **24**, 556–557 (1938)
28. Gödel, K.: Consistency-proof for the generalized continuum-hypothesis. Proc. Natl. Acad. Sci. U.S.A. **25**, 220–224 (1939)
29. Gödel, K.: The Consistency of the Axiom of Choice and of the Generalized Continuum-Hypothesis with the Axioms of Set Theory. Princeton University Press, Princeton (1940)
30. Gödel, K.: What is Cantor's continuum problem? Am. Math. Mon. **54**(9), 515–525 (1947). Errata: ibid., 55 (1948), 151. Revised and expanded version. In: Benacerraf, P., Putnam, H., (eds.) Philosophy of Mathematics, pp. 258–273. Selected Readings. Englewood Cliffs, Prentice Hall (1964)
31. Jech, T.: Set Theory, The Third Millennium Edition. Springer, Berlin (2001/2006)
32. Kanamori, A.: The Higher Infinite. Springer Monographs in Mathematics, 2nd edn. Springer, Berlin (1994/2003/2009)
33. Kennedy, J.: Can the continuum hypothesis be solved?. Inst. Lett., IAS, (Fall, 2011), **1**, 10–11,13 (2011)
34. König, B.: Generic compactness reformulated. Arch. Math. Logic **43**, 311–326 (2004)
35. König, Bernhard, Yoshinobu, Yasuo: Fragments of Martin's maximum in generic extensions. Math. Logic Quarterly **50**(3), 297–302 (2004)

36. Larson, Paul: Separating stationary reflection principles. J. Symb. Logic **65**(1), 247–258 (2000)
37. Magidor, M.: Large cardinals and strong logics, slides of tutorial lectures at the CRM research programme "Large cardinals and strong logics" (2016). http://www.crm.cat/en/Activities/Curs_2016-2017/Documents/Tutorial%20lecture%202.pdf
38. Morgenstern, O.: Tagebücher (diaries). http://gams.uni-graz.at/context:ome
39. Steel, J.: Gödel's program, Chapter 8. In: Kennedy, J. (ed.), Interpreting Gödel, pp. 153–179. Cambridge University Press, Critical Essays (2014)
40. 竹内，外史 (Takeuti, Geishi): 数学基礎論の世界, —ロジックの雑記帳から, (The world of foundations of mathematics — from notebooks of logic), 日本評論社. Nippon Hyoron Sha Co. Ltd (1972)
41. Takeuti, G.: Gödel numbers of product spaces. In: Higher Set Theory (Proceedings of the Conference of Mathematisches Forschungsinstitut, Oberwolfach, 1977). Lecture Notes in Mathematics, vol. 669, pp. 461–471. Springer, Berlin (1978)
42. Takeuti, G.:【新版】 ゲーデル，日本評論社 (Nippon Hyoron Sha Co. Ltd), (1986/1998). English Translation: Memoirs of a Proof Theorist: Gödel and Other Logicians, (translation by: Yasugi, Mariko and Passell, Nicholas). World Scientific, Singapore (2003)
43. Todorčević, S.: Conjectures of Rado and Chang and Cardinal arithmetic. In: Sauer, N.W., Woodrow, R.E., Sands, B. (eds.) Finite and Infinite Combinatorics in Sets and Logic, NATO ASI Series (Series C: Mathematical and Physical Sciences), vol. 411, pp. 385–398. Springer, Berlin (1993)

On Supercompactness of ω_1

Daisuke Ikegami and Nam Trang

Abstract This paper studies structural consequences of supercompactness of ω_1 under ZF. We show that the Axiom of Dependent Choice (DC) follows from "ω_1 is supercompact". "ω_1 is supercompact" also implies that AD^+, a strengthening of the Axiom of Determinacy (AD), is equivalent to $\mathsf{AD}_{\mathbb{R}}$. It is shown that "ω_1 is supercompact" does not imply AD. The most one can hope for is Suslin determinacy. We show that this follows from "ω_1 is supercompact" and Hod Pair Capturing (HPC), an inner-model theoretic hypothesis that imposes certain smallness conditions on the universe of sets. "ω_1 is supercompact" on its own implies that every Suslin set is the projection of a determined (in fact, homogenously Suslin) set. "ω_1 is supercompact" also implies all sets in the Chang model have all the usual regularity properties, like Lebesgue measurability and the Baire property.

Keywords Large cardinal properties · Supercompactness · ω_1 · Axiom of determinacy

Mathematics Subject Classification 03E55 · 03E60 · 03E45

D. Ikegami (✉)
SIT Research Laboratories, Shibaura Institute of Technology, 3-7-5 ,
Toyosu Koto-ward, Tokyo 135-8548, Japan
e-mail: ikegami@shibaura-it.ac.jp
URL: http://www.sic.shibaura-it.ac.jp/ikegami/

N. Trang
Department of Mathematics, University of North Texas, 1155 Union Circle #311430,
Denton, TX 76203-5017, USA
e-mail: Nam.Trang@unt.edu
URL: http://www.math.unt.edu/ntrang/

T. Arai et al. (eds.), *Advances in Mathematical Logic*, Springer Proceedings
in Mathematics & Statistics 369, https://doi.org/10.1007/978-981-16-4173-2_2

1 Introduction

Under ZFC, successor cardinals (like ω_1) are "small". If $\alpha = \beta^+$ is a successor cardinal, then there is an injection from α into $\mathcal{P}(\beta)$.[1] Without the Axiom of Choice, it is possible for successor cardinals like ω_1 to exhibit large cardinal properties. For instance, it has been known since the 1960s that ω_1 can be measurable under ZF; this in particular implies that ω_1 is regular and there is no injection of ω_1 into $\mathcal{P}(\omega)$. We believe this result is independently due to Jech [4] and Takeuti [18]. Furthermore, Takeuti, in the same paper [18], is able to show that "ZF + ω_1 is supercompact" is consistent relative to "ZFC+ there is a supercompact cardinal". Takeuti's model $\mathcal{T}$ in which ω_1 is supercompact is the same as Solovay's model while Takeuti used the method of Boolean valued models to describe his model. Suppose ZFC holds and there is a supercompact cardinal. Let κ be a supercompact cardinal. Let $g \subset Coll(\omega, < \kappa)$ be V-generic for the collapse forcing. Let $\mathbb{R}^* = \mathbb{R}^{V[g]}$. Takeuti's model $\mathcal{T}$ is (in modern terms) the symmetric model $V(\mathbb{R}^*)$.

Another major development started in the 1960s in set theory concerns the theory of infinite games with perfect information. The Axiom of Determinacy (AD) asserts that in an infinite game where players take turns to play integers, one of the players has a winning strategy (see the next section for more detailed discussions on AD and its variations). It is well-known that AD contradicts the Axiom of Choice. Solovay has shown that AD implies ω_1 is measurable and $\mathsf{AD}_{\mathbb{R}}$ implies that there is a supercompact (countably complete, normal, fine) measure on $\mathcal{P}_{\omega_1}\mathbb{R}$. Structural consequences of AD have been extensively investigated, most notably by the Cabal seminar members. Through work of Harrington, Kechris, Neeman, Woodin amongst others, we know that ω_1 is α-supercompact for every ordinal $\alpha < \Theta$ under AD^+, a strengthening of AD.[2] By [21], AD and $\mathsf{AD}_{\mathbb{R}}$ cannot imply ω_1 is supercompact. Woodin (see below) shows that AD is consistent with "ω_1 is supercompact."

It can be shown that the theory "ZF + ω_1 is measurable" is equiconsistent with "ZFC+ there is a measurable cardinal". The question of whether "ZF + ω_1 is supercompact" is equiconsistent with "ZFC+ there is a supercompact cardinal" is much more subtle. Woodin, in an unpublished work in the 1990s, is able to show that the former is much weaker than the latter. Woodin's model is a variation of the Chang model.[3] For each λ, let $\mathcal{F}_\lambda$ be the club filter on $\mathcal{P}_{\omega_1}\lambda^\omega$. Woodin's model is defined as

$$\mathcal{C}^+ = \mathrm{L}(\textstyle\bigcup_{\lambda \in \mathrm{Ord}} \lambda^\omega)[(\mathcal{F}_\lambda \mid \lambda \in Ord)].$$

---

[1] This is equivalent to "there is a surjection from $\mathcal{P}(\beta)$ onto α" under ZFC. Without the Axiom of Choice, the equivalence can fail.

[2] Harrington and Kechris [2] proved that under AD, ω_1 is α-supercompact for all α below a Suslin cardinal. Neeman [11] improved the result to that in $\mathrm{L}(\mathbb{R})$, ω_1 is α-supercompact for all $\alpha < \Theta$. Woodin extended Neeman's result by replacing the assumption $\mathrm{V} = \mathrm{L}(\mathbb{R})$ with AD^+. Unfortunately, Woodin's result is unpublished.

[3] The Chang model is defined as $\mathrm{L}(\bigcup_{\lambda \in Ord} \lambda^\omega)$, the least inner model N of ZF such that for all λ, $\lambda^\omega \in N$.

The model $\mathcal{C}^+$ is the least inner model M of ZF such that for all λ, $\lambda^\omega \in M$ and $M \cap \mathcal{F}_\lambda \in M$.[4]

Woodin shows that if there is a proper class of Woodin cardinals which are limits of Woodin cardinals, then $\mathcal{C}^+$ satisfies AD and ω_1 is supercompact. We note that in Takeuti's model, AD fails.[5] This is because the model $V(\mathbb{R}^*)$ satisfies that $\Theta = \omega_2$ (that is κ^+ in V) while AD implies $\Theta > \omega_2$.[6]

The theory "ω_1 is supercompact" and variations of Woodin's model $\mathcal{C}^+$ are intimately related to determinacy theory as well as modern developments in descriptive inner model theory, cf. [13, Conjecture 1.8]. The following conjecture captures some of these relationships and is an important test question for the future development of descriptive inner model theory and the core model induction.

Conjecture 1 *The following theories are equiconsistent.*

(i) ZF + ω_1 *is supercompact.*
(ii) ZF + AD + ω_1 *is supercompact.*
(iii) ZFC + *there are proper class many Woodin cardinals which are limits of Woodin cardinals.*

References [19–21] made some progress in resolving the conjecture by exploring consistency strength and structural consequences of various fragments of supercompactness of ω_1.

This paper studies structural consequences of (full) supercompactness of ω_1 under ZF. We first show the following basic structural consequences.

Theorem 1 *Assume that ω_1 is supercompact. Then*

1. *the Axiom of Dependent Choices (DC) holds, while*
2. *(Folklore) there is no injection from ω_1 to 2^ω.*

The useful fact that DC holds can be used to derive other determinacy-like consequences such as:

Theorem 2 *Assume ω_1 is supercompact. Then every tree is weakly homogeneous.*

Remark 1 Note that under ZF+DC, every weakly homogeneously Suslin set is co-Suslin. So if ω_1 is supercompact, then every Suslin set is also co-Suslin.

Theorem 3 *Assume ω_1 is supercompact and Hod Pair Capturing (HPC). Then for any A such that A is Suslin, A is determined.*

See Sect. 7 for more detailed discussions on the hypothesis HPC. Under "ω_1 is supercompact", we also show that AD^+ and $\mathsf{AD}_\mathbb{R}$ are equivalent.

---

[4] We need a suitable coding of the sequence $(\mathcal{F}_\lambda \mid \lambda \in Ord)$ to let the model satisfy the above properties of M.

[5] This gives a proof that "ω_1 is supercompact" does not imply AD.

[6] This argument is due to the anonymous referee, which is simpler than what the authors presented in the previous manuscript. We thank the referee for showing us this argument.

Theorem 4 *Assume ω_1 is supercompact. Then the following theories are equivalent:*

1. AD^+.
2. $\mathsf{AD}_{\mathbb{R}}$.

"ω_1 is supercompact" also implies a large collection of sets of reals are determined [21] and perhaps an even larger collection of sets of reals admit ∞-Borel representations.

Theorem 5 *Assume that ω_1 is supercompact. Then every subset of 2^ω in the Chang model $\mathrm{L}(\bigcup_{\lambda \in Ord} \lambda^\omega)$ is ∞-Borel.*

The paper is organized as follows. Section 2 summarizes basic concepts and definitions used in this paper. In Sect. 3, we prove Theorem 1. The proof of Theorem 5 is given in Sect. 4. In Sect. 5, we prove Theorem 2. Section 6 proves Theorem 4. Finally, Sect. 7 explains HPC and proves Theorem 3.

2 Definitions and Basic Concepts

Throughout this paper, we work in ZF without the Axiom of Choice. For a nonempty set A, the axiom DC_A states that for any relation R on A such that for any element x of A there is an element y of A with $(x, y) \in R$, there is a function $f : \omega \to A$ such that for all natural numbers n, $\big(f(n), f(n+1)\big) \in R$. The **Axiom of Dependent Choices** (**DC**) states that for any nonempty set A, DC_A holds.

For a set X, $X^{<\omega}$ denotes the set of all finite sequences of elements of X, and X^ω denotes the set of all functions from ω to X. In particular, 2^ω denotes the set of all function from ω to $2 = \{0, 1\}$, not an ordinal or a cardinal. For a set X, we often consider X^ω as a topological space whose basic open sets are of the form $O_s = \{x \in X^\omega \mid s \subseteq x\}$ for $s \in X^{<\omega}$. For a set X and an infinite cardinal κ, let $\mathcal{P}_\kappa X$ be the set of all subsets σ of X such that σ is well-orderable and its cardinality is less than κ.

Let us review some basic terminology on filters. For a set Z, a **filter on** Z is a collection of subsets of Z closed under supersets and finite intersections. A filter on Z is σ **-complete** if it is closed under countable intersections. A filter on Z is **non-trivial** if the empty set $\emptyset$ does not belong to the filter. A filter on Z is an **ultrafilter** (or a **measure**) if it is non-trivial and for any subset A of Z, either A or $Z \setminus A$ is in the filter. Given a formula ϕ and an ultrafilter μ on Z, if the set $A = \{\sigma \in Z \mid \phi(\sigma)\}$ is in μ, then we say "for μ-measure one many σ, $\phi(\sigma)$ holds".

Let us introduce fineness and normality of ultrafilters on $\mathcal{P}_\kappa X$. An ultrafilter μ on $\mathcal{P}_\kappa X$ is **fine** if for any element x of X, for μ-measure one many σ, x is in σ. An ultrafilter μ on $\mathcal{P}_\kappa X$ is **normal** if for any set A in μ and $f : A \to \mathcal{P}_\kappa X$ with $\emptyset \neq f(\sigma) \subseteq \sigma$ for all $\sigma \in A$, there is an $x_0 \in X$ such that for μ-measure one many σ in A, $x_0 \in f(\sigma)$. Notice that this definition of normality is equivalent to the closure under diagonal intersections in ZF while it may not be equivalent to the standard definition of normality with regressive functions $f : A \to X$ without the axiom of choice. An ultrafilter on $\mathcal{P}_\kappa X$ is a **fine measure on** $\mathcal{P}_\kappa X$ if it is σ-complete and fine. A fine measure on $\mathcal{P}_\kappa X$ is a **normal measure on** $\mathcal{P}_\kappa X$ if it is normal.

We now introduce the main definitions of this paper:

Definition 1 Let κ be an infinite cardinal.

1. Let X be a set.
   (a) κ is **X-strongly compact** if there is a fine measure on $\mathcal{P}_\kappa X$.
   (b) κ is **X-supercompact** if there is a normal measure on $\mathcal{P}_\kappa X$.
2. κ is **strongly compact** if for any set X, κ is X-strongly compact.
3. κ is **supercompact** if for any set X, κ is X-supercompact.

We now review basic notions on determinacy axioms. For a nonempty set X, the **Axiom of Determinacy in** X^ω (AD_X) states that for any subset A of X^ω, in the Gale-Stewart game with the payoff set A, one of the players must have a winning strategy. We write AD for AD_ω. The ordinal Θ is defined as the supremum of ordinals which are surjective images of $\mathbb{R}$. Under ZF+AD, Θ is very big, e.g., it is a limit of measurable cardinals while under ZFC, Θ is equal to the successor cardinal of the continuum $|\mathbb{R}|$. **Ordinal Determinacy** states that for any $\lambda < \Theta$, any continuous function $\pi : \lambda^\omega \to \omega^\omega$, and any $A \subseteq \omega^\omega$, in the Gale-Stewart game with the payoff set $\pi^{-1}(A)$, one of the players must have a winning strategy. In particular, Ordinal Determinacy implies AD while it is still open whether the converse holds under ZF+DC.

We will introduce the notion of ∞-Borel codes. Before that, we review some terminology on trees. Given a set X, a **tree on** X is a collection of finite sequences of elements of X closed under initial segments. Given an element t of $X^{<\omega}$, $\mathrm{lh}(t)$ denotes its length, i.e., the domain or the cardinality of t. Given a tree T on X and elements s and t of T, s is an **immediate successor of** t **in** T if s is an extension of t and $\mathrm{lh}(s) = \mathrm{lh}(t) + 1$. Given a tree T on X and an element t of T, $\mathrm{Succ}_T(t)$ denotes the collection of all immediate successors of t in T. An element t of a tree T on X is **terminal** if $\mathrm{Succ}_T(t) = \emptyset$. For an element t of a tree T on X, $\mathrm{term}(T)$ denotes the collection of all terminal elements of T. Given a tree T on X, $[\mathbf{T}]$ denotes the collection of all $x \in X^\omega$ such that for all natural numbers n, $x \restriction n$ is in T. A tree T on X is **well-founded** if $[T] = \emptyset$. We often identify a tree T on $X \times Y$ with a subset of the set $\{(s, t) \in X^{<\omega} \times Y^{<\omega} \mid \mathrm{lh}(s) = \mathrm{lh}(t)\}$, and $\mathrm{p}[T]$ denotes the collection of all $x \in X^\omega$ such that there is a $y \in Y^\omega$ with $(x, y) \in [T]$.

Definition 2 Let λ be a non-zero ordinal.

1. An **∞-Borel code in** λ^ω is a pair (T, ρ) where T is a well-founded tree on some ordinal γ, and ρ is a function from $\mathrm{term}(T)$ to $\lambda^{<\omega}$.
2. Given an ∞-Borel code $c = (T, \rho)$ in λ^ω, to each element t of T, we assign a subset $B_{c,t}$ of λ^ω by induction on t using the well-foundedness of the tree T as follows:

(a) If t is a terminal element of T, let $B_{c,t}$ be the basic open set $O_{\rho(t)}$ in λ^ω.
(b) If $\mathrm{Succ}_T(t)$ is a singleton of the form $\{s\}$, let $B_{c,t}$ be the complement of $B_{c,s}$.
(c) If $\mathrm{Succ}_T(t)$ has more than one element, then let $B_{c,t}$ be the union of all sets of the form $B_{c,s}$ where s is in $\mathrm{Succ}_T(t)$.

We write B_c for $B_{c,\emptyset}$.
3. A subset A of λ^ω is **∞-Borel** if there is an ∞-Borel code c in λ^ω such that $A = B_c$.

Usually, we use ∞-Borel codes and ∞-Borel sets only in the spaces ω^ω or 2^ω. We use them for general spaces λ^ω in Sect. 4.

In Sect. 4, we will use the following characterization of ∞-Borelness in the space λ^ω:

Fact 1 *Let λ be a non-zero ordinal and A be a subset of λ^ω. Then the following are equivalent:*

1. *A is ∞-Borel, and*
2. *for some formula ϕ and some set S of ordinals, for all elements x of λ^ω, x is in A if and only if $\mathbf{L}[S, x] \models$ "$\phi(S, x)$".*

Proof For the case $\lambda = 2$, one can refer to [8, Theorem 9.0.4]. The general case can be proved in the same way. □

Remark 2 In fact, the second item of Fact 1 is equivalent to the following using Lévy's Reflection Principle:

- for some $\gamma > \lambda$, some formula ϕ, and some set S of ordinals, for all elements x of λ^ω, x is in A if and only if $\mathbf{L}_\gamma[S, x] \models$ "$\phi(S, x)$".

Throughout this paper, we will freely use either of the equivalent conditions of ∞-Borelness.

We now introduce the axiom AD^+, and review some notions on Suslin sets. The axiom AD^+ states that (a) $\mathsf{DC}_\mathbb{R}$ holds, (b) Ordinal Determinacy holds, and (c) every subset of 2^ω is ∞-Borel. Since AD^+ demands Ordinal Determinacy, AD^+ implies AD while it is open whether the converse holds in $\mathsf{ZF{+}DC}$. A subset A of 2^ω (ot ω^ω) is **Suslin** if there are some ordinal λ and a tree T on $2 \times \lambda$ ($\omega \times \lambda$ respectively) such that $A = \mathrm{p}[T]$. A is **co-Suslin** if the complement of A is Suslin. An infinite cardinal λ is a **Suslin cardinal** if there is a subset A of 2^ω (ω^ω) such that there is a tree on $2 \times \lambda$ ($\omega \times \lambda$) such that $A = \mathrm{p}[T]$ while there are no $\gamma < \lambda$ and a tree S on $2 \times \gamma$ ($\omega \times \lambda$) such that $A = \mathrm{p}[S]$. Under $\mathsf{ZF{+}DC}_\mathbb{R}$, AD^+ is equivalent to the assertion that Suslin cardinals are closed below Θ in the order topology of $(\Theta, <)$.

3 Choice Principles and Supercompactness of ω_1

In this section, we prove Theorem 1.

Proof (Theorem 1) 1. Let A be any nonempty set and R be any relation on A such that for any $x \in A$ there is a $y \in A$ such that $(x, y) \in R$. We will show that there is a function $f : \omega \to A$ such that for all natural numbers n, $\big(f(n), f(n+1)\big) \in R$.

Since ω_1 is supercompact, there is a fine normal measure on $\mathcal{P}_{\omega_1} A$. We fix such a measure μ.

Claim 1 *For μ-measure one many elements σ of $\mathcal{P}_{\omega_1} A$, the following holds:*

$$(\forall x \in \sigma)\ (\exists y \in \sigma)\ (x, y) \in R$$

Proof (Claim 1) Suppose not. We will derive a contradiction using μ. Since μ is an ultrafilter on $\mathcal{P}_{\omega_1} A$, for μ-measure one many elements σ of $\mathcal{P}_{\omega_1} A$, the following holds:

$$(\exists x \in \sigma)\ (\forall y \in \sigma)\ (x, y) \notin R$$

By normality of μ, there is an $x_0 \in A$ such that for μ-measure one many elements σ of $\mathcal{P}_{\omega_1} A$ with $x_0 \in \sigma$, for all $y \in \sigma$, $(x_0, y) \notin R$.

On the other hand, by the assumption on R, there is a $y_0 \in A$ such that $(x_0, y_0) \in R$. By fineness of μ, for μ-measure one many elements σ of $\mathcal{P}_{\omega_1} A$, both x_0 and y_0 are elements of σ.

Since μ is a filter, for μ-measure one many elements σ of $\mathcal{P}_{\omega_1} A$ with $x_0 \in \sigma$, for all $y \in \sigma$, $(x_0, y) \notin R$ while both x_0 and y_0 are elements of σ and $(x_0, y_0) \in R$. This gives us both $(x_0, y_0) \notin R$ and $(x_0, y_0) \in R$, a contradiction. This finishes the proof of the claim. □

We now know that for μ-measure one many elements σ of $\mathcal{P}_{\omega_1} A$, the following holds:

$$(\forall x \in \sigma)\ (\exists y \in \sigma)\ (x, y) \in R$$

Let us pick such a σ. Then for any $x \in \sigma$, there is a $y \in \sigma$ such that $(x, y) \in R$. Since σ is an element of $\mathcal{P}_{\omega_1} A$, it is countable, so we can fix a surjection $\pi : \omega \to \sigma$. Using this π, the above property of σ, and the well-orderedness of $(\omega, <)$, one can easily construct a desired $f : \omega \to A$. This finishes the proof of 1.

2. This is a well-known fact to the experts. Nevertheless, we will give a proof for the sake of completeness. Suppose that there was an injection $i : \omega_1 \to 2^\omega$. We will derive a contradiction using supercompactness of ω_1. For each $\alpha < \omega_1$, we write x_α for $i(\alpha)$.

We first note that there is a non-principal σ-complete ultrafilter on ω_1, i.e., ω_1 is measurable. Since ω_1 is supercompact, we can fix a fine normal measure μ on $\mathcal{P}_{\omega_1} \omega_1$. Let ν be as follows:

$$\nu = \{A \subseteq \omega_1 \mid \text{ for } \mu\text{-measure one many elements } \sigma \text{ of } \mathcal{P}_{\omega_1}\omega_1, \sup\sigma \in A\}$$

Then it is easy to see that ν is a non-principal σ-complete ultrafilter on ω_1.

Using this ν, we will derive a contradiction as follows. Since ν is an ultrafilter on ω_1, for any natural number n, there is an $k_n \in \{0, 1\}$ such that the set $A_n = \{\alpha < \omega_1 \mid x_\alpha(n) = k_n\}$ is of ν-measure one. Since ν is σ-complete, the set $A = \bigcap_{n \in \omega} A_n$ is of ν-measure one. By the property of each A_n, for any α in A, for all natural numbers n, $x_\alpha(n) = k_n$. But since i is injective, A has at most one element. This contradicts that A is of ν-measure one and ν is non-principal. This finishes the proof of 2. This completes the proof of Theorem 1. □

Remark 3 (2) of Theorem 1 is the best one can hope for. "ω_1 is supercompact" does not imply "there is no injection $f : \omega_2 \to \mathcal{P}(\omega_1)$". To see this, assume ZFC and there is a supercompact cardinal κ. Let $f : \kappa^+ \to \mathcal{P}(\kappa)$ be an injection in V. Let $\mathcal{T}$ be the Takeuti model defined at κ. Then clearly $f \in \mathcal{T}$ and in $\mathcal{T}$, $\kappa = \omega_1$ and $(\kappa^+)^V = \omega_2$.

4 Chang Model and Supercompactness of ω_1

In this section, we prove Theorem 5. As a corollary, one can obtain usual regularity properties for sets of reals in the Chang model:

Corollary 1 *Assume that ω_1 is supercompact. Then every subset of 2^ω in the Chang model is Lebesgue measurable and has the Baire property.*

Corollary 1 directly follows from Theorem 1, Theorem 5, and the following fact:

Fact 2 *(Essentially Solovay) Assume that there is no injection from ω_1 to 2^ω. Let A be a subset of 2^ω which is ∞-Borel. Then A is Lebesgue measurable and has the Baire property.*

For the proof of Fact 2, one can refer to e.g., [3, Theorem 2.4.2 and Proposition 3.2.13].

To prove Theorem 5, we use the following lemma:

Lemma 1 $\mathrm{L}(\bigcup_{\lambda \in Ord} \lambda^\omega) = \bigcup_{\lambda \in Ord} \mathrm{L}(\lambda^\omega)$.

Proof Given a set X, let $J(X)$ be the rudimental closure of $X \cup \{X\}$. Let $(C_\alpha \mid \alpha \in \mathrm{Ord})$ be the following sequence: $C_0 = \mathrm{L}_\omega$, $C_{\alpha+1} = J(C_\alpha \cup \alpha^\omega)$, and $C_\beta = \bigcup_{\alpha<\beta} C_\alpha$ when β is a limit ordinal. Set $C = \bigcup_{\alpha \in \mathrm{Ord}} C_\alpha$.

We first argue that C is equal to the Chang model $\mathrm{L}(\bigcup_{\lambda \in \mathrm{Ord}} \lambda^\omega)$. It is easy to see that C is contained in the Chang model because the construction of the sequence $(C_\alpha \mid \alpha \in \mathrm{Ord})$ is absolute between the Chang model and V. So it is enough to prove that C contains the Chang model. For that it is enough to show that C is an inner model of ZF containing all sets in Ord^ω. By the construction of $(C_\alpha \mid \alpha \in \mathrm{Ord})$, it is easy to see that C contains all the sets in Ord^ω, is rudimentarily closed, satisfies Comprehension Scheme, and for any subset X of C in V, there is a set Y in C such that $X \subseteq Y$ (namely C_α for some big α). Therefore, C is an inner model of ZF containing all the sets in Ord^ω, as desired.

Next, we claim that for all ordinals λ, $C_\lambda \in \mathrm{L}(\lambda^\omega)$. For this, it is enough to see that the construction of the sequence $(C_\alpha \mid \alpha \leq \lambda)$ is absolute between $\mathrm{L}(\lambda^\omega)$ and V, which follows by observing that λ^ω is in $\mathrm{L}(\lambda^\omega)$.

We now argue that the Chang model $\mathrm{L}(\bigcup_{\lambda \in \mathrm{Ord}} \lambda^\omega)$ is equal to $\bigcup_{\lambda \in \mathrm{Ord}} \mathrm{L}(\lambda^\omega)$. The inclusion $\bigcup_{\lambda \in \mathrm{Ord}} \mathrm{L}(\lambda^\omega) \subseteq \mathrm{L}(\bigcup_{\lambda \in \mathrm{Ord}} \lambda^\omega)$ is clear. We will see the other inclusion. Let A be any set in the Chang model. By the second to last paragraph, A is in C and hence there is an ordinal λ such that A is in C_λ. By the last paragraph, C_λ is in $\mathrm{L}(\lambda^\omega)$. Therefore, A is in $\mathrm{L}(\lambda^\omega)$, as desired. This completes the proof of Lemma 1. □

We now come to the proof of Theorem 5.

Proof (Theorem 5)

We assume that ω_1 is supercompact and will show that every subset of 2^ω in the Chang model is ∞-Borel.

By Lemma 1, to obtain Theorem 5, it is enough to prove that for all λ, every subset of 2^ω in $\mathrm{L}(\lambda^\omega)$ is ∞-Borel.

Throughout the rest of this section, we fix an infinite ordinal λ and a fine measure μ on $\mathcal{P}_{\omega_1}\lambda^\omega$ whose existence is ensured by the supercompactness of ω_1. We will show that every subset of λ^ω in $\mathrm{L}(\lambda^\omega)$ is ∞-Borel using μ, which will give us that every subset of 2^ω in $\mathrm{L}(\lambda^\omega)$ is ∞-Borel. The arguments are a generalization of the proof of Woodin's theorem in [1, Theorem 1.9].

By Fact 1, it is enough to show that for any subset A of λ^ω in $\mathrm{L}(\lambda^\omega)$, there are some formula ϕ and a set S of ordinals such that for all elements x of λ^ω,

$$x \in A \iff \mathrm{L}[S, x] \vDash \phi[S, x]. \qquad (\dagger)$$

If (†) holds for all elements x of λ^ω, then we say that A is **defined from the pair** (ϕ, S) and we write $B_{(\phi,S)}$ for A.

The following is the key claim in this section:

Claim 2 *There is a function F which is OD from μ such that*

1. *F is defined for all pairs (ϕ, S) where ϕ is a formula and S is a set of ordinals,*
2. *$F(\phi, S)$ is of the form (ψ, T) such that if A is a subset of $(\lambda^\omega)^{n+1}$ defined from (ϕ, S), then $pA = \{\boldsymbol{x} \in (\lambda^\omega)^n \mid (\exists y)\ (\boldsymbol{x}, y) \in A\}$ is defined from (ψ, T), i.e., if $A = B_{(\phi,S)}$, then $pA = B_{F(\phi,S)}$.*

To prove Claim 2, we use a variant of Vopěnka algebra: Let S be a set of ordinals and σ be an element of $\mathcal{P}_{\omega_1}\lambda^\omega$. We fix an injection $\iota\colon \mathrm{OD}_{S,\sigma} \cap \mathcal{P}(\sigma) \to \mathrm{HOD}_{S,\sigma}$ which is OD from S and σ such that for all $t \in \lambda^{<\omega}$, $\iota(O_t) = t$ where $O_t = \{x \in \sigma \mid$

$t \subseteq x\}$.[7] Set $B_\sigma = \{\iota(A) \mid A \in \mathrm{OD}_{S,\sigma} \cap \mathcal{P}(\sigma)\}$. For $p, q \in B_\sigma$, $p \leq q$ if $\iota^{-1}(p) \subseteq \iota^{-1}(q)$. Note that the structure $(B_\sigma, \leq)$ is in $\mathrm{HOD}_{S,\sigma}$. For an element x of σ, set $G_x = \{p \in B_\sigma \mid x \in \iota^{-1}(p)\}$.

Fact 3 (Vopěnka)

1. *In $\mathrm{HOD}_{S,\sigma}$, B_σ is a complete Boolean algebra, and*
2. *for any element x of σ, G_x is B_σ-generic over $\mathrm{HOD}_{S,\sigma}$, and x is in* $\mathrm{L}[S, B_\sigma][G_x]$, *which is a subclass of $\mathrm{HOD}_{S,\sigma}[G_x]$.*

Recall that we have fixed the fine measure μ on $\mathcal{P}_{\omega_1}\lambda^\omega$. For each $\sigma \in \lambda^\omega$, let $Q_\sigma = (B_\sigma)^{\mathrm{L}(S,\sigma)}$. We will consider the ultraproducts $\prod_\sigma \mathrm{L}[S, Q_\sigma][x]/\mu$ for $x \in \lambda^\omega$. Using the fineness of μ, one can prove Łos' theorem for these ultraproducts (the proof is essentially the same as the one given in [19, Lemma 2.3]).[8] By DC from Theorem 1, the above ultraproducts are all well-founded and we identify them with their transitive collapses. For each $y \in \bigcap_\sigma \mathrm{L}[S, Q_\sigma]$, let $y_\infty = \prod_\sigma y/\mu$. In particular, $S_\infty = \prod_\sigma S/\mu$. Let $h\colon \lambda \to \mathrm{Ord}$ be such that $h(\alpha) = \alpha_\infty$ for all $\alpha < \lambda$. We also set $Q_\infty = \prod_\sigma Q_\sigma/\mu$.

We are now ready to prove Claim 2.

Proof (Claim 2)
For simplicity, we will assume $n = 1$ (the general case is treated in the same way). Let $A \subseteq (\lambda^\omega)^2$ be defined from (ϕ, S), i.e., $A = B_{(\phi,S)}$. Then for all $x \in \lambda^\omega$,

$$
\begin{aligned}
& x \in \mathrm{p}A \\
\iff & (\exists y \in \lambda^\omega)\ (x, y) \in B_{(\phi,S)} \\
\iff & \text{for } \mu\text{-measure one many } \sigma,\ \mathrm{L}(S, \sigma) \models \text{“}(\exists y)\ (x, y) \in B_{(\phi,S)}\text{”} \\
\iff & \text{for } \mu\text{-measure one many } \sigma, \\
& \mathrm{L}[S, Q_\sigma, x] \models \text{“}\Big(\exists p \in \mathrm{Coll}\big(\omega, |\mathcal{P}(Q_\sigma)|\big)\Big) p \Vdash (\exists y)\ (\check{x}, y) \in B_{(\phi,S)}\text{”} \\
\iff & \prod_\sigma \mathrm{L}[S, Q_\sigma, x]/\mu \models \text{“}\Big(\exists p \in \mathrm{Coll}\big(\omega, |\mathcal{P}(Q_\infty)|\big)\Big)\ p \Vdash (\exists y)\ (\check{x_\infty}, y) \in B_{(\phi,S_\infty)}\text{”} \\
\iff & \mathrm{L}[S_\infty, Q_\infty, x_\infty] \models \text{“}\Big(\exists p \in \mathrm{Coll}\big(\omega, |\mathcal{P}(Q_\infty)|\big)\Big)\ p \Vdash (\exists y)\ (\check{x_\infty}, y) \in B_{(\phi,S_\infty)}\text{”}
\end{aligned}
$$

The first equivalence follows from the assumption that A is defined from (ϕ, S). The second equivalence follows from the fineness of μ. The forward direction of the third equivalence follows from the property of the Vopěnka algebra Q_σ given in Fact 3: Given a y in σ with $(x, y) \in B_{\phi,S}$, letting z code x and y in a simple way, $z \in \mathrm{L}[S, Q_\sigma][G_z]$ by Fact 3. Hence y is in a set generic extension of $\mathrm{L}[S, Q_\sigma, x]$ whose poset is of size at most $|Q_\sigma|$ in $\mathrm{L}[S, Q_\sigma, x]$. In particular, in $\mathrm{L}[S, Q_\sigma, x]$,

---

[7] We are demanding that for all $t \in \lambda^{<\omega}$, $\iota(O_t) = t$ to have that "x is in $\mathrm{L}[S, B_\sigma][G_x]$" in Fact 3. Namely, x should be easily computable from B_σ and G_x without referring to ι because ι may not be definable in $\mathrm{L}[S, B_\sigma][G_x]$.

[8] For the proof of Łos' theorem, we do not need the normality of μ because for each σ, one can define a well-order on $\mathrm{L}[S, Q_\sigma][x]$ uniformly in σ.

one can force to add such a y over $\mathrm{Coll}(\omega, |\mathcal{P}(Q_\sigma)|)$. The backward direction of the third equivalence follows from the fact that $\mathcal{P}(\mathcal{P}(Q_\sigma))^{\mathrm{L}[S,Q_\sigma,x]}$ is countable in V because Q_σ is countable by the fact that $\mathrm{OD}^{\mathrm{L}(S,\sigma)} \cap \mathcal{P}(\sigma)$ is well-orderable and σ is countable in V, and because $\mathrm{L}[S, Q_\sigma, x]$ is a transitive model of ZFC. The fourth & fifth equivalences follow from Łos' theorem for the ultraproduct $\prod_\sigma \mathrm{L}[S, Q_\sigma][x]/\mu$ and the definitions of S_∞, Q_∞, and x_∞.

Now let T be the set of ordinals simply coding S_∞, Q_∞, and h. Then for each $x \in \lambda^\omega$, $x_\infty \in \mathrm{L}[T, x]$ because $x_\infty = h[x]$. Let ψ be the formula stating "$\mathrm{L}[S_\infty, Q_\infty, x_\infty] \models \left(\exists p \in \mathrm{Coll}(\omega, |\mathcal{P}(Q_\infty)|)\right)\ p \Vdash (\exists y)\ (\check{x}, y) \in B_{(\phi, S_\infty)}$". Then for each $x \in \lambda^\omega$,

$$
\begin{aligned}
& x \in \mathrm{p}A \\
\iff & \mathrm{L}[S_\infty, Q_\infty, x_\infty] \models \text{“}\left(\exists p \in \mathrm{Coll}(\omega, |\mathcal{P}(Q_\infty)|)\right)\ p \Vdash (\exists y)\ (\check{x_\infty}, y) \in B_{(\phi, S_\infty)}\text{”} \\
\iff & \mathrm{L}[T, x] \models \text{“}\psi[T, x]\text{”}
\end{aligned}
$$

Therefore, $F(\phi, S) = (\psi, T)$ satisfies the desired equivalence. This completes the proof of Claim 2. □

As is mentioned before Claim 2, we shall prove that for all subsets A of λ^ω in $\mathrm{L}(\lambda^\omega)$, A is defined from some pair (ϕ, S) as in (†), which gives us Theorem 5. The idea is to look at the hierarchy $\left(\mathrm{L}_\alpha(\lambda^\omega) \mid \alpha \in \mathrm{Ord}, \alpha \geq \omega\right)$, and by induction on α, to each definition of an element A of $\mathrm{L}_\alpha(\lambda^\omega)$, we assign certain ϕ and S such that A is defined from (ϕ, S). We fix an F from Claim 2.

Definition 3 The hierarchy $\left(\mathrm{L}_\alpha(\lambda^\omega) \mid \alpha \in \mathrm{Ord}, \alpha \geq \omega\right)$ is defined as follows:

1. $\mathrm{L}_\omega(\lambda^\omega) = \lambda \cup \lambda^\omega$,
2. $\mathrm{L}_{\alpha+1}(\lambda^\omega) = \mathrm{Def}\left(\mathrm{L}_\alpha(\lambda^\omega), \in\right)$, and
3. $\mathrm{L}_\beta(\lambda^\omega) = \bigcup_{\alpha<\beta} \mathrm{L}_\alpha(\lambda^\omega)$ when β is a limit ordinal bigger than ω.

Note that the above definition is not standard in the following two senses: One is that the indexing of the hierarchy starts with $\alpha = \omega$, not $\alpha = 0$. The other is that the first stage $\mathrm{L}_\omega(\lambda^\omega)$ is not transitive. However, the above definition satisfies that for all $\alpha \geq \omega + \omega$, $\mathrm{L}_\alpha(\lambda^\omega)$ is transitive and that the union of all $\mathrm{L}_\alpha(\lambda^\omega)$ $(\alpha \geq \omega)$ coincides with $\mathrm{L}(\lambda^\omega)$, the least inner model of ZF containing λ^ω as an element. These two properties are enough for us to prove Theorem 5. We start the indexing with $\alpha = \omega$ to ensure that one can code any formula with a natural number smaller than α in the arguments below. Also we use the above definition of $\mathrm{L}_\omega(\lambda^\omega)$ for convenience of proving Claim 3 below.

Remark 4 There is a sequence of partial surjections $\left(\pi_\alpha \colon \alpha^{<\omega} \times \lambda^\omega \to \mathrm{L}_\alpha(\lambda^\omega) \mid \alpha \in \mathrm{Ord}, \alpha \geq \omega\right)$ which is OD such that

1. for all $\alpha \geq \omega$, $\pi_\alpha(\emptyset, x) = x$,

2. $\pi_\omega(\boldsymbol{\beta}, x) = x(0)$ when $\boldsymbol{\beta} \neq \emptyset$,
3. if $\omega \leq \beta < \alpha$, then $\pi_\beta = \pi_\alpha \restriction \beta^{<\omega} \times \lambda^\omega$, and
4. if $\alpha > \omega$, $(\boldsymbol{\beta}, x) \in \alpha^{<\omega} \times \lambda^\omega$, $\pi_\alpha(\boldsymbol{\beta}, x)$ is defined, and $\boldsymbol{\beta} = (\beta_0, \beta_1, \cdots, \beta_k)$, then $\pi_\alpha(\boldsymbol{\beta}, x)$ is an element of $\mathrm{L}_{\beta_0+1}(\lambda^\omega)$ which is defined in the structure $(\mathrm{L}_{\beta_0}(\lambda^\omega), \in)$ via a formula coded by β_1 with some parameters of the form $\pi_{\beta_0}(\boldsymbol{\gamma}, y)$ where $\boldsymbol{\gamma}$ here depends only on $\boldsymbol{\beta}$, not on x.

Definition 4 For an ordinal $\alpha \geq \omega$, a formula ϕ, and $\boldsymbol{\beta}^0, \cdots, \boldsymbol{\beta}^{n-1} \in \alpha^{<\omega}$, let

$$T^\alpha_{(\phi, \boldsymbol{\beta}^0, \cdots, \boldsymbol{\beta}^{n-1})} = \{(x_0, \cdots, x_{n-1}) \in (\lambda^\omega)^n \mid \mathrm{L}_\alpha(\lambda^\omega) \vDash \phi[\pi_\alpha(\boldsymbol{\beta}^0, x_0), \cdots, \pi_\alpha(\boldsymbol{\beta}^{n-1}, x_{n-1})]\}$$

Claim 3 *There is a function which is OD from μ sending (α, p) to an ∞-Borel code $q^\alpha_p = (\psi, S)$, where α is an ordinal with $\alpha \geq \omega$ and $p = (\phi, \boldsymbol{\beta}^0, \cdots, \boldsymbol{\beta}^{n-1})$ is as in Definition 4, such that T^α_p is defined from q^α_p.*

Proof (Claim 3)

We prove the claim by induction on α. Let us fix α. Then we prove the statement by induction on the complexity of ϕ.

Case 1: When ϕ is of the form $v \in w$ or $v = w$.

Suppose that $\alpha = \omega$. This is the base case of the double induction. Let $\boldsymbol{\beta}^0, \boldsymbol{\beta}^1 \in \alpha^{<\omega}$. Then the set $T^0_{\phi, \boldsymbol{\beta}^0, \boldsymbol{\beta}^1}$ is of the form $\emptyset$, $\{(x_0, x_1) \mid x_0(0) \in x_1(0)\}$, $\{(x_0, x_1) \mid x_1(0) \in x_0(0)\}$, $\{(x_0, x_1) \mid x_0(0) = x_1(0)\}$, or $\{(x_0, x_1) \mid x_0 = x_1\}$. In each case, one can assign a suitable code q^α_p in a simple way.

Suppose that $\alpha > \omega$. Let $\beta_* = \max\{\beta^0_0, \beta^1_0\}$. Then $\beta^* < \alpha$ and by Remark 4, both $\pi_\alpha(\boldsymbol{\beta}^0, x_0)$ and $\pi_\alpha(\boldsymbol{\beta}^1, x_1)$ are definable in the structure $(\mathrm{L}_{\beta_*}(\lambda^\omega), \in)$ with some parameters of the form $\pi_{\beta_*}(\boldsymbol{\gamma}, y)$ where $\boldsymbol{\gamma}$ here depends only on $\boldsymbol{\beta}^0$ and $\boldsymbol{\beta}^1$. Then one can find a formula ϕ' and some $\boldsymbol{\gamma}^0, \boldsymbol{\gamma}^1$ such that $T^\alpha_{\phi, \boldsymbol{\beta}^0, \boldsymbol{\beta}^1} = T^{\beta_*}_{\phi', \boldsymbol{\gamma}^0, \boldsymbol{\gamma}^1}$. By induction hypothesis, one can find a desired code q^α_p.

Case 2: When ϕ is of the form $\neg\, \phi'$.

In this case, by induction hypothesis, letting $p' = (\phi', \boldsymbol{\beta}^0, \cdots, \boldsymbol{\beta}^{n-1})$, we have $q^\alpha_{p'} = (\psi, S)$. Then $q^\alpha_p = (\neg\, \psi, S)$ is the desired code.

Case 3: When ϕ is of the form $\phi_1 \wedge \phi_2$.

In this case, by induction hypothesis, letting $p_1 = (\phi_1, \boldsymbol{\beta}^0, \cdots, \boldsymbol{\beta}^{n-1})$ and $p_2 = (\phi_2, \boldsymbol{\beta}^0, \cdots, \boldsymbol{\beta}^{n-1})$, we have $q^\alpha_{p_1} = (\psi_1, S_1)$ and $q^\alpha_{p_2} = (\psi_2, S_2)$. Then let $q^\alpha_p = (\psi, S)$ where S is a set of ordinals simply coding S_1 and S_2, and $\psi(S, x)$ states that both "$\mathrm{L}[S_1, x] \vDash \psi_1[S_1, x]$" and "$\mathrm{L}[S_2, x] \vDash \psi_2[S_2, x]$" hold. Then q^α_p is the desired code.

Case 4: When ϕ is of the form $\exists v\, \phi'$.

In this case, by induction hypothesis, for each $\boldsymbol{\beta} \in \alpha^{<\omega}$, setting $p_{\boldsymbol{\beta}} = (\phi', \boldsymbol{\beta}, \boldsymbol{\beta}^0, \cdots, \boldsymbol{\beta}^{n-1})$, we have the code $q^\alpha_{p_{\boldsymbol{\beta}}}$. We write $q_{\boldsymbol{\beta}}$ for $q^\alpha_{p_{\boldsymbol{\beta}}}$. Note that

$$
\begin{aligned}
&T^{\alpha}_{\phi,\beta^0,\cdots,\beta^{n-1}}\\
=&\{(x_0,\cdots x_{n-1}) \mid \big(\exists y \in \mathrm{L}_\alpha(\lambda^\omega)\big)\\
&\qquad \mathrm{L}_\alpha(\lambda^\omega) \models \text{“}\phi'[y, \pi_\alpha(\boldsymbol{\beta}^0, x_0), \cdots, \pi_\alpha(\boldsymbol{\beta}^{n-1}, x_{n-1})]\text{"}\}\\
=&\{(x_0,\cdots x_{n-1}) \mid (\exists \boldsymbol{\beta} \in \alpha^{<\omega})\ (\exists x \in \lambda^\omega)\\
&\qquad \mathrm{L}_\alpha(\lambda^\omega) \models \text{“}\phi'[\pi_\alpha(\boldsymbol{\beta}, x), \pi_\alpha(\boldsymbol{\beta}^0, x_0), \cdots, \pi_\alpha(\boldsymbol{\beta}^{n-1}, x_{n-1})]\text{"}\}\\
=&\bigcup_{x\in\lambda^\omega}\ \bigcup_{\boldsymbol{\beta}\in\alpha^{<\omega}} T^{\alpha}_{\phi',\boldsymbol{\beta},\boldsymbol{\beta}^0,\cdots,\boldsymbol{\beta}^{n-1}}\\
=&\bigcup_{x\in\lambda^\omega}\ \bigcup_{\boldsymbol{\beta}\in\alpha^{<\omega}} B_{q_{\boldsymbol{\beta}}}\\
=&\bigcup_{x\in\lambda^\omega} B_{\bigvee_{\boldsymbol{\beta}\in\alpha^{<\omega}} q_{\boldsymbol{\beta}}}\\
=&B_{F\left(\bigvee_{\boldsymbol{\beta}\in\alpha^{<\omega}} q_{\boldsymbol{\beta}}\right)},
\end{aligned}
$$

where $B_{q_{\boldsymbol{\beta}}}$ is the subset of $(\lambda^\omega)^n$ defined from the code $q_{\boldsymbol{\beta}}$ as in (†), $\bigvee_{\boldsymbol{\beta}\in\alpha^{<\omega}} q_{\boldsymbol{\beta}}$ is the pair (ψ, S) defining the union $\bigcup_{\boldsymbol{\beta}\in\alpha^{<\omega}} B_{q_{\boldsymbol{\beta}}}$ in a similar way as Case 3, and F is from Claim 2. Therefore, $q^\alpha_p = F\big(\bigvee_{\boldsymbol{\beta}\in\alpha^{<\omega}} q_{\boldsymbol{\beta}}\big)$ is the desired code. This completes the proof of the claim. □

We are now ready to finish the proof of Theorem 5.

Let A be a subset of λ^ω in $\mathrm{L}(\lambda^\omega)$. By Fact 1, it is enough to find a pair (ϕ, S) which defines A as in (†). Since A is in $\mathrm{L}(\lambda^\omega)$, there is an ordinal α such that $A \in \mathrm{L}_{\alpha+1}(\lambda^\omega) \setminus \mathrm{L}_\alpha(\lambda^\omega)$. Let ψ be a formula defining A in the structure $\big(\mathrm{L}_\alpha(\lambda^\omega), \in\big)$ with some parameters $\pi_\alpha(\boldsymbol{\beta}^0, x_0), \cdots, \pi_\alpha(\boldsymbol{\beta}^{n-1}, x_{n-1})$. By Remark 4, $\pi_\alpha(\emptyset, x) = x$ for all $x \in \lambda^\omega$. Hence

$$
\begin{aligned}
A =&\{x \mid \mathrm{L}_\alpha(\lambda^\omega) \models \text{“}\psi[\pi_\alpha(\emptyset, x), \pi_\alpha(\boldsymbol{\beta}^0, x_0), \cdots, \pi_\alpha(\boldsymbol{\beta}^{n-1}, x_{n-1})]\text{"}\}\\
=&\{x \mid (x, x_0, \cdots, x_{n-1}) \in T^{\alpha}_{\psi,\emptyset,\boldsymbol{\beta}^0,\cdots,\boldsymbol{\beta}^{n-1}}\}\\
=&\{x \mid (x, x_0, \cdots, x_{n-1}) \in B_{q^\alpha_p}\},
\end{aligned}
$$

where $p = (\psi, \emptyset, \boldsymbol{\beta}^0, \cdots, \boldsymbol{\beta}^{n-1})$ and q^α_p is from Claim 3. This shows that A is defined from q^α_p with parameters $x_0, \cdots, x_{n-1}$, which easily gives us that A is defined from (ϕ, S) for some ϕ and S.

This completes the proof of Theorem 5. □

5 Weak Homogeneity and Supercompactness of ω_1

In this section, we prove Theorem 2. A tree T is said to be on $\omega \times \kappa$ if $T \subset (\omega \times \kappa)^{<\omega}$. For a tree T on $\omega \times \kappa$, for $s \in \omega^{<\omega}$, let $T_s = \{t \in \kappa^{lh(s)} \mid (s, t) \in T\}$. Let also $p_0[T] = \{x \in \omega^\omega \mid \exists f \forall n (x \restriction n, f \restriction n) \in T\}$, $p_1[T] = \{f \in \kappa^\omega \mid \exists x \forall n (x \restriction n, f \restriction n) \in T\}$. Every tree T considered in the following will be on $\omega \times \kappa$ for some κ.

Following [9], we define what it means for a tree T on $\omega \times \kappa$ to be **weakly homogeneous**. First, for $n < \omega$, κ an infinite cardinal, λ a nonzero ordinal, let $\mathrm{MEAS}_n^{\kappa,\lambda}$ be the set of all κ-complete measures on λ^n. For $m < n < \omega$, for $X \subseteq \lambda^m$, let $ext_n(X) = \{t \in \lambda^n \mid t \restriction m \in X\}$. A **$\lambda$-tower of measures** is a sequence $(\mu_n \mid n < \omega)$ such that

(i) for each n, $\mu_n \in \mathrm{MEAS}_n^{\omega_1,\lambda}$, and
(ii) for $m < n$, $\mu_m = proj_m(\mu_n)$, where $proj_m(\mu_n) = \{X \subseteq \lambda^m \mid ext_n(X) \in \mu_n\}$.

A tower $(\mu_n \mid n < \omega)$ is **countably complete** if for every sequence $(X_n \mid n < \omega)$ such that $X_n \in \mu_n$ for all $n < \omega$, there is a function $f : \omega \to \lambda$ such that $f \restriction n \in X_n$ for all n.

Definition 5 Let T be a tree on $\omega \times \lambda$. T is **weakly homogeneous** if there is a sequence $(M_s \mid s \in \omega^{<\omega})$ such that

(i) for each s, M_s is a countable subset of $\mathrm{MEAS}_{lh(s)}^{\omega_1,\lambda}$ and for each $\mu \in M_s$, $T_s \in \mu_s$.
(ii) for all $x \in p_0[T]$, there is a countably complete λ-tower of measures $(\mu_n \mid n < \omega)$ such that for each n, $\mu_n \in M_{x \restriction n}$.

Remark 5 We will not work directly with weakly homogenous trees in the proof of Theorem 2. Rather, the conclusion that T is weakly homogeneous is reached by verifying that the hypotheses needed to run the proof in [9] follow from our hypothesis that ω_1 is supercompact. Theorem 2 is similar to one of the main results of [9], which states that "every tree is weakly homogeneous" follows from $\mathsf{AD}_{\mathbb{R}}$.

Proof (Theorem 2) Let T be a tree on $\omega \times \lambda$. Reference [9] shows that T is weakly homogeneous provided the following conditions hold:

(A) There is a countably complete, normal fine measure on $\mathcal{P}_{\omega_1}(\bigcup_n(\mathcal{P}(\lambda^n) \cup \mathrm{MEAS}_n^{\omega_1,\lambda}))$.
(B) The Axiom of Dependent Choice holds for relations on $\mathcal{P}(\lambda)$.
(C) There is a wellorder on $\bigcup_n \mathrm{MEAS}_n^{\omega_1,\lambda}$.

We need to verify (A), (B), (C) follow from the supercompactness of ω_1. (A) is obvious. (B) follows from Theorem 1. Now we verify (C). Let $X = \bigcup_n \mathrm{MEAS}_n^{\omega_1,\lambda}$. We need to show that X is wellorderable.[9]

It is enough to prove that $\mathrm{MEAS}_1^{\omega_1,\lambda}$ is well-orderable. This is because for each $n > 1$, there is a bijection from $\mathrm{MEAS}_1^{\omega_1,\lambda}$ onto $\mathrm{MEAS}_n^{\omega_1,\lambda}$. Such a bijection is induced by a bijection between λ and λ^n. Hence, $\mathrm{MEAS}_n^{\omega_1,\lambda}$ is well-orderable. Using a definable bijection from λ onto $\lambda^{<\omega}$, we conclude that X is well-orderable.

Let $Z = \mathcal{P}(\lambda)$. Let U be a countably complete, normal fine measure on $\mathcal{P}_{\omega_1} Z$. Given $\mu \in \mathrm{MEAS}_1^{\omega_1,\lambda}$ and $\sigma \in \mathcal{P}_{\omega_1} Z$, let

$$f_\mu(\sigma) = min \bigcap(\sigma \cap \mu).$$

So f_μ is a function from $\mathcal{P}_{\omega_1} Z$ into the ordinals.

[9] This is similar to Kunen's proof that under AD, every countably complete measure on an ordinal is OD. See [5, Corollary 28.21].

Claim 4 *Suppose $\mu \neq \nu$ are in $MEAS_1^{\omega_1,\lambda}$. Then $\forall^*_U \sigma\ f_\mu(\sigma) \neq f_\nu(\sigma)$; here "$\forall^*_U \sigma\varphi(\sigma)$" abbreviates the statement "the set of σ such that $\varphi(\sigma)$ is in U".*

Proof Let A witness $\mu \neq \nu$. Without loss of generality, assume $A \in \mu$ and $\neg A \in \nu$. By fineness of U, $\forall^*_U \sigma, \{A, \neg A\} \subset \sigma$. Fix such a σ. Then $f_\mu(\sigma) \in A$ and $f_\nu(\sigma) \in \neg A$. Since $A, \neg A$ are disjoint, $f_\mu(\sigma) \neq f_\nu(\sigma)$. □

Let $\pi : X \to \prod_{\sigma \in \mathcal{P}_{\omega_1} Z} Ord/U$ be defined as: $\pi(\mu) = [f_\mu]_U$. The claim gives us that π is an injection. By DC, $\prod_{\sigma \in \mathcal{P}_{\omega_1} Z} Ord/U$ is well-founded and furthermore is well-ordered. Therefore, X is well-ordered as desired. □

6 AD$^+$, AD$_\mathbb{R}$, and Supercompactness of ω_1

In this section, we prove Theorem 4. The following fundamental fact about AD$^+$ is due to W.H. Woodin (cf. [7]).

Theorem 6 (Woodin) *The following are equivalent.*

1. *AD$^+$.*
2. *AD + the class of Suslin cardinals is closed below Θ.*

We will also need the following results due to D. A. Martin and Woodin.

Theorem 7 *Assume ZF + DC. The following are equivalent.*

1. *AD$_\mathbb{R}$.*
2. *AD$^+$ + every set is Suslin.*

We now use Theorems 6 and 7 to prove Theorem 4.

Proof (Theorem 4) First, note that by Theorem 1, DC follows from supercompactness of ω_1. The ($\Leftarrow$) direction follows immediately from Theorem 7. For the ($\Rightarrow$) direction, suppose AD$_\mathbb{R}$ fails. Let $\kappa < \Theta$ be the largest Suslin cardinal and $\Gamma = S(\kappa)$. The existence of κ follows from Theorems 6 and 7. By [6, Theorem 1.3], there is a universal Γ-set. Let A be such a universal set. Then by results of Sect. 5, A is weakly homogeneously Suslin. By the Martin-Solovay construction, $\neg A$ is Suslin. But $\neg A \in \check{\Gamma} \backslash \Gamma$. This contradicts the fact that Γ is the largest Suslin pointclass.

Remark 6 Wilson's methods, cf. [22], using the theory of envelopes of pointclasses can be used to show that ZF + DC + ω_1 is strongly compact implies every Suslin set of reals is co-Suslin directly, without using weak homogeneity.

The following may be a more approachable version of a well-known conjecture that AD is equivalent to AD$^+$.

Conjecture 2 *Assume ω_1 is supercompact.* AD *is equivalent to* AD$^+$.

7 HPC and Supercompactness of ω_1

In this section, we will prove Theorem 3. First, we note that we do not need the full "ω_1 is supercompact" hypothesis in the proof of the theorem; one just needs:

- ω_1 is $\mathbb{R}$-supercompact, and
- $\neg\Box_{\omega_1}$.

Both of these are consequences of ω_1 is supercompact, cf. [21, Sect. 1].

Now we explain *Hod Pair Capturing* (HPC). This hypothesis and the notion of *least branch hod pair* (lbr hod pair) are formulated by John Steel. The reader can see [16] for a detailed discussion regarding topics concerning least-branch hod premise, lbr hod pairs, and HPC. The main thing one needs from HPC are the facts given by Theorem 8. For basic facts about inner model theoretic notions such as iteration strategies, see [15]. In particular, a complete strategy for $\mathcal{P}$ is an iteration strategy Σ that acts on all finite stacks (of normal trees) on $\mathcal{P}$ that are according to Σ.

Definition 6 (*lbr hod pair,* [16]) $(\mathcal{P}, \Sigma)$ is an lbr hod pair if $\mathcal{P}$ is an lpm (least-branch hod premouse) and Σ is a complete strategy for $\mathcal{P}$ that normalizes well and has strong hull condensation.

Definition 7 (HPC, [16]) Suppose A is Suslin co-Suslin. Then there is an lbr hod pair $(\mathcal{P}, \Sigma)$ such that A is Wadge reducible to $Code(\Sigma)$.

Remark 7 We caution the reader that the formulation of HPC here in Definition 7 is slightly different from Steel's formulation of HPC. The difference is that we do not work under AD^+ in this section. In applications using the core model induction, we are proving that HPC or its variations holds in a universe where AD typically fails (assuming certain smallness hypotheses). From our hypotheses and Steel's results in [16, 17], we get that such a Σ as in Definition 7 is Suslin co-Suslin.

It is conjectured that AD^+ implies HPC. HPC and its variations have been shown to hold in very strong models of determinacy, cf. [12].

In the above, a complete strategy acts on all countable stacks of countable normal trees. The reader can consult [16] for more details on lbr hod pairs. The basic theory of lbr hod pairs has been worked out in [16]. What we need are a couple of facts about them. In the following, we fix a canonical coding $Code$ of subsets of HC by subsets of $\mathbb{R}$.[10] Given an lbr hod pair $(\mathcal{P}, \Sigma)$, for $n < \omega$, $\mathcal{M}_n^{\Sigma,\sharp}$ is the minimal, active Σ-mouse that has n Woodin cardinals. See for instance [14] for a precise definition. The following facts are relevant for us.

Lemma 2 *Let $(\mathcal{P}, \Sigma)$ be an lbr hod pair. Let $M = \mathcal{M}_n^{\Sigma,\sharp}$ and Λ be M's canonical strategy. Let λ be the largest Woodin cardinal of M. There is a term $\tau_\Sigma \in M^{Coll(\omega,\lambda)}$ such that whenever $i : M \to N$ is an iteration embedding via an iteration according to λ, and $g \subseteq Coll(\omega, i(\lambda))$ is N-generic, then*

[10] One way of defining $Code$ is as follows. Let $\pi : \mathbb{R} \to HC$ be a surjection defined as: for any x that codes a well-founded relation E_x on ω, let $\pi(x)$ be the transitive collapse of the structure (ω, E_x). Then for any $A \subseteq HC$, $Code(A)$ is defined to be $\pi^{-1}[A]$.

$$[i(\tau_\Sigma)]_g = Code(\Sigma) \cap N[g].$$

Theorem 8 *Suppose ω_1 is $\mathbb{R}$-supercompact, $\neg\square_{\omega_1}$. Suppose $(\mathcal{P}, \Sigma)$ is an lbr hod pair. Then*

1. *[17, Sect. 2] $Code(\Sigma)$ is Suslin co-Suslin.*
2. *[21, Sect. 3] $\mathcal{M}_2^{\Sigma,\sharp}$ exists.*

Remark 8 We note that in the above theorem, the hypothesis ω_1 is $\mathbb{R}$-supercompact is used in (1) to extend Σ to act on all stacks $\mathcal{W}$ such that there is a surjection of $\mathbb{R}$ onto $\mathcal{W}$. The proof of (2) just needs ω_1 is $\mathbb{R}$-strongly compact and $\neg\square_{\omega_1}$.

The following theorem, due to Neeman, is our main tool for proving determinacy.

Theorem 9 (Neeman, [10]) *Suppose $A \subseteq \mathbb{R}$. Suppose (M, Λ, δ) is such that*

1. *M is a countable, transitive model of* ZFC*;*
2. *$M \vDash \delta$ is Woodin;*
3. *Λ is an $\omega_1 + 1$-iteration strategy for M*[11]*;*
4. *there is a term $\tau \in M^{Coll(\omega,\delta)}$ such that whenever $i : M \to N$ is an iteration map according to Λ, $g \subseteq Coll(\omega, i(\delta))$ is N-generic, then $i(\tau)_g = A \cap N[g]$.*

Then A is determined.

Proof (Theorem 3) Let A be Suslin. Then A is also co-Suslin by Remark 1. By HPC, let $(\mathcal{P}, \Sigma)$ be an lbr hod pair such that A is Wadge reducible to $Code(\Sigma)$. Let $x \in \mathbb{R}$ witness this; we let τ_x be the continuous function given by x such that $\tau_x^{-1}[Code(\Sigma)] = A$.[12] By Theorem 8, $Code(\Sigma)$ is Suslin co-Suslin and $\mathcal{M}_2^{\Sigma,\sharp}$ exists. Let Λ be the canonical iteration strategy for $\mathcal{M}_2^{\Sigma,\sharp}$ and $\delta_0 < \delta_1$ be the Woodin cardinals of $\mathcal{M}_2^{\Sigma,\sharp}$. Let $\mathcal{T}$ be an iteration tree with the following properties:

- $\mathcal{T}$ is according to Λ.
- Letting $i : \mathcal{M}_2^{\Sigma,\sharp} \to \mathcal{N}$ be the corresponding iteration embedding, then x is $\mathcal{N}$-generic for the extender algebra at $i(\delta_0)$.[13]

Now we can construe $\mathcal{N}[x]$ as a Σ-mouse over x, which we will call $\mathcal{M}$. Note that $i(\delta_1)$ is a Woodin cardinal of $\mathcal{M}$ and Λ induces a strategy Ψ on $\mathcal{M}$.

We note that $(\mathcal{M}, \Psi, i(\delta_1))$ satisfies the hypothesis of Theorem 9 for A. Let τ_Σ be given as in Lemma 2 for $(\mathcal{M}_2^{\Sigma,\sharp}, \Lambda, \delta_1)$, then $i(\tau_\Sigma)$ induces a term $\sigma_\Sigma \in M^{Coll(\omega,i(\delta_1))}$ satisfying (3) of Theorem 9. Let $\mathbb{P} = Coll(\omega, i(\delta_1))$. The term τ consists of $(1_\mathbb{P}, \sigma)$ where $1_\mathbb{P} \Vdash_\mathbb{P}$ "σ is a real and $\sigma \in \tau_x^{-1}[\sigma_\Sigma]$".

By Theorem 9, A is determined. This completes the proof of Theorem 3. □

---

[11] Neeman [10] needs less iterability than $\omega_1 + 1$-iterability. $\omega_1 + 1$-iterability is available to us in this context.

[12] We can construe x as a function $\omega^{<\omega} \to \omega^{<\omega}$ and this naturally gives rise to a continuous (in fact Lipschitz) function $\sigma_x : \omega^\omega \to \omega^\omega$.

[13] The reader can see [15] for more detailed discussions about the extender algebra and genericity iteration trees. All we need here is the fact there is some forcing $\mathbb{Q} \in \mathcal{N}$ such that $\text{card}^{\mathcal{N}}(\mathbb{Q}) < i(\delta_1)$ and that there is some $\mathcal{N}$-generic filter $h \subset \mathbb{Q}$ with $x \in \mathcal{N}[h]$.

We conjecture that HPC is not needed in Theorem 3. Reference [21] has shown that ω_1 is supercompact implies that all sets in $L(\mathbb{R})$ are Suslin and co-Suslin and are determined and much more.[14] One may hope to prove Conjecture 3 by showing that every Suslin set is homogeneously Suslin. Theorem 2 shows that every Suslin set is a projection of a homogenously Suslin, hence determined, set.

Conjecture 3 *Assume ω_1 is supercompact. For any Suslin set A, A is determined.*

Acknowledgements We would like to thank Hugh Woodin for communicating his insights on this subject as well as his results concerning the model $\mathcal{C}^+$. We are grateful to the anonymous referee for numerous helpful comments. The first author would like to thank the Japan Society for the Promotion of Science (JSPS) for its generous support through the grant with JSPS KAKENHI Grant Number 15K17586 and 19K03604. He is also grateful to the Sumitomo Foundation for its generous support through Grant for Basic Science Research. The second author would like to thank the National Science Foundation (NSF) for its generous support through Grants DMS-1565808 and DMS-1849295.

References

1. Caicedo, A.E., Ketchersid, R.: A trichotomy theorem in natural models of AD^+. In: Set theory and its applications. Contemp. Math. **533**, 227–258. Am. Math. Soc., Providence (2011). https://doi.org/10.1090/conm/533/10510
2. Harrington, L.A., Kechris, A.S.: On the determinacy of games on ordinals. Ann. Math. Logic **20**(2), 109–154 (1981). https://doi.org/10.1016/0003-4843(81)90001-2
3. Ikegami, D.: Games in set theory and logic. Ph.D. thesis, Institute for Logic, Language and Computation, Universiteit van Amsterdam (2010). http://www.sic.shibaura-it.ac.jp/%7Eikegami/thesis-latest.pdf
4. Jech, T.: ω_1 can be measurable. Israel J. Math. **6**, 363–367 (1969) (1968)
5. Kanamori, A.: The Higher Infinite: Large Cardinals in Set Theory from Their Beginnings. Springer Science & Business Media, Berlin (2008)
6. Kechris, A.S.: Souslin cardinals, κ-Souslin sets and the scale property in the hyperprojective hierarchy. In: Cabal Seminar 77–79 (Proc. Caltech-UCLA Logic Sem., 1977–79). Lecture Notes in Mathematics, vol. 839, pp. 127–146. Springer, Berlin (1981)
7. Ketchersid, R.: More structural consequences of AD. Contemp. Math. **533**, 71–105 (2011)
8. Larson, P.: Extensions of the Axion of Determinacy. http://www.users.miamioh.edu/larsonpb/adplusbook_public.pdf. It is the version as of June 26, 2019
9. Martin, D.A., Woodin, W.H.: Weakly homogeneous trees. In: Games, Scales, and Suslin Cardinals: The Cabal Seminar, vol. 1, pp. 421–438 (2008)
10. Neeman, I.: Optimal proofs of determinacy. Bull. Symb. Logic **1**(3), 327–339 (1995)
11. Neeman, I.: Inner models and ultrafilters in $L(\mathbb{R})$. Bull. Symb. Logic **13**(1), 31–53 (2007). http://projecteuclid.org/euclid.bsl/1174668217
12. Sargsyan, G., Trang, N.: The largest Suslin axiom. Submitted. http://sites.math.rutgers.edu/%7Egs481/lsa.pdf
13. Sargsyan, G., Trang, N.: The exact consistency strength of the generic absoluteness for the universally Baire sets (2019). In preparation
14. Schlutzenberg, F., Trang, N.: Scales in hybrid mice over $\mathbb{R}$ (2016). https://www.math.uci.edu/%7Entrang. Submitted

[14] Reference [21] shows that there is a transitive class model M containing all reals such that $M \models \mathsf{AD}_{\mathbb{R}} + \mathsf{DC}$.

15. Steel, J.R.: An outline of inner model theory. In: Handbook of Set Theory, pp. 1595–1684 (2010)
16. Steel, J.R.: Normalizing iteration trees and comparing iteration strategies (2016).https://math.berkeley.edu/%7Esteel/papers/Publications.html
17. Steel, J.R.: Mouse pairs and suslin cardinals (2018). https://math.berkeley.edu/%7Esteel/papers/Publications.html
18. Takeuti, G.: A relativization of axioms of strong infinity to ω_1. Ann. Japan Assoc. Philos. Sci. **3**, 191–204 (1970)
19. Trang, N.: Determinacy in $L(\mathbb{R}, \mu)$. J. Math. Log. **14**(1), 1450,006, 23 (2014). https://doi.org/10.1142/S0219061314500068
20. Trang, N.: Structure theory of $L(\mathbb{R}, \mu)$ and its applications. J. Symb. Logic **80**(01), 29–55 (2015)
21. Trang, N., Wilson, T.: Determinacy from strong compactness of ω_1. arXiv preprint arXiv:1609.05411 (2016)
22. Wilson, T.M.: Contributions to descriptive inner model theory. Ph.D. thesis, University of California (2012). Available at author's website

Interpolation Properties for Sacchetti's Logics

Sohei Iwata

Abstract In this paper we investigate interpolation properties for Sacchetti's logics $\mathbf{wGL}_n$, which are fragments of Gödel-Löb logic **GL**. We prove the effective Lyndon interpolation property for Sacchetti's logics $\mathbf{wGL}_n$ using Shamkanov's circular proof argument.

Keywords Sacchetti's logic · Lyndon interpolation property · Effective fixed-point theorem

1 Introduction

A system **L** is said to have Craig interpolation property (CIP) if for any implication $\varphi \to \psi$ which is provable in **L**, there exists a formula θ (called an interpolant of $\varphi \to \psi$) such that θ consists of common variables of φ and ψ, and satisfies $\mathbf{L} \vdash \varphi \to \theta$ and $\mathbf{L} \vdash \theta \to \psi$. A system **L** is said to have Lyndon interpolation property (LIP) if for any provable implication $\varphi \to \psi$, it has a stronger interpolant θ which preserves positivity of variables. Gödel-Löb logic **GL** consists of axioms and rules of the modal logic **K** and Löb axiom $\Box(\Box p \to p) \to \Box p$. The logic **GL** analyzes properties of formal provability in Peano arithmetic. There are several classical investigations of CIP for **GL** (see Smoryński [10] or Boolos [1]), however it had been open that whether **GL** posses LIP until Shamkanov solved in [8, 9].

Shamkanov proved LIP for **GL** by two ways. In the former paper [8] he gave a semantical proof by using Kripke models. The later one [9] he introduced a circular proof system which is equivalent to **GL**, and proved LIP for **GL** syntactically.

Let **L** be a system. A circular proof system $^\circ\mathbf{L}$ of **L** is one which admits a "circular proof" that is a derivation tree of **L** whose leaves are either axioms of **L** or equal to a sequent below that leaf. Shamkanov showed that **GL** is equivalent to the circular proof system of the modal logic **K4**. He also established a method of constructing

S. Iwata (✉)
Graduate School of System Informatics, Kobe University, 1-1 Rokkodai-cho, 657-8501 Nada-ku, Kobe, Japan
e-mail: soh.iwata@people.kobe-u.ac.jp

T. Arai et al. (eds.), *Advances in Mathematical Logic*, Springer Proceedings in Mathematics & Statistics 369, https://doi.org/10.1007/978-981-16-4173-2_3

a Lyndon interpolant in the circular proof system with the aid of the fixed-point theorem for **GL**.

In the present paper we try to generalize Shamkanov's results into fragments of **GL**, which are developed by Sacchetti [7]. For a non-zero natural number n, Sacchetti's logic $\mathbf{wGL}_n$ is obtained from **GL** by replacing Löb axiom $\Box(\Box p \to p) \to \Box p$ by a weaker one $\Box(\Box^n p \to p) \to \Box p$. Sacchetti [7] proved CIP for $\mathbf{wGL}_n$ $(n > 1)$ via Kripke semantics, and the fixed-point theorem for $\mathbf{wGL}_n$. It seems to be open whether $\mathbf{wGL}_n$ posses LIP for $n \geq 2 \to n > 1$. In this paper we will answer this question positively.

This paper is organized as follows: First, in Sect. 2 we give precise definitions and basic facts of Sacchetti's logics $\mathbf{wGL}_n$. In Sect. 3 we develop one-sided sequent calcululi $\mathbf{wGL}_n^{\mathbf{G}}$, and their cut-admissibility via Kripke semantics. A proof search procedure in $\mathbf{wGL}_n^{\mathbf{G}}$ will be introduced. Using $\mathbf{wGL}_n^{\mathbf{G}}$, we give a new proof of CIP for $\mathbf{wGL}_n$ in Sect. 4. Second, in Sect. 5 we introduce a circular proof system which is equivalent to $\mathbf{wGL}_n^{\mathbf{G}}$. Finally, in Sect. 6, we prove LIP for $\mathbf{wGL}_n$ by applying Shamkanov's argument. We also show that there is an effective way to construct an interpolant of a given formula $\varphi \to \psi$.

2 Sacchetti's Logics $\mathbf{wGL}_n$

Throughout this paper, *propositional variables* are denoted by $p, q, \ldots$ and their complements by $\overline{p}, \overline{q}, \ldots$. We call them *literals*.

Formulas, denoted by $\varphi, \psi, \ldots$, are build up using the following grammar:

$$\varphi ::= p \mid \overline{p} \mid \top \mid \bot \mid \varphi \wedge \varphi \mid \varphi \vee \varphi \mid \Box\varphi \mid \Diamond\varphi.$$

We inductively define the *negation* $\overline{\varphi}$ of a formula φ in the usual way:

$$\overline{\top} :\equiv \bot,\quad \overline{\bot} \equiv \top,\quad \overline{(p)} :\equiv \overline{p},\quad \overline{(\overline{p})} :\equiv p,\quad \overline{\varphi \vee \psi} :\equiv \overline{\varphi} \wedge \overline{\psi},\quad \overline{\varphi \wedge \psi} :\equiv \overline{\varphi} \vee \overline{\psi},$$
$$\overline{\Box\varphi} :\equiv \Diamond\overline{\varphi},\quad \overline{\Diamond\varphi} :\equiv \Box\overline{\varphi}.$$

We put $(\varphi \to \psi) :\equiv (\overline{\varphi} \vee \psi)$, and $(\varphi \leftrightarrow \psi) :\equiv (\varphi \to \psi) \wedge (\psi \to \varphi)$.

Formulas $\Box^n\varphi$ and $\Diamond^n\varphi$ stand for $\overbrace{\Box \cdots \Box}^{n}\varphi$ and $\overbrace{\Diamond \cdots \Diamond}^{n}\varphi$, respectively.

Definition 1 Let n be a non-zero natural number. The Hilbert-style axiomatization for Sacchetti's logic ($\mathbf{wGL}_n$) is defined as follows:
Axioms:

1. Propositional tautologies;
2. $\Box(p \to q) \to (\Box p \to \Box q)$;
3. $\Box(\Box^n p \to p) \to \Box p$,

Rules: *modus ponens* $\dfrac{\varphi \quad \varphi \to \psi}{\psi}$, *necessitation* $\dfrac{\varphi}{\Box\varphi}$, and *substitution*.

Note that $\mathbf{wGL}_1$ coincides with the standard provability logic $\mathbf{GL}$, and for $n \geq 2$, $\mathbf{wGL}_n$ is a proper fragment of $\mathbf{GL}$.

The next Propositions 1 and 2 state basic properties which will be used in Sects. 4 and 5.

Proposition 1 *Assume $n \geq 1$. For any φ, $\mathbf{wGL}_n \vdash \Box\varphi \to \Box^{n+1}\varphi$.*

Proof See Sacchetti [7] or Kurahashi and Okawa [4]. □

We give some notations. For $n \geq 1$ and φ, we put:

$$[n]\varphi :\equiv \Box\varphi \wedge \Box^2\varphi \wedge \cdots \wedge \Box^n\varphi, \quad [n]^+\varphi :\equiv \varphi \wedge [n]\varphi.$$

Let Γ be a set of formulas. The sets $\Box^n\Gamma$, $[n]\Gamma$, and $[n]^+\Gamma$ denote the ones obtained from Γ by replacing every formula φ in Γ by $\Box^n\varphi$, $[n]\varphi$, and $[n]^+\varphi$, respectively.

Lemma 1 *Assume $n \geq 1$. For any formula φ,*

1. $\mathbf{wGL}_n \vdash [n]\varphi \leftrightarrow [n+1]\varphi$.
2. $\mathbf{wGL}_n \vdash [n]\varphi \to \Box[n]\varphi$.

Proof Clearly follows from Proposition 1. □

Proposition 2 *Assume $n \geq 1$. For any sets of formulas Γ and Δ, and any formula φ,*

1. $\mathbf{wGL}_n \vdash [n]^+\varphi \to \varphi$.
2. *If*
$$\mathbf{wGL}_n \vdash \bigwedge [n]^+\Gamma \wedge \bigwedge [n]\Delta \to \varphi,$$
then
$$\mathbf{wGL}_n \vdash \bigwedge [n](\Gamma \cup \Delta) \to \Box\varphi.$$
3. *If*
$$\mathbf{wGL}_n \vdash \bigwedge [n]^+\Gamma \wedge \bigwedge [n]\Delta \wedge [n]\varphi \to \varphi,$$
then
$$\mathbf{wGL}_n \vdash \bigwedge [n]^+\Gamma \wedge \bigwedge [n]\Delta \to \varphi.$$

Proof

1. Trivial.
2. Suppose that $\mathbf{wGL}_n \vdash \bigwedge [n]^+\Gamma \wedge \bigwedge [n]\Delta \to \varphi$. Since $\mathbf{wGL}_n$ is normal,
$$\mathbf{wGL}_n \vdash \bigwedge [n+1]\Gamma \wedge \bigwedge \Box[n]\Delta \to \Box\varphi.$$

By Lemma 1, the premises can be simplified as follows:

$$\mathbf{wGL}_n \vdash \bigwedge [n]\Gamma \wedge \bigwedge [n]\Delta \to \Box\varphi.$$

That is,

$$\mathbf{wGL}_n \vdash \bigwedge [n]\,(\Gamma \cup \Delta) \to \Box\varphi.$$

3. The argument is based on Kurahashi and Okawa [4] Proposition 3.4. Suppose that $\mathbf{wGL}_n \vdash \bigwedge [n]^+\Gamma \wedge \bigwedge [n]\Delta \wedge [n]\varphi \to \varphi$. We claim that for any k ($0 \leq k \leq n-1$),

$$\mathbf{wGL}_n \vdash \bigwedge [n]^+\Gamma \wedge \bigwedge [n]\Delta \wedge (\Box^{k+1}\varphi \wedge \cdots \wedge \Box^n\varphi) \to [n]^+\varphi.$$

We prove the claim by induction on k.
Base case ($k = 0$). It is clear that $\mathbf{wGL}_n \vdash \bigwedge [n]^+\Gamma \wedge \bigwedge [n]\Delta \wedge [n]\varphi \to [n]\varphi$. Combining with the supposition we have

$$\mathbf{wGL}_n \vdash \bigwedge [n]^+\Gamma \wedge \bigwedge [n]\Delta \wedge [n]\varphi \to [n]^+\varphi. \tag{1}$$

Inductive case. Suppose that the claim holds for k, i.e.,

$$\mathbf{wGL}_n \vdash \bigwedge [n]^+\Gamma \wedge \bigwedge [n]\Delta \wedge (\Box^{k+1}\varphi \wedge \cdots \wedge \Box^n\varphi) \to [n]^+\varphi.$$

Since $\mathbf{wGL}_n$ is normal,

$$\mathbf{wGL}_n \vdash \bigwedge [n+1]\Gamma \wedge \bigwedge \Box[n]\Delta \wedge (\Box^{k+2}\varphi \wedge \cdots \wedge \Box^{n+1}\varphi) \to [n+1]\varphi.$$

By Lemma 1,

$$\mathbf{wGL}_n \vdash \bigwedge [n]\Gamma \wedge \bigwedge [n]\Delta \wedge (\Box^{k+2}\varphi \wedge \cdots \wedge \Box^{n+1}\varphi) \to [n]\varphi. \tag{2}$$

On the other hand, by the inductive hypothesis,

$$\begin{aligned}
\mathbf{wGL}_n &\vdash \bigwedge [n]^+\Gamma \wedge \bigwedge [n]\Delta \wedge (\Box^{k+1}\varphi \wedge \cdots \wedge \Box^n\varphi) \to \varphi, \\
&\vdash \bigwedge [n]^+\Gamma \wedge \bigwedge [n]\Delta \wedge (\Box^{k+1}\varphi \wedge \cdots \wedge \Box^{n-1}\varphi) \to (\Box^n\varphi \to \varphi), \\
&\vdash \bigwedge [n+1]\Gamma \wedge \bigwedge \Box[n]\Delta \wedge (\Box^{k+2}\varphi \wedge \cdots \wedge \Box^n\varphi) \to \Box(\Box^n\varphi \to \varphi), \\
&\vdash \bigwedge [n+1]\Gamma \wedge \bigwedge \Box[n]\Delta \wedge (\Box^{k+2}\varphi \wedge \cdots \wedge \Box^n\varphi) \to \Box\varphi.
\end{aligned}$$

By Proposition 1 and Lemma 1,

$$\begin{aligned}\mathbf{wGL}_n &\vdash \bigwedge[n+1]\Gamma \wedge \bigwedge\Box[n]\Delta \wedge \left(\Box^{k+2}\varphi \wedge \cdots \wedge \Box^n\varphi\right) \to \Box^{n+1}\varphi,\\ &\vdash \bigwedge[n]\Gamma \wedge \bigwedge[n]\Delta \wedge \left(\Box^{k+2}\varphi \wedge \cdots \wedge \Box^n\varphi\right) \to \Box^{n+1}\varphi.\end{aligned}$$

Combining with (2), we have

$$\begin{aligned}\mathbf{wGL}_n &\vdash \bigwedge[n]\Gamma \wedge \bigwedge[n]\Delta \wedge \left(\Box^{k+2}\varphi \wedge \cdots \wedge \Box^n\varphi\right) \to [n]\varphi,\\ &\vdash \bigwedge[n]^+\Gamma \wedge \bigwedge[n]\Delta \wedge \left(\Box^{k+2}\varphi \wedge \cdots \wedge \Box^n\varphi\right) \to [n]\varphi.\end{aligned}$$

From this and (1), we obtain

$$\mathbf{wGL}_n \vdash \bigwedge[n]^+\Gamma \wedge \bigwedge[n]\Delta \wedge \left(\Box^{k+2}\varphi \wedge \cdots \wedge \Box^n\varphi\right) \to [n]^+\varphi.$$

The proof of the claim is completed.

We return to the proof of Proposition 2.3. By the claim, If $k = n - 1$, then

$$\begin{aligned}\mathbf{wGL}_n &\vdash \bigwedge[n]^+\Gamma \wedge \bigwedge[n]\Delta \wedge \Box^n\varphi \to [n]^+\varphi,\\ &\vdash \bigwedge[n]^+\Gamma \wedge \bigwedge[n]\Delta \to \left(\Box^n\varphi \to [n]^+\varphi\right).\end{aligned}$$

Since $\mathbf{wGL}_n \vdash [n]^+\varphi \to \varphi$,

$$\mathbf{wGL}_n \vdash \bigwedge[n]^+\Gamma \wedge \bigwedge[n]\Delta \to \left(\Box^n\varphi \to \varphi\right). \tag{3}$$

Then we have

$$\begin{aligned}\mathbf{wGL}_n &\vdash \bigwedge[n+1]\Gamma \wedge \bigwedge\Box[n]\Delta \to \Box\left(\Box^n\varphi \to \varphi\right), && \text{by the normality,}\\ &\vdash \bigwedge[n+1]\Gamma \wedge \bigwedge\Box[n]\Delta \to \Box\varphi,\\ &\vdash \bigwedge[n]\Gamma \wedge \bigwedge[n]\Delta \to \Box\varphi, && \text{by Lemma 1.}\end{aligned}$$

Moreover,

$$\begin{aligned}\mathbf{wGL}_n &\vdash \bigwedge\Box[n]\Gamma \wedge \bigwedge\Box[n]\Delta \to \Box^2\varphi,\\ &\vdash \bigwedge[n]\Gamma \wedge \bigwedge[n]\Delta \to \Box^2\varphi,\\ &\vdots\\ &\vdash \bigwedge[n]\Gamma \wedge \bigwedge[n]\Delta \to \Box^n\varphi,\\ &\vdash \bigwedge[n]^+\Gamma \wedge \bigwedge[n]\Delta \to \Box^n\varphi.\end{aligned}$$

From this and (3), we conclude $\mathbf{wGL}_n \vdash \bigwedge[n]^+\Gamma \wedge \bigwedge[n]\Delta \to \varphi$. □

3 Sequent Calculi for Sacchetti's Logics

In this section we present one-sided sequent calculi $\mathbf{wGL}_n^{\mathbf{G}}$ for Sacchetti's logics. *Sequents*, denoted by $\Gamma, \Delta, \ldots$ etc., are defined as *sets* of formulas. Let Γ and Δ be sequents and φ be a formula. We define the sequent (Γ, Δ) as the union of Γ and Δ, and (Γ, φ) as the set $\Gamma \cup \{\varphi\}$. As mentioned in Sect. 2, for a given Γ, $\Box^n \Gamma$ (and $\Diamond^n \Gamma$) denotes the sequents obtained from Γ by replacing every φ in Γ by $\Box^n \varphi$ (resp. $\Diamond^n \varphi$). A *derivation* in a calculus is a finite tree whose nodes are assigned by sequents that is constructed according to the rules of the calculus. A *proof* in a calculus is a derivation such that every leaf is labeled with axioms.

Definition 2 Assume $n \geq 1$. The one-sided sequent calculus $\mathbf{wGL}_n^{\mathbf{G}}$ consists of the following axioms and rules.

Axioms:

$$(p, \overline{p}), \qquad \top$$

Structural Rule:

$$\frac{\Gamma}{\Gamma, \Delta}\ (weak)$$

Propositional Rules:

$$\frac{\Gamma, \varphi, \psi}{\Gamma, \varphi \vee \psi}\ (\vee) \quad \frac{\Gamma, \varphi \quad \Gamma, \psi}{\Gamma, \varphi \wedge \psi}\ (\wedge)$$

Modal rule:

$$\frac{\Diamond^n \Gamma, \Gamma, \Diamond^n \overline{\varphi}, \varphi}{\Diamond \Gamma, \Box \varphi}\ (\Box_n)$$

The aim of this section is showing the following facts:

- For any sequent Γ, if $\mathbf{wGL}_n^{\mathbf{G}} \vdash \Gamma$, then we can construct a proof of Γ in $\mathbf{wGL}_n^{\mathbf{G}}$ effectively;
- The following rules are admissible in $\mathbf{wGL}_n^{\mathbf{G}}$:

$$\frac{\Gamma, \varphi \quad \Gamma, \overline{\varphi}}{\Gamma}\ (cut), \quad \text{and} \quad \frac{\Diamond^n \Gamma, \Gamma, \Diamond^n \overline{\varphi}, \varphi}{\Diamond^n \Gamma, \Gamma, \varphi}\ (L\ddot{o}b).$$

The argument is based on Sambin and Valentini [6].

3.1 Kripke Semantics

We present Kripke semantics for $\mathbf{wGL}_n$ developed by Sacchetti.

A *frame* F is a pair $\langle W, \prec \rangle$, where W is a non-empty set, and $\prec$ is a binary relation on W. A *model* M is a triple $\langle W, \prec, V \rangle$, where $\langle W, \prec \rangle$ is a frame and V is a valuation function from the set of propositional variables to $\mathcal{P}(W)$. For a model M, we say a propositional variable p is *true* in w (write $M, w \models p$) if $w \in V(p)$. The valuation of formulas in a model M is uniquely determined by V in a usual way. In particular, $M, w \models \Box\varphi$ iff for any $x \in W$, $w \prec x$ implies $M, x \models \varphi$. We say a formula φ is *valid in a model* $M = \langle W, \prec, V \rangle$ (write $M \models \varphi$) if $M, w \models \varphi$ for any $w \in W$. We say a formula φ is *valid in a frame* $F = \langle W, \prec \rangle$ (write $F \models \varphi$) if for any valuation V on F, φ is valid in the model $\langle W, \prec, V \rangle$.

We inductively define the binary relation $\prec^n$ on W: $x \prec^0 y$ iff $x = y$, and $x \prec^{n+1} y$ iff $\exists z \in W$ s.t. $x \prec^n z$ and $z \prec y$. A binary relation $\prec$ on W is said to be *n-transitive* if for any $x, y \in W$, $x \prec^n y$ implies $x \prec y$. A frame $F = \langle W, \prec \rangle$ (resp. a model $M = \langle W, \prec, V \rangle$) is said to be a $\mathbf{wGL}_n$ *-frame* (resp. a $\mathbf{wGL}_n$*-model*) if $\prec$ is $(n+1)$-transitive and conversely well-founded.

The following lemma will be needed in our proof of Theorem 1.

Lemma 2 *Let $\prec$ be a binary relation on a set W and suppose that $\prec$ is $(n+1)$-transitive. Then for any $x, y \in W$ and $k \geq 1$, if $x \prec^{kn} y$ then $x \prec^n y$.*

Proof Induction on k. The case for $k = 1$ is trivial. Assume Lemma holds for $\leq k$, and $x \prec^{(k+1)n} y$. Then there exist $x_1, \ldots, x_{(k+1)n} \in W$ such that $x \prec x_1 \prec \cdots \prec x_{(k+1)n}$ and $y = x_{(k+1)n}$. In particular, $x \prec^{n+1} x_{n+1} \prec^{kn-1} y$. Since $\prec$ is $(n+1)$-transitive, $x \prec x_{n+1}$. Hence we get $x \prec^{kn} y$. By the induction hypothesis, we obtain $x \prec^n y$. □

3.2 Proof Search Procedure

We give an effective way of constructing a proof of Γ in $\mathbf{wGL}_n^{\mathbf{G}}$, for every sequent Γ with $\mathbf{wGL}_n^{\mathbf{G}} \vdash \Gamma$ (see Proposition 4 in this subsection).

Lemma 3 *Let Γ be a sequent and φ be a formula. Then $\mathbf{wGL}_n^{\mathbf{G}} \vdash (\Gamma, \varphi, \overline{\varphi})$. Moreover, we can construct a proof of $(\Gamma, \varphi, \overline{\varphi})$ in $\mathbf{wGL}_n^{\mathbf{G}}$ effectively from Γ and φ.*

Proof Induction on the construction of φ.

- Suppose that φ is one of the form $\langle p, \overline{p}, \top, \bot \rangle$. In this case, a proof of $(\Gamma, \varphi, \overline{\varphi})$ in $\mathbf{wGL}_n^{\mathbf{G}}$ is given as follows:

$$\frac{\top}{\Gamma, \top, \bot}\,(weak), \quad \frac{p, \overline{p}}{\Gamma, p, \overline{p}}\,(weak).$$

- Assume $\varphi \equiv \psi \vee \theta$ or $\varphi \equiv \psi \wedge \theta$. Consider the following derivations:

$$\dfrac{\dfrac{\Gamma, \psi, \theta, \overline{\psi}}{\Gamma, \psi \vee \theta, \overline{\psi}}(\vee) \quad \dfrac{\Gamma, \psi, \theta, \overline{\theta}}{\Gamma, \psi \vee \theta, \overline{\theta}}(\vee)}{\Gamma, \psi \vee \theta, \overline{\psi} \wedge \overline{\theta}}(\wedge), \quad \dfrac{\dfrac{\Gamma, \psi, \overline{\psi}, \overline{\theta}}{\Gamma, \psi, \overline{\psi} \vee \overline{\theta}}(\vee) \quad \dfrac{\Gamma, \theta, \overline{\psi}, \overline{\theta}}{\Gamma, \theta, \overline{\psi} \vee \overline{\theta}}(\vee)}{\Gamma, \psi \wedge \theta, \overline{\psi} \vee \overline{\theta}}(\wedge).$$

By the induction hypothesis, we can effectively construct a proof of each assumption in the derivations. Thus we can also effectively construct a proof of $(\Gamma, \varphi, \overline{\varphi})$.

– Assume $\varphi \equiv \Box\psi$ or $\varphi \equiv \Diamond\psi$. Consider the following derivations:

$$\dfrac{\dfrac{\Diamond^n\overline{\psi}, \psi, \overline{\psi}}{\Box\psi, \Diamond\overline{\psi}}(\Box_n)}{\Gamma, \Box\psi, \Diamond\overline{\psi}}(weak), \quad \dfrac{\dfrac{\Diamond^n\psi, \psi, \overline{\psi}}{\Diamond\psi, \Box\overline{\psi}}(\Box_n)}{\Gamma, \Diamond\psi, \Box\overline{\psi}}(weak)$$

By the induction hypothesis, we can effectively construct a proof of each assumption in the derivations. Thus we can also effectively construct a proof of $(\Gamma, \varphi, \overline{\varphi})$.

□

Next, we describe the proof search procedure $\mathcal{P}$. For a given Γ, $\mathcal{P}$ generates a derivation of Γ in the following variant of $\mathbf{wGL}_n^{\mathbf{G}}$.

Definition 3 Assume $n \geq 1$. The calculus $\mathbf{wGL}_n^{\mathbf{G}'}$ consists of the following axioms and rules:

Axioms:

$$\Gamma, \varphi, \overline{\varphi} \qquad \Gamma, \top$$

Rules: Propositional rules of $\mathbf{wGL}_n^{\mathbf{G}}$, and

$$\dfrac{\Diamond^n\Gamma, \Gamma, \Diamond^n\overline{\varphi_1}, \varphi_1 \quad \cdots \quad \Diamond^n\Gamma, \Gamma, \Diamond^n\overline{\varphi_m}, \varphi_m}{L, \Diamond\Gamma, \Box\varphi_1, \ldots, \Box\varphi_m}\Box'_n,$$

where L is a set of literals and the constant $\bot$.

The rule $(\Box'_n)$ has m assumptions of the form $(\Diamond^n\Gamma, \Gamma, \Diamond^n\overline{\varphi_i}, \varphi_i)$. The meaning of $(\Box'_n)$ is that the conclusion is provable if at least one of the assumptions is provable.

For a given sequent, the proof search procedure $\mathcal{P}$ tries to apply all applicable rules of $\mathbf{wGL}_n^{\mathbf{G}'}$ until every leaf is decomposed into an axiom of $\mathbf{wGL}_n^{\mathbf{G}'}$ or a sequent to which no more rules are applicable.

The following proposition asserts that the procedure $\mathcal{P}$ works soundly.

Proposition 3 *For any input Γ, the proof search procedure $\mathcal{P}$ of Γ always halts.*

Proof It suffices to show that $\mathcal{P}$ never generates an infinite branch. Since any propositional rule lowers the numbers of connectives, $\mathcal{P}$ always halts as long as it applies only propositional rules. Therefore we have to show that $\mathcal{P}$ never generates infinitely many applications of $(\Box_n')$. Suppose, for contradiction, that $\mathcal{P}$ produces an infinitely many applications of $\Box_n'$ for some sequent Γ.

$$\dfrac{\dfrac{\dfrac{\vdots}{\cdots \quad \Diamond^n \Delta_2, \Delta_2, \Diamond^n \overline{\psi_2}, \psi_2 \quad \cdots}}{L_2, \Diamond \Delta_2, \Box \Sigma_2 \ni \Box \psi_2}\,(\Box_n')}{\vdots}$$

$$\dfrac{\cdots \quad \Diamond^n \Delta_1, \Delta_1, \Diamond^n \overline{\psi_1}, \psi_1 \quad \cdots}{L_1, \Diamond \Delta_1, \Box \Sigma_1 \ni \Box \psi_1}\,(\Box_n')$$

$$\vdots$$

$$\Gamma$$

Note that every sequent in this infinite branch is non-axiomatic. Each time $\Box_n'$ is applied, we need at least one formula of the form $\Box \psi$. Since every propositional rule does not affect to any sequents of the form $\Diamond \Delta$, every $\Box \psi_{i+1}$ is obtained from either ψ_i or a formula in Δ_i by applying some propositional rules.

It is impossible that for all $i \geq 1$, $\Box \psi_{i+1}$ are obtained from ψ_i. Hence, for some natural number k, $\Box \psi_{k+1}$ is obtained from a formula θ in Δ_k by applying some propositional rules. The following table describes some formulas generated by $\mathcal{P}$ in each application of $(\Box_n')$ from the kth application of $(\Box_n')$.

k	$k+1$	$k+2$	$\cdots$	$k+(n-1)$	$k+n$
	$\Diamond^n \overline{\psi_{k+1}}, \psi_{k+1}$	$\Diamond^{n-1} \overline{\psi_{k+1}}$	$\cdots$	$\Diamond^2 \overline{\psi_{k+1}}$	$\Diamond \overline{\psi_{k+1}}$
$\Diamond^n \theta, \theta$	$\Diamond^{n-1} \theta$	$\Diamond^{n-2} \theta$	$\cdots$	$\Diamond \theta$	$\Diamond^n \theta, \theta$

When the $(k+n)$th $(\Box_n')$ is applied, the resulting sequent contains $\Diamond \overline{\psi_{k+1}}$ and θ, and $\mathcal{P}$ will decompose θ into $\Box \psi_{k+1}$ by applying propositional rules. Hence this branch contains a sequent containing $\Diamond \overline{\psi_{k+1}}$ and $\Box \psi_{k+1}$, i.e., an axiom of $\mathbf{wGL}_n^{\mathbf{G}'}$. This contradicts that the infinite branch consists of non-axiomatic sequents. □

Let π be a derivation of a sequent Γ in $\mathbf{wGL}_n^{\mathbf{G}'}$ generated by $\mathcal{P}$. A *search* of π is a subtree obtained by choosing a particular branch at each application of $(\Box_n)'$. A search π' of π is said to be *successful* if every branch of π' terminates in axioms of $\mathbf{wGL}_n^{\mathbf{G}'}$.

Proposition 4 *For an input Γ, let π be a derivation of Γ in $\mathbf{wGL}_n^{\mathbf{G}'}$ generated by $\mathcal{P}$. If π contains a successful search, then we can construct a proof of Γ in $\mathbf{wGL}_n^{\mathbf{G}}$ from π.*

Proof Let π' be a successful search of π. Then from π' we can construct a proof of Γ in $\mathbf{wGL}_n^{\mathbf{G}}$ by Lemma 3 and transforming each application of $(\Box_n')$ as follows:

$$\dfrac{\begin{array}{c}\vdots\\ \Diamond^n\Gamma,\Gamma,\Diamond^n\overline{\varphi_i},\varphi_i\end{array}}{\begin{array}{c}L,\Diamond\Gamma,\Box\Sigma\\ \vdots\end{array}}\,(\Box'_n) \longmapsto \dfrac{\dfrac{\begin{array}{c}\vdots\\ \Diamond^n\Gamma,\Gamma,\Diamond^n\overline{\varphi_i},\varphi_i\end{array}}{\Diamond\Gamma,\Box\varphi_i}\,(\Box_n)}{\begin{array}{c}L,\Diamond\Gamma,\Box\Sigma\\ \vdots\end{array}}\,(weak).$$

□

3.3 Cut-Admissibility

For a sequent Γ, let $\Gamma^\#$ be the formula $\bigvee\{\varphi \mid \varphi \in \Gamma\}$. We prove the following theorem.

Theorem 1 *For any sequent Γ, the following are equivalent:*

1. $\mathbf{wGL}_n \vdash \Gamma^\#$;
2. $F \models \Gamma^\#$ *for any finite* $\mathbf{wGL}_n$*-frame* F;
3. *For the input* Γ, *the proof search procedure* $\mathcal{P}$ *generates a derivation which has a successful search of* Γ;
4. $\mathbf{wGL}_n^{\mathbf{G}} \vdash \Gamma$;
5. $\mathbf{wGL}_n^{\mathbf{G}} + (cut) \vdash \Gamma$.

Proof $(3 \Rightarrow 4)$: Clearly follows from Proposition 4.

$(4 \Rightarrow 5)$: Trivial.

$(5 \Rightarrow 1)$: By induction on the length of proofs in $\mathbf{wGL}_n^{\mathbf{G}} + (cut)$.

$(1 \Rightarrow 2)$: Due to Sacchetti [7].

We prove $(2 \Rightarrow 3)$. We construct a finite countermodel of Γ from the derivation of Γ which is generated by $\mathcal{P}$ and has no successful searches.

Definition 4 Let π be a derivation of a sequent Γ in $\mathbf{wGL}_n^{\mathbf{G}'}$. We define that π is unsuccessful inductively as follows:

- If π consists of a single sequent Γ, then π is *unsuccessful* iff $\Gamma = (L, \Diamond\Pi)$ where L is a set of literals and the constant $\bot$ satisfying that there is no propositional variable p such that $p, \overline{p} \in L$;
- If the last application of π is $(\vee)$ or $(\wedge)$, then π is unsuccessful iff for some sub-derivation of the assumption sequent is unsuccessful;
- If the last application of π is $\Box'_n$, then π is unsuccessful iff every sub-derivation of the assumption sequent is unsuccessful.

Clearly if π has no successful searches then π is unsuccessful. It suffices to show the following lemma.

Lemma 4 *For any unsuccessful derivation in* $\mathbf{wGL}_n^{\mathbf{G}'}$, *if* π *is a derivation of* Γ, *then there is a finite* $\mathbf{wGL}_n$*-model* $\mathcal{M}$ *such that* $\mathcal{M} \not\models \Gamma^\#$.

Proof Induction on the height of unsuccessful π.

- Suppose that π consists of a single sequent $\Gamma = (L, \Diamond\Pi)$. Define $\mathcal{M} = \langle W, \prec, V \rangle$ as follows:
  - $W := \{w\}$, and $\prec := \emptyset$;
  - $M, w \models p :\iff \overline{p} \in L$.

  Then it is clear that for any $\varphi \in \Gamma$, $\mathcal{M}, w \not\models \varphi$, and hence $\mathcal{M} \not\models \Gamma^\#$.
- Suppose that the last application of π is $(\vee)$ or $(\wedge)$. By Definition 4 and the induction hypothesis, for some assumption sequent Δ, there is a model $\mathcal{M}$ such that $\mathcal{M} \not\models \Delta^\#$. It is clear that $\mathcal{M}$ also falsifies $\Gamma^\#$.
- Suppose that $\Gamma = (L, \Diamond\Pi, \Box\Sigma)$ and the last application of π is $(\Box'_n)$.

$$\dfrac{\begin{array}{c}\pi_1\\ \vdots \\ \Diamond^n\Pi, \Pi, \Diamond^n\overline{\varphi_1}, \varphi_1\end{array} \quad \cdots \quad \begin{array}{c}\pi_m\\ \vdots \\ \Diamond^n\Pi, \Pi, \Diamond^n\overline{\varphi_m}, \varphi_m\end{array}}{L, \Diamond\Pi, \Box\Sigma}\ \Box'_n$$

  By Definition 4 and the induction hypothesis, each π_i $(1 \le i \le m)$ has a model $\mathcal{M}_i = \langle W_i, \prec_i, V_i \rangle$ such that $\mathcal{M}_i \not\models (\Diamond^n\Pi, \Pi, \Diamond^n\overline{\varphi_i}, \varphi_i)^\#$. We assume for any $1 \le i, j \le m$, if $i \ne j$ then the sets W_i and W_j are disjoint. Moreover, we may assume each $\mathcal{M}_i$ has the root $w_i \in W_i$ and $\mathcal{M}_i, w_i \not\models (\Diamond^n\Pi, \Pi, \Diamond^n\overline{\varphi_i}, \varphi_i)^\#$. Define a model $\mathcal{M} = \langle W, \prec, V \rangle$ as follows:
  - $W := \bigcup W_i \cup \{w\}$ where w is a new object not contained in $\bigcup W_i$;
  - $x \prec y :\Leftrightarrow \begin{cases} x \in W_i \text{ and } x \prec_i y \text{ for some } 1 \le i \le m, \text{ or} \\ x = w \text{ and } w_i \prec_i^{kn} y \text{ for some } 1 \le i \le m \text{ and } k \ge 0 \end{cases};$
  - If $x \in W_i$ for some i, then $M, x \models p :\Leftrightarrow M_i, x \models p$;
  - $\mathcal{M}, w \models p :\Leftrightarrow \overline{p} \in L$.

  Clearly the relation $\prec$ is irreflexive and acyclic. We show that $\prec$ is $(n+1)$-transitive. Suppose that $x \prec^{n+1} y$. If $x \in W_i$ for some i, then $x \prec y$ immediately follows from the $(n+1)$-transitivity of $\prec_i$. Assume $x = w$. Then there are $x_1, \ldots, x_n \in W$ such that $x \prec x_1 \prec \cdots \prec x_n \prec y$. By the definition of $\prec$, we have $w_i \prec_i^{kn} x_1$ and $x_1 \prec_i \cdots \prec_i x_n \prec_i y$ for some $1 \le i \le m$ and $k \ge 0$. Hence we obtain $w_i \prec_i^{(k+1)n} y$, i.e., $x \prec y$. Thus, our $\mathcal{M}$ is a finite $\mathbf{wGL}_n$-frame.
  It suffices to show that for any formula $\varphi \in \Gamma$, $\mathcal{M}, w \not\models \varphi$. If $\varphi \in L$, then $\mathcal{M}, w \not\models \varphi$ clearly follows from the definition of V. Assume $\varphi \equiv \Box\varphi_i \in \Box\Sigma$ for some $1 \le i \le m$. Then we have $w \prec w_i$. By the induction hypothesis, $\mathcal{M}_i, w_i \not\models \varphi_i$. Note that if $x \in W_i$, then $\mathcal{M}_i, x \models \varphi \iff \mathcal{M}, x \models \varphi$. Therefore $\mathcal{M}, w_i \not\models \varphi_i$, i.e., $\mathcal{M}, w \not\models \Box\varphi_i$. Assume $\varphi \equiv \Diamond\psi \in \Diamond\Pi$. We have to show that for every x, if $w \prec x$, then $\mathcal{M}, x \not\models \psi$. By the definition of $\prec$, we have $w_i \prec_i^{kn} x$ for some $1 \le i \le m$ and $k \ge 0$. If $k = 0$, then $x = w_i$. In this case, $\mathcal{M}_i, w_i \not\models \psi$ by the induction hypothesis, i.e., $\mathcal{M}, w_i \not\models \psi$. If $k \ge 1$, then by Lemma 2, $w_i \prec_i^n x$. By the induction hypothesis, $\mathcal{M}_i, w_i \not\models \Diamond^n\psi$, and hence $\mathcal{M}_i, x \not\models \psi$, i.e., $\mathcal{M}, x \not\models \psi$.

Now the proof of Lemma 4 is completed. □

By Lemma 4, (2 $\Rightarrow$ 3) is verified. □

Corollary 1 *The rules* (cut) *and* (Löb) *are admissible in* $\mathbf{wGL}_n^{\mathbf{G}}$.

Proof The (cut)-admissibility immediately holds from Theorem 1. Suppose that $\mathbf{wGL}_n^{\mathbf{G}} \vdash (\Diamond^n \Gamma, \Gamma, \Diamond^n \overline{\varphi}, \varphi)$. Let π be a proof of this sequent. Then we can infer $(\Diamond^n \Gamma, \Gamma, \varphi)$ in $\mathbf{wGL}_n^{\mathbf{G}} + (cut)$ as follows:

$$\dfrac{\dfrac{\dfrac{\dfrac{\begin{matrix}\pi \\ \vdots \\ \Diamond^n \Gamma, \Gamma, \Diamond^n \overline{\varphi}, \varphi\end{matrix}}{\Diamond \Gamma, \Box \varphi}\,(\Box_n)}{\Diamond^n \Diamond \Gamma, \Diamond \Gamma, \Diamond^n \overline{\Box \varphi}, \Box \varphi}\,(weak)}{\Diamond^2 \Gamma, \Box^2 \varphi}\,(\Box_n)}{}$$

$$\dfrac{\begin{matrix}\pi \\ \vdots \\ \Diamond^n \Gamma, \Gamma, \varphi, \Diamond^n \overline{\varphi}\end{matrix} \qquad \dfrac{\begin{matrix}\vdots \\ \Diamond^n \Gamma, \Box^n \varphi\end{matrix}}{\Diamond^n \Gamma, \Gamma, \varphi, \Box^n \varphi}\,(weak)}{\Diamond^n \Gamma, \Gamma, \varphi}\,(cut)$$

By Theorem 1, $(\Diamond^n \Gamma, \Gamma, \varphi)$ is provable in $\mathbf{wGL}_n^{\mathbf{G}}$. Thus the rule (Löb) is admissible. □

4 Craig Interpolation Property for $\mathbf{wGL}_n$

In this section, we give a new proof of the Craig interpolation theorem for Sacchetti's logics via $\mathbf{wGL}_n^{\mathbf{G}}$. For a formula φ, we define $var(\varphi) := \{p \mid p \text{ occurs in } \varphi\} \cup \{q \mid \overline{q} \text{ occurs in } \varphi\}$. For a sequent Γ, we also define $var(\Gamma) := \bigcup\{var(\varphi) \mid \varphi \in \Gamma\}$. Put $\overline{\Gamma} := \{\overline{\varphi} \mid \varphi \in \Gamma\}$.

Theorem 2 (Craig interpolation theorem for $\mathbf{wGL}_n$) *Assume* $n \geq 1$. *If* $\mathbf{wGL}_n \vdash \varphi \to \psi$, *then there is a formula* θ *(called a Craig interpolant of* $\varphi \to \psi$*) such that:*

1. $\mathbf{wGL}_n \vdash \varphi \to \theta$ *and* $\mathbf{wGL}_n \vdash \theta \to \psi$;
2. $var(\theta) \subseteq var(\varphi) \cap var(\psi)$.

Moreover, such a θ *is effectively constructable from* φ *and* ψ.

In order to prove Theorem 2, we introduce a *split derivation system*. A *split sequent* is one of the form $[\theta]\ \Gamma_1 \mid \Gamma_2$ where θ is a formula and Γ_1 and Γ_2 are sequents. The natural meaning of $(\Gamma_1 \mid \Gamma_2)$ is the formula $\overline{\Gamma_1^{\#}} \to \Gamma_2^{\#}$. The formula θ in the bracket is a corresponding interpolant of $(\Gamma_1 \mid \Gamma_2)$.

Definition 5 The split derivation system $\mathbf{wGL}_n^{\mathbf{Sp}}$ consists of the following axioms and rules.

Axioms:

$$[\bot]\ p, \overline{p} \mid \emptyset \qquad [p]\ \overline{p} \mid p \qquad [\overline{p}]\ p \mid \overline{p} \qquad [\top]\ \emptyset \mid p, \overline{p}$$
$$[\bot]\ \top \mid \emptyset \qquad [\top]\ \emptyset \mid \top$$

Structural rule:

$$\frac{[\theta]\ \Gamma_1 \mid \Gamma_2}{[\theta]\ \Gamma_1, \Delta_1 \mid \Gamma_2, \Delta_2}\ (weak)$$

Propositional Rules:

$$\frac{[\theta]\ \Gamma_1, \varphi, \psi \mid \Gamma_2}{[\theta]\ \Gamma_1, \varphi \vee \psi \mid \Gamma_2}\ (\vee^l) \qquad \frac{[\theta]\ \Gamma_1 \mid \Gamma_2, \varphi, \psi}{[\theta]\ \Gamma_1 \mid \Gamma_2, \varphi \vee \psi}\ (\vee^r)$$

$$\frac{[\theta_1]\ \Gamma_1, \varphi \mid \Gamma_2 \quad [\theta_2]\ \Gamma_1, \psi \mid \Gamma_2}{[\theta_1 \vee \theta_2]\ \Gamma_1, \varphi \wedge \psi \mid \Gamma_2}\ (\wedge^l) \qquad \frac{[\theta_1]\ \Gamma_1 \mid \Gamma_2, \varphi \quad [\theta_2]\ \Gamma_1 \mid \Gamma_2, \psi}{[\theta_1 \wedge \theta_2]\ \Gamma_1 \mid \Gamma_2, \varphi \wedge \psi}\ (\wedge^r)$$

Modal rules:

$$\frac{[\theta]\ \Diamond^n \Gamma_1, \Gamma_1, \Diamond^n \overline{\varphi}, \varphi \mid \Diamond^n \Gamma_2, \Gamma_2}{[\Diamond\theta]\ \Diamond\Gamma_1, \Box\varphi \mid \Diamond\Gamma_2}\ (\Box_n^l) \qquad \frac{[\theta]\ \Diamond^n \Gamma_1, \Gamma_1 \mid \Diamond^n \Gamma_2, \Gamma_2, \Diamond^n \overline{\varphi}, \varphi}{[\Box\theta]\ \Diamond\Gamma_1 \mid \Diamond\Gamma_2, \Box\varphi}\ (\Box_n^r)$$

Lemma 5 *Assume $n \geq 1$, and let Γ_1 and Γ_2 be sequents. Then the following statements are equivalent:*

1. $\mathbf{wGL}_n \vdash \overline{\Gamma_1^\#} \to \Gamma_2^\#$;
2. $\mathbf{wGL}_n^{\mathbf{G}} \vdash (\Gamma_1, \Gamma_2)$;
3. $\mathbf{wGL}_n^{\mathbf{Sp}} \vdash (\Gamma_1 \mid \Gamma_2)$.

Proof (1⇔2): Clearly follows from Theorem 1.

(2⇒3): Let π be a proof of (Γ_1, Γ_2) in $\mathbf{wGL}_n^{\mathbf{G}}$. By bottom-up splitting of sequents, we obtain a proof of $(\Gamma_1 \mid \Gamma_2)$ in $\mathbf{wGL}_n^{\mathbf{Sp}}$.

(3⇒2): Let π' be a proof of $(\Gamma_1 \mid \Gamma_2)$. We can obtain a proof of (Γ_1, Γ_2) in $\mathbf{wGL}_n^{\mathbf{G}}$ by removing all splittings in π'. □

Note that if a proof π of (Γ_1, Γ_2) is given, then a proof π' of $(\Gamma_1 \mid \Gamma_2)$ in $\mathbf{wGL}_n^{\mathbf{Sp}}$ is effectively constructible from π.

Lemma 6 *Assume $n \geq 1$. Suppose that $\mathbf{wGL}_n^{\mathbf{Sp}} \vdash (\Gamma_1 \mid \Gamma_2)$, and let π be a proof of $[\theta]\ (\Gamma_1 \mid \Gamma_2)$ in $\mathbf{wGL}_n^{\mathbf{Sp}}$. Then:*

1. $var(\theta) \subseteq var\left(\overline{\Gamma_1}\right) \cap var(\Gamma_2)$;
2. $\mathbf{wGL}_n^{\mathbf{G}} \vdash (\Gamma_1, \theta)$ *and* $\mathbf{wGL}_n^{\mathbf{G}} \vdash \left(\Gamma_2, \overline{\theta}\right)$.

Proof By induction on the length of π.

– Suppose that π consists of an axiom $[\theta]\ \Gamma_1 \mid \Gamma_2$. Then it is clear that the corresponding formula θ satisfies the conditions 1 and 2.

– Suppose that π is one of the following derivations:

$$\dfrac{\begin{matrix}\pi_1\\ \vdots\\ [\theta_1]\ \Delta_{11} \mid \Delta_{12}\end{matrix}}{[\theta]\ \Gamma_1 \mid \Gamma_2}\,(R), \qquad \dfrac{\begin{matrix}\pi_1\\ \vdots\\ [\theta_1]\ \Delta_{11} \mid \Delta_{12}\end{matrix} \quad \begin{matrix}\pi_2\\ \vdots\\ [\theta_2]\ \Delta_{21} \mid \Delta_{22}\end{matrix}}{[\theta]\ \Gamma_1 \mid \Gamma_2}\,(R)$$

where π_i $(i = 1, 2)$ is the subproof of each hypothesis, and θ_i is the formula according to the rules in π_i. By the induction hypothesis for π_i, θ_i satisfies the following conditions:

- $var\,(\theta_i) \subseteq var\,\left(\overline{\Delta_{i1}}\right) \cup var\,(\Delta_{i2})$;
- $\mathbf{wGL}_n^{\mathbf{G}} \vdash (\Delta_{i1}, \theta)$ and $\mathbf{wGL}_n^{\mathbf{G}} \vdash \left(\Delta_{i2}, \overline{\theta}\right)$.

Then we can easily deduce from the above facts that θ enjoys the conditions 1 and 2. We only describe the case for $(\Box_n^r)$. Assume the last application of π is $(\Box_n^r)$.

$$\dfrac{\begin{matrix}\pi_1\\ \vdots\\ [\theta]\ \Diamond^n \Delta_1, \Delta_1 \mid \Diamond^n \Delta_2, \Delta_2, \Diamond^n \overline{\varphi}, \varphi\end{matrix}}{[\Box\theta]\ \Diamond \Delta_1 \mid \Diamond \Delta_2, \Box\varphi}\,(\Box_n^r) \qquad (*)$$

By the induction hypothesis,

- $var\,(\theta) \subseteq var\,\left(\overline{\Diamond^n \Delta_1, \Delta_1}\right) \cap var\,(\Diamond^n \Delta_2, \Delta_2, \Diamond^n \overline{\varphi}, \varphi)$;
- $\mathbf{wGL}_n^{\mathbf{G}} \vdash (\Diamond^n \Delta_1, \Delta_1, \theta)$ and $\mathbf{wGL}_n^{\mathbf{G}} \vdash \left(\Diamond^n \Delta_2, \Delta_2, \Diamond^n \overline{\varphi}, \varphi, \overline{\theta}\right)$.

1. We have

$$\begin{aligned} var\,(\Box\theta) = var\,(\theta) &\subseteq var\,\left(\overline{\Diamond^n \Delta_1, \Delta_1}\right) \cap var\,\left(\Diamond^n \Delta_2, \Delta_2, \Diamond^n \overline{\varphi}, \varphi\right), \\ &= var\,\left(\overline{\Diamond \Delta_1}\right) \cap var\,(\Diamond \Delta_2, \Box\varphi)\,. \end{aligned}$$

2. Consider the following derivations:

$$\dfrac{\dfrac{\begin{matrix}\rho_1\\ \vdots\\ \Diamond^n \Delta_1, \Delta_1, \theta\end{matrix}}{\Diamond^n \Delta_1, \Delta_1, \Diamond^n \overline{\theta}, \theta}\,(weak)}{\Diamond \Delta_1, \Box\theta}\,(\Box_n), \qquad \dfrac{\dfrac{\begin{matrix}\rho_2\\ \vdots\\ \Diamond^n \Delta_2, \Delta_2, \Diamond^n \overline{\varphi}, \varphi, \overline{\theta}\end{matrix}}{\Diamond^n \Delta_2, \Delta_2, \Diamond^n \overline{\varphi}, \varphi, \Diamond^n \overline{\theta}, \overline{\theta}}\,(weak)}{\Diamond \Delta_2, \Box\varphi, \Diamond\overline{\theta}}\,(\Box_n) \qquad \Box$$

Now we prove Theorem 2. Assume $\mathbf{wGL}_n \vdash \varphi \rightarrow \psi$. By Theorem 1 and Lemma 5, we can effectively obtain a proof π of $[\theta]\ (\overline{\varphi} \mid \psi)$ in $\mathbf{wGL}_n^{\mathbf{Sp}}$. By Lemma 6, we have $var(\theta) \subseteq var(\varphi) \cap var(\psi)$, $\mathbf{wGL}_n^{\mathbf{G}} \vdash \overline{\varphi}, \theta$ and $\mathbf{wGL}_n^{\mathbf{G}} \vdash \psi, \overline{\theta}$. Thus θ is indeed a Craig interpolant of $\varphi \rightarrow \psi$. □

Remark 1 Let φ be a formula. We define $var^+(\varphi)$ (resp. $var^-(\varphi)$) as the set of literals occurring positively (resp. negatively) in φ. Suppose that $\mathbf{wGL}_n \vdash \varphi \rightarrow \psi$. A *Lyndon interpolant* of $\varphi \rightarrow \psi$ is a formula θ satisfying the following conditions:

- $\mathbf{wGL}_n \vdash \varphi \rightarrow \theta$ and $\mathbf{wGL}_n \vdash \theta \rightarrow \psi$;
- $var^+(\theta) \subseteq var^+(\varphi) \cap var^+(\psi)$ and $var^-(\theta) \subseteq var^-(\varphi) \cap var^-(\psi)$.

It is not always true that $\mathbf{wGL}_n^{\mathbf{Sp}}$ supplies a Lyndon interpolant of $\varphi \rightarrow \psi$. Consider the derivation π as in (∗), and suppose that θ is a Lyndon interpolant of $\overline{(\Diamond^n \Delta_1, \Delta_1)^\#} \rightarrow (\Diamond^n \Delta_2, \Delta_2, \Diamond^n \overline{\varphi}, \varphi)^\#$. Then θ may contain a literal l such that:

- l occurs in both $\overline{\Delta_1}$ and $\overline{\varphi}$;
- l does not occur in φ nor any formula in Δ_2.

Then the formula $\Box\theta$ also contains l, however, the assumption $\Diamond^n\overline{\varphi}$ is eliminated in the conclusion sequent of $(\Box_n^r)$. Thus, l is not a common literal of the conclusion sequent, and $\Box\theta$ is no longer a Lyndon interpolant of $\overline{\Diamond \Delta_1{}^\#} \rightarrow \Box\varphi$.

To avoid this problem, in the next section we will develop a system which preserves the positiveness of formulas, and is equivalent to $\mathbf{wGL}_n^{\mathbf{Sp}}$.

5 Circular Proof System

In this section, we describe the calculus which admits *circular proofs*, and is equivalent to $\mathbf{wGL}_n^{\mathbf{G}}$.

Definition 6 The sequent calculus $\mathbf{wK4}_n^{\mathbf{G}}$ is obtained from $\mathbf{wGL}_n^{\mathbf{G}}$ by replacing the modal rule $\Box_n$ by

$$\frac{\Diamond^n \Gamma, \Gamma, \varphi}{\Diamond \Gamma, \Box\varphi}\ (\blacksquare_n).$$

A *circular derivation* of a calculus is a pair $\pi = (\kappa, d)$ where κ is a derivation in the calculus and d is a back-link function from some leaf x to an interior node y with an identical sequent, such that y lies on the path from the root of κ to x, and there exists at least one application of $\blacksquare_n$ between x and y. We call such an (x, y) a *circular pair*. (In other words, d is the set of circular pairs in κ.) A *circular proof* is a circular derivation such that every leaf is either marked by an axiom or connected by the back-link function. The circular proof system $^\circ\mathbf{wK4}_n^{\mathbf{G}}$ is obtained from $\mathbf{wK4}_n^{\mathbf{G}}$ by admitting circular proofs.

The following diagram is an example of a circular proof in $^\circ\mathbf{wK4}_2^{\mathbf{G}}$.

$$
\dfrac{
  \dfrac{
    \dfrac{
      \dfrac{\Diamond(\Box^2 p \wedge \overline{p}), \Box p}{\Diamond^3(\Box^2 p \wedge \overline{p}), \Diamond(\Box^2 p \wedge \overline{p}), \Box p}\,(weak)
    }{\Diamond^2(\Box^2 p \wedge \overline{p}), \Box^2 p}\,(\blacksquare_2)
  }{\Diamond^2(\Box^2 p \wedge \overline{p}), \Box^2 p, p}\,(weak)
  \qquad
  \dfrac{\overline{p}, p}{\Diamond^2(\Box^2 p \wedge \overline{p}), \overline{p}, p}\,(weak)
}{
  \dfrac{\Diamond^2(\Box^2 p \wedge \overline{p}), \Box^2 p \wedge \overline{p}, p}{\Diamond(\Box^2 p \wedge \overline{p}), \Box p}\,(\blacksquare_2)
}\,(\wedge)
$$

Let n be an arbitrary natural number, and consider the following diagram:

$$
\dfrac{\dfrac{p \vee q}{p \vee q, p, q}\,(weak)}{p \vee q}\,(\vee)
$$

This diagram contains a pair of nodes labeled by the same sequent $(p \vee q)$, however, there is no application of the rule $\blacksquare_n$. Therefore this diagram is *not* a circular proof in $^\circ\mathbf{wK4}_n^{\mathbf{G}}$.

In the rest of this section we prove the following theorem.

Theorem 3 *Assume $n \geq 1$. For any sequent Γ,*

$$
\mathbf{wGL}_n^{\mathbf{G}} \vdash \Gamma \iff {}^\circ\mathbf{wK4}_n^{\mathbf{G}} \vdash \Gamma.
$$

Lemma 7 *Assume $n \geq 1$. For any sequent Γ, $\mathbf{wGL}_n^{\mathbf{G}} \vdash \Gamma \Longrightarrow {}^\circ\mathbf{wK4}_n^{\mathbf{G}} \vdash \Gamma$. Moreover, if π is a proof of Γ in $\mathbf{wGL}_n^{\mathbf{G}}$, then we can construct a proof $\pi' = (\kappa, d)$ of Γ in $^\circ\mathbf{wK4}_n^{\mathbf{G}}$ from π in an effective way.*

Proof Assume $\mathbf{wGL}_n^{\mathbf{G}} \vdash \Gamma$, and let κ be a proof of Γ in $\mathbf{wGL}_n^{\mathbf{G}}$. First we introduce a complexity of formulas in κ. We denote by $Sub(\Gamma)$ the set of all subformulas of formulas contained in Γ. We put $\overline{Sub}(\Gamma) := Sub(\Gamma) \cup \{\overline{\varphi} \mid \varphi \in Sub(\Gamma)\}$. Let φ be a formula occurring in some sequent in κ. By the definition of $\mathbf{wGL}_n^{\mathbf{G}}$, φ is either an element of $\overline{Sub}(\Gamma)$ or of the form $\Diamond^k \psi$ where $\psi \in \overline{Sub}(\Gamma)$. The *complexity* of φ (write $c(\varphi)$) is defined as the least natural number k such that $\varphi \equiv \Diamond^k \psi$ for some $\psi \in \overline{Sub}(\Gamma)$.

In order to obtain a circular proof of Γ in $^\circ\mathbf{wK4}_n^{\mathbf{G}}$, we construct the sequence $\{\pi_i = (\kappa_i, d_i)\}_{i \leq j}$ of circular derivation satisfying the following properties:

- Each π_i is a circular derivation in the system $\mathbf{wGL}_n^{\mathbf{G}} \cup \mathbf{wK4}_n^{\mathbf{G}}$;
- For any κ_i and application of $\blacksquare_n$ in κ_i, there is no application of $\Box_n$ in the path from the root of κ_i to the conclusion of $\blacksquare_n$;
- π_j is a circular proof of Γ in $^\circ\mathbf{wK4}_n^{\mathbf{G}}$.

We construct such a sequence $\{\pi_i = (\kappa_i, d_i)\}_{i \leq j}$ from a given κ in the following steps.

1. Let $\pi_0 := (\kappa_0, d_0)$, where $\kappa_0 := \kappa$ and $d_0 := \emptyset$.

2. If π_i contains no application of $\Box_n$, then the sequence stops. Otherwise, consider the lowest application of $\Box_n$ in κ_i:

$$\dfrac{\begin{matrix}\vdots\\ \Diamond^n\Delta, \Delta, \Diamond^n\overline{\varphi}, \varphi\end{matrix}}{\begin{matrix}\Diamond\Delta, \Box\varphi\\ \vdots\end{matrix}}\ (\Box_n)$$

3. Search the path below $(\Diamond\Delta, \Box\varphi)$ for a pair (x, y) such that x and y are marked by an identical sequent and there is an application of $\blacksquare_n$ between x and y. If we find such a pair (x, y), then cut away all nodes higher than x. Let κ_{i+1} be the obtained derivation and $d_{i+1} := d_i \cup (x, y)$, and return to Step 2. Otherwise, go to Step 4.
4. Let ρ be the subproof of the sequent $(\Diamond^n\Delta, \Delta, \Diamond^n\overline{\varphi}, \varphi)$. Note that ρ contains no application of $\blacksquare_n$, and hence is a proof in $\mathbf{wGL}_n^{\mathbf{G}}$. By Corollary 1, there is a proof ρ' of $(\Diamond^n\Delta, \Delta, \varphi)$ in $\mathbf{wGL}_n^{\mathbf{G}}$. Replace κ_i by:

$$\dfrac{\begin{matrix}\rho\\ \vdots\\ \Diamond^n\Delta, \Delta, \Diamond^n\overline{\varphi}, \varphi\end{matrix}}{\begin{matrix}\Diamond\Delta, \Box\varphi\\ \vdots\end{matrix}}\ (\Box_n) \quad\longmapsto\quad \dfrac{\begin{matrix}\rho'\\ \vdots\\ \Diamond^n\Delta, \Delta, \varphi\end{matrix}}{\begin{matrix}\Diamond\Delta, \Box\varphi\\ \vdots\end{matrix}}\ (\blacksquare_n)$$

If $\Diamond^n\Delta$ contains no formula ψ such that $c(\psi) > n$, then let π_{i+1} be the obtained derivation, and return to Step 2. Otherwise, go to Step 5.
5. Let $\Sigma \subseteq \Delta$ be the set of all formulas ψ such that $c(\Diamond^n\psi) > n$, and put $\Pi := \Delta\backslash\Sigma$. Recall that for any $\psi \in \Sigma$, $\mathbf{wGL}_n^{\mathbf{G}} \vdash \psi, \Box^n\overline{\psi}$. By Corollary 1, the assumptions $\Diamond^n\Sigma$ can be eliminated. Let ρ'' be a proof of $(\Sigma, \Diamond^n\Pi, \Pi, \varphi)$ in $\mathbf{wGL}_n^{\mathbf{G}}$. Replace the derivation by:

$$\dfrac{\begin{matrix}\rho'\\ \vdots\\ \Diamond^n\Sigma, \Sigma, \Diamond^n\Pi, \Pi, \varphi\end{matrix}}{\begin{matrix}\Diamond\Delta, \Box\varphi\\ \vdots\end{matrix}}\ (\blacksquare_n) \quad\longmapsto\quad \dfrac{\dfrac{\begin{matrix}\rho''\\ \vdots\\ \Sigma, \Diamond^n\Pi, \Pi, \varphi\end{matrix}}{\Diamond^n\Sigma, \Sigma, \Diamond^n\Pi, \Pi, \varphi}\ (weak)}{\begin{matrix}\Diamond\Delta, \Box\varphi\\ \vdots\end{matrix}}\ (\blacksquare_n)$$

(The sequent $(\Sigma, \Diamond^n\Pi, \Pi, \varphi)$ consists of formulas of which complexities are $\leq n$.) Let π_{i+1} be the obtained derivation, and return to Step 2.

In Steps 4–5 the procedure always generates a new sequent which consists of formulas having complexity $\leq n$. Since the number of such sequents is finite, the sequence must stop at some j. Suppose that the construction terminates at π_j. This is our desired circular proof of Γ in $^\circ\mathbf{wK4}^{\mathbf{G}}_n$. □

Lemma 8 $^\circ\mathbf{wK4}^{\mathbf{G}}_n \vdash \Gamma \Longrightarrow \mathbf{wGL}^{\mathbf{G}}_n \vdash \Gamma$.

Proof For a circular derivation $\pi = (\kappa, d)$ in $^\circ\mathbf{wK4}^{\mathbf{G}}_n$ and a leaf a of π, a is called an *assumption leaf* if a is non-axiomatic and not connected by the back-link function d. An assumption leaf a is *boxed* if there is an application of $\blacksquare_n$ on the path from the root to a. Let $BH(\pi)$ and $H(\pi)$ be the sets of boxed, respectively not boxed assumption leaves of π. The sequent of a is denoted by Δ_a.

Claim 1 *For any circular derivation $\pi = (\kappa, d)$ of Γ,*

$$\mathbf{wGL}_n \vdash \bigwedge [n]^+\{\Delta_a^\# \mid a \in H(\pi)\} \wedge \bigwedge [n]\{\Delta_a^\# \mid a \in BH(\pi)\} \to \Gamma^\#.$$

Proof Induction on the construction of κ.

- Assume κ consists of a single node a. The claim clearly holds if Δ_a is an axiom of $\mathbf{wK4}^{\mathbf{G}}_n$. Otherwise, since $a \in H(\pi)$ and by Proposition 2.1, we have $\mathbf{wGL}_n \vdash [n]^+\Delta_a^\# \to \Delta_a^\#$.
- Consider π is one of the following derivations:

$$\dfrac{\begin{array}{c}\pi_1\\ \vdots\\ \Delta\end{array}}{\Delta, \Sigma}\,(weak), \quad \dfrac{\begin{array}{c}\pi_1\\ \vdots\\ \Delta, \varphi, \psi\end{array}}{\Delta, \varphi \vee \psi}\,(\vee), \quad \dfrac{\begin{array}{cc}\pi_1 & \pi_2\\ \vdots & \vdots\\ \Delta, \varphi & \Delta, \psi\end{array}}{\Delta, \varphi \wedge \psi}\,(\wedge).$$

  Suppose that Γ is not connected by d. In this case,

  $$H(\pi) = \bigcup_{i=1,2} H(\pi_i), \text{ and } BH(\pi) = \bigcup_{i=1,2} BH(\pi_i).$$

  Note that

  $$\mathbf{wGL}_n \vdash \Delta^\# \to (\Delta, \Sigma)^\#, \quad \mathbf{wGL}_n \vdash (\Delta, \varphi, \psi)^\# \to (\Delta, \varphi \vee \psi)^\#,$$
  $$\mathbf{wGL}_n \vdash (\Delta, \varphi)^\# \wedge (\Delta, \psi)^\# \to (\Delta, \varphi \wedge \psi)^\#.$$

  By the induction hypotheses for π_1 and π_2, the claim holds.
  Suppose that Γ is connected by d. Let b be such a leaf connecting with the root. In this case,

  $$H(\pi) = \bigcup_{i=1,2} H(\pi_i) \text{ and } BH(\pi) = \bigcup_{i=1,2} BH(\pi_i)\backslash\{b\}.$$

  From a similar argument as above, we obtain

  $$\mathbf{wGL}_n \vdash \bigwedge [n]^+\{\Delta_a^\# \mid a \in H(\pi)\} \wedge \bigwedge [n]\{\Delta_a^\# \mid a \in BH(\pi)\} \wedge [n]\Delta_b^\# \to \Gamma^\#,$$

  where Δ_b is exactly Γ. By Proposition 2.3,

$$\mathbf{wGL}_n \vdash \bigwedge [n]^+\{\Delta_a^\# \mid a \in H(\pi)\} \wedge \bigwedge [n]\{\Delta_a^\# \mid a \in BH(\pi)\} \to \Gamma^\#.$$

– Assume the last application of π is $\blacksquare_n$:

$$\cfrac{\begin{array}{c}\pi_1\\ \vdots\\ \Diamond^n\Delta, \Delta, \varphi\end{array}}{\Diamond\Delta, \Box\varphi}\ (\blacksquare_n).$$

Suppose that Γ is not connected by d. In this case,

$$H(\pi) = \emptyset \text{ and } BH(\pi) = H(\pi_1) \cup BH(\pi_1).$$

By the induction hypothesis for π_1,

$$\mathbf{wGL}_n \vdash \bigwedge [n]^+\{\Delta_a^\# \mid a \in H(\pi_1)\} \wedge \bigwedge [n]\{\Delta_a^\# \mid a \in BH(\pi_1)\} \\ \to \left(\Diamond^n\Delta, \Delta, \varphi\right)^\#.$$

By Proposition 2.2,

$$\mathbf{wGL}_n \vdash \bigwedge [n]\{\Delta_a^\# \mid a \in BH(\pi)\} \to \Box\left(\Diamond^n\Delta, \Delta, \varphi\right)^\#.$$

Note that $\mathbf{wGL}_n \vdash \Box(\Diamond^n\Delta, \Delta, \varphi)^\# \to \Box\left(\overline{(\Diamond^n\Delta, \Delta)^\#} \to \varphi\right)$. By Proposition 1, $\mathbf{wGL}_n \vdash \overline{\Diamond\Delta^\#} \to \Box\overline{(\Diamond^n\Delta, \Delta)^\#}$. Hence $\mathbf{wGL}_n \vdash \Box(\Diamond^n\Delta, \Delta, \varphi)^\# \to \left(\overline{\Diamond\Delta^\#} \to \Box\varphi\right)$, i.e., $\mathbf{wGL}_n \vdash \Box\,(\Diamond^n\Delta, \Delta, \varphi)^\# \to (\Diamond\Delta, \Box\varphi)^\#$. Thus we obtain

$$\mathbf{wGL}_n \vdash \bigwedge [n]\{\Delta_a^\# \mid a \in BH(\pi)\} \to \Gamma^\#.$$

Suppose that Γ is connected with b by d. In this case,

$$H(\pi) = \emptyset \text{ and } BH(\pi) = H(\pi_1) \cup BH(\pi_1)\backslash\{b\}.$$

Again by the induction hypothesis for π_1 and Proposition 2.2, we obtain

$$\mathbf{wGL}_n \vdash \bigwedge [n]\{\Delta_a^\# \mid a \in BH(\pi)\} \wedge [n]\Gamma^\# \to \Gamma^\#.$$

By Proposition 2.3, we conclude

$$\mathbf{wGL}_n \vdash \bigwedge [n]\{\Delta_a^\# \mid a \in BH(\pi)\} \to \Gamma^\#.$$

The proof of the claim is completed. □

Now if π is a circular proof of ${}^\circ\mathbf{wK4}_n^{\mathbf{G}}$, then $H(\pi) = BH(\pi) = \emptyset$, and hence $\mathbf{wGL}_n \vdash \Gamma^\#$. By Theorem 1, we conclude $\mathbf{wGL}_n^{\mathbf{G}} \vdash \Gamma$. □

Theorem 3 immediately follows from Lemmas 7 and 8.

Remark 2 Iemhoff [2] studies some sufficient conditions for a type of modal sequent calculus to have an equivalent circular proof system. The calculus $\mathbf{wGL}_n^{\mathbf{G}}$ does not enjoy Iemhoff's conditions, however, has an equivalent circular proof counterpart.

6 Lyndon Interpolation Property for $\mathbf{wGL_n}$

Shamkanov [8] originally showed that the standard provability logic **GL** enjoys the Lyndon interpolation property. In [9] he also gave a syntactical proof of the Lyndon interpolation theorem for **GL** by using the circular proof argument.

In this section, we show that Shamkanov's argument can be applied to the case for $\mathbf{wGL}_n$, i.e., for $n \geq 2$, if $\mathbf{wGL}_n \vdash \varphi \to \psi$, then we can construct a Lyndon interpolant of $\varphi \to \psi$ effectively. Before proving, we give some terminology. Definitions and notations are according to Shamkanov [9].

For any formula φ, we define $u(\varphi)$ as the set of literals l occurring in φ out of the scope of all modal operators. We use new symbols of the form p° and $\overline{p}^\circ$ (we call them *marked literals*.) to specify literals within the scope of modal operators. We define $v(\varphi)$ as the set of marked literals l° such that l occurs in φ within the scope of a modal operator. Let $w(\varphi) := u(\varphi) \cup v(\varphi)$, and $w(\Gamma) := \bigcup\{w(\varphi) \mid \varphi \in \Gamma\}$.

Theorem 4 (Lyndon interpolation theorem for $\mathbf{wGL}_n$) *Assume $n \geq 2$. If $\mathbf{wGL}_n \vdash \varphi \to \psi$, then there is a formula θ (called a Lyndon interpolant of $\varphi \to \psi$) such that:*

1. $\mathbf{wGL}_n \vdash \varphi \to \theta$ *and* $\mathbf{wGL}_n \vdash \theta \to \psi$;
2. $w(\theta) \subseteq w(\varphi) \cap w(\psi)$.

Moreover, such a θ is effectively constructible from φ and ψ.

First, we develop a split derivation system based on $\mathbf{wK4}_n^{\mathbf{G}}$.

Definition 7 The split derivation system $\mathbf{wK4}_n^{\mathbf{Sp}}$ is obtained from $\mathbf{wGL}_n^{\mathbf{Sp}}$ by replacing the modal rules $(\Box_n^l)$ and $(\Box_n^r)$ by:

$$\frac{[\theta]\ \Diamond^n\Gamma_1, \Gamma_1, \varphi \mid \Diamond^n\Gamma_2, \Gamma_2}{[\Diamond\theta]\ \Diamond\Gamma_1, \Box\varphi \mid \Diamond\Gamma_2}\ (\blacksquare_n^l),$$

and

$$\frac{[\theta]\ \Diamond^n\Gamma_1, \Gamma_1 \mid \Diamond^n\Gamma_2, \Gamma_2, \varphi}{[\Box\theta]\ \Diamond\Gamma_1 \mid \Diamond\Gamma_2, \Box\varphi}\ (\blacksquare_n^r).$$

Similarly, the split circular proof system $^\circ\mathbf{wK4}_n^{\mathbf{Sp}}$ is obtained from $\mathbf{wK4}_n^{\mathbf{Sp}}$ by admitting circular proofs.

Proposition 5 *For any sequents Γ_1 and Γ_2,*

$$^{\circ}\mathbf{wK4}^{\mathbf{G}}_n \vdash \Gamma_1, \Gamma_2 \iff {}^{\circ}\mathbf{wK4}^{\mathbf{Sp}}_n \vdash \Gamma_1 \mid \Gamma_2.$$

Proof ($\Longrightarrow$): Let π be a proof of (Γ_1, Γ_2) in $^{\circ}\mathbf{wK4}^{\mathbf{G}}_n$. First we expand π to an infinite derivation by adding subproofs to each leaf a connected by d. By the bottom-up splitting of sequents, we obtain an infinite derivation π_∞ of $(\Gamma_1 \mid \Gamma_2)$ in $\mathbf{wK4}^{\mathbf{Sp}}_n$. Since π consists of only finitely many different sequents, each sequent can be split into only finitely many split sequents. Therefore, π_∞ consists of finitely many different split sequents.

($\Longleftarrow$): For a given split circular proof π in $^{\circ}\mathbf{wK4}^{\mathbf{Sp}}_n$, we obtain a circular proof in $^{\circ}\mathbf{wK4}^{\mathbf{G}}_n$ by removing all splittings in π. □

Notice that for a given circular proof π of (Γ_1, Γ_2) in $^{\circ}\mathbf{wK4}^{\mathbf{G}}_n$, we can effectively construct a split circular proof of $(\Gamma_1 \mid \Gamma_2)$ in $^{\circ}\mathbf{wK4}^{\mathbf{Sp}}_n$.

We describe several facts that will be used in the proof of Theorem 4. Let φ be a formula and $w(\varphi) := u(\varphi) \cup v(\varphi)$ as before. For a set S of literals and marked literals, we define $S^{\circ} := \{l^{\circ} \mid l \in S \text{ or } l^{\circ} \in S\}$, and $w^*(\varphi) := w(\varphi) \cup w(\varphi)^{\circ}$. The following theorem states that $\mathbf{wGL}_n$ has effective fixed-points.

Theorem 5 (Fixed-point theorem. Kurahashi and Okawa [4]) *Let $\varphi(p)$ be a formula in which p occurs only within the scope of a modal operator. Then there is a formula ψ satisfying the following conditions:*

1. $w(\psi) \subseteq w^*(\varphi) \cup w^*(\overline{\varphi}) \setminus \{p^{\circ}, \overline{p}^{\circ}\}$;
2. $\mathbf{wGL}_n \vdash \psi \leftrightarrow \varphi(\psi)$.

Moreover, if φ does not contain $\overline{p}$, then $w(\psi) \subseteq w^*(\varphi) \setminus \{p\}$.

The last property "$w(\psi) \subseteq w^*(\varphi) \setminus \{p\}$" will be essentially needed in our proof. Kurahashi and Okawa [4] gave an effective procedure which produces a fixed-point ψ of a given formula $\varphi(p)$. Moreover, by the construction of fixed-points in [4], such a ψ also satisfies the conditions in Theorem 5, and the procedure does not use any kind of interpolation. Briefly, a fixed-point of $\varphi(p)$ is obtained by multi-substituting formulas containing only literals which occur in A, for each occurrence of p. Therefore ψ enjoys the condition 1. Moreover, if $\varphi(p)$ contains no occurrences of $\overline{p}$, then the procedure preserves the positiveness of literals in every substitution (see Lindström [5], and Kurahashi and Okawa [4]).

Lemma 9 *For each rule of* $\mathbf{wK4}^{\mathbf{Sp}}_n$*, the following corresponding statement holds:*

1. $\mathbf{wGL}_n \vdash \left(\overline{\Gamma_1^{\#}} \to \theta\right) \wedge \left(\theta \to \Gamma_2^{\#}\right) \to \left(\overline{(\Gamma_1, \Delta_1)^{\#}} \to \theta\right) \wedge \left(\theta \to (\Gamma_2, \Delta_2)^{\#}\right)$;
2. $\mathbf{wGL}_n \vdash \left(\overline{(\Gamma_1, \varphi, \psi)^{\#}} \to \theta\right) \wedge \left(\theta \to \Gamma_2^{\#}\right)$

$$\to \left(\overline{(\Gamma_1, \varphi \vee \psi)^{\#}} \to \theta\right) \wedge \left(\theta \to \Gamma_2^{\#}\right);$$

3. $\mathbf{wGL}_n \vdash \left(\overline{\Gamma_1^{\#}} \to \theta\right) \wedge \left(\theta \to (\Gamma_2, \varphi, \psi)^{\#}\right)$

$$\to \left(\overline{\Gamma_1^\#} \to \theta\right) \wedge \left(\theta \to (\Gamma_2, \varphi \vee \psi)^\#\right);$$

4. $\mathbf{wGL}_n \vdash \left(\overline{(\Gamma_1, \varphi)^\#} \to \theta_1\right) \wedge \left(\theta_1 \to \Gamma_2^\#\right) \wedge \left(\overline{(\Gamma_1, \psi)^\#} \to \theta_2\right) \wedge \left(\theta_2 \to \Gamma_2^\#\right)$
$$\to \left(\overline{(\Gamma_1, \varphi \wedge \psi)^\#} \to \theta_1 \vee \theta_2\right) \wedge \left(\theta_1 \vee \theta_2 \to \Gamma_2^\#\right);$$
5. $\mathbf{wGL}_n \vdash \left(\overline{\Gamma_1^\#} \to \theta_1\right) \wedge \left(\theta_1 \to (\Gamma_2, \varphi)^\#\right) \wedge \left(\overline{\Gamma_1^\#} \to \theta_2\right) \wedge \left(\theta_2 \to (\Gamma_2, \psi)^\#\right)$
$$\to \left(\overline{\Gamma_1^\#} \to \theta_1 \wedge \theta_2\right) \wedge \left(\theta_1 \wedge \theta_2 \to (\Gamma_2, \varphi \wedge \psi)^\#\right);$$
6. $\mathbf{wGL}_n \vdash \Box\left[\left(\overline{(\Diamond^n \Gamma_1, \Gamma_1, \varphi)^\#} \to \theta\right) \wedge \left(\theta \to (\Diamond^n \Gamma_2, \Gamma_2)^\#\right)\right]$
$$\to \left(\overline{(\Diamond \Gamma_1, \Box\varphi)^\#} \to \Diamond\theta\right) \wedge \left(\Diamond\theta \to (\Diamond \Gamma_2)^\#\right);$$
7. $\mathbf{wGL}_n \vdash \Box\left[\left(\overline{(\Diamond^n \Gamma_1, \Gamma_1)^\#} \to \theta\right) \wedge \left(\theta \to (\Diamond^n \Gamma_2, \Gamma_2, \varphi)^\#\right)\right]$
$$\to \left(\overline{(\Diamond \Gamma_1)^\#} \to \Box\theta\right) \wedge \left(\Box\theta \to (\Diamond \Gamma_2, \Box\varphi)^\#\right).$$

Suppose that $^\circ\mathbf{wK4}_n^{\mathbf{Sp}} \vdash \Gamma_1 \mid \Gamma_2$. We define an *interpolant* of $\Gamma_1 \mid \Gamma_2$ as a Lyndon interpolant of $\overline{\Gamma_1^\#} \to \Gamma_2^\#$.

For a given split circular derivation $\pi = (\kappa, d)$ of $(\Gamma_1 \mid \Gamma_2)$, we define two sets of leaves $BH(\pi)$ and $H(\pi)$ as in Sect. 4. For each non-axiomatic leaf a of κ, we fix two variables x_a and w_a. The first variable x_a plays a role of the provisional interpolant of a. The second variable w_a ranges over sets of literals and marked literals. We interpret the second variable w_a as $w(x_a)$. Let $(\Delta_1 \mid \Delta_2)$ be the split sequent of a. Define the formula I_a and the statement I'_a as follows:

$$I_a :\equiv \left(\overline{\Delta_1^\#} \to x_a\right) \wedge \left(x_a \to \Delta_2^\#\right);$$
$$I'_a :\Leftrightarrow w_a \subseteq w(\overline{\Delta_1}) \cap w(\Delta_2).$$

Lemma 10 *Let π be a split circular derivation of $(\Gamma_1 \mid \Gamma_2)$. Then there is a formula θ satisfying the following conditions:*

1. *θ does not contain literals of the form $\overline{x_a}$;*
2. *If $a \in BH(\pi)$, then x_a occurs in θ only within the scope of modal operators;*
3. $\mathbf{wGL}_n \vdash \bigwedge\{[n]I_a \mid a \in BH(\pi)\} \wedge \bigwedge\{[n]^+ I_a \mid a \in H(\pi)\}$
$$\to \left(\overline{\Gamma_1^\#} \to \theta\right) \wedge \left(\theta \to \Gamma_2^\#\right);$$
4. *$T(\pi) \Rightarrow w_X(\theta) \cup \bigcup\{w_a^\circ \mid a \in BH(\pi)\} \cup \bigcup\{w_a \mid a \in H(\pi)\} \subseteq w(\overline{\Gamma_1}) \cap w(\Gamma_2)$, where $T(\pi)$ is the statement $\bigwedge\{I'_a \mid a \in BH(\pi) \cup H(\pi)\}$, and $w_X(\theta) := w(\theta) \backslash \{x_a, x_a^\circ \mid x_a$ occurs in $\theta\}$.*

Proof Induction on the construction of κ. We argue five cases.

Case 1 Assume κ consists of a single node a. If $(\Gamma_1 \mid \Gamma_2)$ is an axiom of $\mathbf{wK4}_n^{\mathbf{Sp}}$, then we take θ as the formula bracketed in the corresponding axiom. Otherwise, we put $\theta :\equiv x_a$ (Note that $H(\pi) = \{a\}$). In both cases, θ clearly satisfies the conditions 1–4.

Case 2 Assume the last application of κ is *weak* or one of the propositional rules, and $(\Gamma_1 \mid \Gamma_2)$ is not connected by d. We show that the formula θ bracketed in each conclusion satisfies Conditions 1–4.

$$\dfrac{\begin{array}{c}\pi_1\\ \vdots\\ [\theta_1]\, \Delta_{11} \mid \Delta_{12}\end{array}}{[\theta]\, \Gamma_1 \mid \Gamma_2} \qquad \dfrac{\begin{array}{c}\pi_1\\ \vdots\\ [\theta_1]\, \Delta_{11} \mid \Delta_{12}\end{array} \quad \begin{array}{c}\pi_2\\ \vdots\\ [\theta_2]\, \Delta_{21} \mid \Delta_{22}\end{array}}{[\theta]\, \Gamma_1 \mid \Gamma_2}$$

Conditions 1 and 2 are clear.

Condition 3 By the induction hypotheses, we have for $i = 1, 2$,

$$\mathbf{wGL}_n \vdash \bigwedge\{[n]I_a \mid a \in BH(\pi_i)\} \wedge \bigwedge\{[n]^+ I_a \mid a \in H(\pi_i)\} \\ \to \left(\overline{\Delta_{i1}^{\#}} \to \theta_i\right) \wedge \left(\theta_i \to \Delta_{i2}^{\#}\right).$$

Recall that $BH(\pi) = \bigcup_{i=1,2} BH(\pi_i)$ and $H(\pi) = \bigcup_{i=1,2} H(\pi_i)$. By Lemma 9.1–5, we obtain

$$\mathbf{wGL}_n \vdash \bigwedge\{[n]I_a \mid a \in BH(\pi)\} \wedge \bigwedge\{[n]^+ I_a \mid a \in H(\pi)\} \\ \to \left(\overline{\Gamma_1^{\#}} \to \theta\right) \wedge \left(\theta \to \Gamma_2^{\#}\right).$$

Thus θ satisfies the condition *3*.

Condition 4 By the induction hypothesis, we have for $i = 1, 2$,

$$T(\pi_i) \Rightarrow w_X(\theta_i) \cup \bigcup\{w_a^\circ \mid a \in BH(\pi_i)\} \cup \bigcup\{w_a \mid a \in H(\pi_i)\} \\ \subseteq w(\overline{\Delta_{i1}}) \cap w(\Delta_{i2}).$$

Suppose $T(\pi)$ holds. (Note that $T(\pi) \Leftrightarrow T(\pi_1) \wedge T(\pi_2)$.) By $w_X(\theta) = w_X(\theta_1) \cup w_X(\theta_2)$ and $w(\overline{\Delta_{i1}}) \cap w(\Delta_{i2}) \subseteq w(\overline{\Gamma_1}) \cap w(\Gamma_2)$, we conclude

$$w_X(\theta) \cup \bigcup\{w_a^\circ \mid a \in BH(\pi)\} \cup \bigcup\{w_a \mid a \in H(\pi)\} \subseteq w(\overline{\Gamma_1}) \cap w(\Gamma_2).$$

Case 3 Assume the last application of κ is $\blacksquare_n^l$ or $\blacksquare_n^r$, and $(\Gamma_1 \mid \Gamma_2)$ is not connected by d. Again we show that the formula θ bracketed in the corresponding conclusion satisfies the conditions 1–4.

$$\dfrac{\begin{array}{c}\pi_1\\ \vdots\\ [\theta_1]\ \Delta_1 \mid \Delta_2\end{array}}{[\theta]\ \Gamma_1 \mid \Gamma_2}$$

The conditions 1 and 2 are clear.

Condition 3 By the induction hypotheses, we have

$$\mathbf{wGL}_n \vdash \bigwedge\{[n]I_a \mid a \in BH(\pi_1)\} \wedge \bigwedge\{[n]^+ I_a \mid a \in H(\pi_1)\}$$
$$\to \left(\overline{\Delta_1^\#} \to \theta_1\right) \wedge \left(\theta_1 \to \Delta_2^\#\right).$$

By Proposition 2.2,

$$\mathbf{wGL}_n \vdash \bigwedge\{[n]I_a \mid a \in BH(\pi_1)\} \wedge \bigwedge\{[n]I_a \mid a \in H(\pi_1)\}$$
$$\to \Box\left[\left(\overline{\Delta_1^\#} \to \theta_1\right) \wedge \left(\theta_1 \to \Delta_2^\#\right)\right].$$

Note that $BH(\pi) = BH(\pi_1) \cup H(\pi_1)$ and $H(\pi) = \emptyset$. By Lemma 9.6–7, we obtain

$$\mathbf{wGL}_n \vdash \bigwedge\{[n]I_a \mid a \in BH(\pi)\} \to \left(\overline{\Gamma_1^\#} \to \theta\right) \wedge \left(\theta \to \Gamma_2^\#\right).$$

Condition 4 By the induction hypothesis,

$$T(\pi_1) \Rightarrow w_X(\theta_1) \cup \bigcup\{w_a^\circ \mid a \in BH(\pi_1)\} \cup \bigcup\{w_a \mid a \in H(\pi_1)\}$$
$$\subseteq w(\overline{\Delta_1}) \cap w(\Delta_2). \quad (1)$$

Suppose that $T(\pi)$ is true. Since $T(\pi) \Leftrightarrow T(\pi_1)$, the consequence of (1) is also true. It suffices to show that: (i) $w_X(\theta) \subseteq w(\overline{\Gamma_1}) \cap w(\Gamma_2)$ and (ii) if $a \in BH(\pi)$, then $w_a^\circ \subseteq w(\overline{\Gamma_1}) \cap w(\Gamma_2)$.

(i): By the conclusion of (1), $w_X(\theta_1) \subseteq w(\overline{\Delta_1}) \cap w(\Delta_2)$. Then $w_X(\theta_1)^\circ \subseteq w(\overline{\Delta_1})^\circ \cap w(\Delta_2)^\circ$. In this case, we have $w(\overline{\Delta_1})^\circ \subseteq w(\overline{\Gamma_1})$ and $w(\Delta_2)^\circ \subseteq w(\Gamma_2)$. Hence $w_X(\theta) = w_X(\theta_1)^\circ \subseteq w(\overline{\Gamma_1}) \cap w(\Gamma_2)$.

(ii): Let $a \in BH(\pi) = BH(\pi_1) \cup H(\pi_1)$. If $a \in BH(\pi_1)$, then $w_a^\circ \subseteq w(\overline{\Delta_1}) \cap w(\Delta_2)$ by (1). If $a \in H(\pi_1)$, then $w_a \subseteq w(\overline{\Delta_1}) \cap w(\Delta_2)$ by (1). In either case, we have $w_a^\circ \subseteq w(\overline{\Delta_1})^\circ \cap w(\Delta_2)^\circ \subseteq w(\overline{\Gamma_1}) \cap w(\Gamma_2)$.

Case 4 Assume that the last application of κ is *weak* or one of the propositional rules, and $(\Gamma_1 \mid \Gamma_2)$ is connected with b by d.

$$\dfrac{\begin{matrix}\pi_1\\ \vdots\\ [\theta_1]\,\Delta_{11} \mid \Delta_{12}\end{matrix}}{[\theta]\,\Gamma_1 \mid \Gamma_2} \qquad \dfrac{\begin{matrix}\pi_1\\ \vdots\\ [\theta_1]\,\Delta_{11} \mid \Delta_{12}\end{matrix} \quad \begin{matrix}\pi_2\\ \vdots\\ [\theta_2]\,\Delta_{21} \mid \Delta_{22}\end{matrix}}{[\theta]\,\Gamma_1 \mid \Gamma_2}$$

(By the construction of θ, θ contains x_b.) There is at least one application of modal rules between b and the root of κ, and hence $b \in BH(\pi_1)$ or $b \in BH(\pi_2)$. By the induction hypothesis, for $i = 1, 2$,

$$\mathbf{wGL}_n \vdash \bigwedge\{[n]I_a : a \in BH(\pi_i)\} \wedge \bigwedge\{[n]^+ I_a : a \in H(\pi_i)\} \\ \rightarrow \left(\overline{\Delta_{i1}^{\#}} \rightarrow \theta_i\right) \wedge \left(\theta_i \rightarrow \Delta_{i2}^{\#}\right).$$

Note that $BH(\pi) = \bigcup_{i=1,2} BH(\pi_i)\backslash\{b\}$, $H(\pi) = \bigcup_{i=1,2} H(\pi_i)$. By Lemma 9.1–5,

$$\mathbf{wGL}_n \vdash \bigwedge\{[n]I_a : a \in BH(\pi)\} \wedge \bigwedge\{[n]^+ I_a : a \in H(\pi)\} \wedge [n]I_b \\ \rightarrow \left(\overline{\Gamma_1^{\#}} \rightarrow \theta\right) \wedge \left(\theta \rightarrow \Gamma_2^{\#}\right).$$

By the construction of θ, x_b only occurs in θ positively and within the scope of modal operators. By Theorem 5, we can construct a formula ψ satisfying $w(\psi) \subseteq w^*(\theta)\backslash\{x_b^\circ\}$, and $\mathbf{wGL}_n \vdash \theta(\psi) \leftrightarrow \psi$. (Here $\theta(\psi)$ is the formula obtained from θ by substituting ψ for all occurrences of x_b.) Thus we have

$$\mathbf{wGL}_n \vdash \bigwedge\{[n]I_a : a \in BH(\pi)\} \wedge \bigwedge\{[n]^+ I_a : a \in H(\pi)\} \wedge [n]I_b(\psi) \\ \rightarrow \left(\overline{\Gamma_1^{\#}} \rightarrow \theta(\psi)\right) \wedge \left(\theta(\psi) \rightarrow \Gamma_2^{\#}\right). \quad (2)$$

By $\mathbf{wGL}_n^{\mathbf{G}} \vdash \theta(\psi) \leftrightarrow \psi$ and the definition of I_b,

$$\mathbf{wGL}_n \vdash \left(\overline{\Gamma_1^{\#}} \rightarrow \theta(\psi)\right) \wedge \left(\theta(\psi) \rightarrow \Gamma_2^{\#}\right) \leftrightarrow I_b(\psi). \quad (3)$$

From (2) and (3),

$$\mathbf{wGL}_n \vdash \bigwedge\{[n]I_a : a \in BH(\pi)\} \wedge \bigwedge\{[n]^+ I_a : a \in H(\pi)\} \wedge [n]I_b(\psi) \rightarrow I_b(\psi).$$

By Proposition 2.3,

$$\mathbf{wGL}_n \vdash \bigwedge\{[n]I_a : a \in BH(\pi)\} \wedge \bigwedge\{[n]^+ I_a : a \in H(\pi)\} \rightarrow I_b(\psi).$$

Again by (3),

$$\mathbf{wGL}_n \vdash \bigwedge\{[n]I_a : a \in BH(\pi)\} \wedge \bigwedge\{[n]^+ I_a : a \in H(\pi)\}$$
$$\to \left(\overline{\Gamma_1^\#} \to \theta(\psi)\right) \wedge \left(\theta(\psi) \to \Gamma_2^\#\right),$$

i.e., $\theta(\psi)$ satisfies the condition 3.
Moreover, by the constructions of θ and ψ, $\theta(\psi)$ does not contain literals of the form $\overline{x_a}$, and if $a \in BH(\pi)$, then x_a occurs only within the scope of modal operators. Thus $\theta(\psi)$ enjoys the conditions 1–2.

Condition 4 By the induction hypothesis, for $i = 1, 2$,

$$T(\pi_i) \Rightarrow w_X(\theta_i) \cup \bigcup\{w_a^\circ \mid a \in BH(\pi_i)\} \cup \bigcup\{w_a \mid a \in H(\pi_i)\}$$
$$\subseteq w(\overline{\Delta_{i1}}) \cap w(\Delta_{i2}).$$

Suppose that $T(\pi)$ is true. Since $T(\pi) \wedge I_b'$ implies $T(\pi_1) \wedge T(\pi_2)$,

$$I_b' \Rightarrow w_X(\theta_i) \cup \bigcup\{w_a^\circ \mid a \in BH(\pi_i)\} \cup \bigcup\{w_a \mid a \in H(\pi_i)\}$$
$$\subseteq w(\overline{\Delta_{i1}}) \cap w(\Delta_{i2}).$$

Note that $w_X(\theta) = w_X(\theta_1) \cup w_X(\theta_2)$ and $w(\overline{\Delta_{i1}}) \cap w(\Delta_{i2}) \subseteq w(\overline{\Gamma_1}) \cap w(\Gamma_2)$. We have

$$I_b' \Rightarrow w_X(\theta) \cup \bigcup\{w_a^\circ \mid a \in BH(\pi)\} \cup w_b^\circ \cup \bigcup\{w_a \mid a \in H(\pi)\}$$
$$\subseteq w(\overline{\Gamma_1}) \cap w(\Gamma_2). \quad (4)$$

We show that $w_X(\theta(\psi)) \subseteq w(\overline{\Gamma_1}) \cap w(\Gamma_2)$. From (4),

$$I_b' \Rightarrow w_X(\theta) \cup w_b^\circ \subseteq w(\overline{\Gamma_1}) \cap w(\Gamma_2). \quad (5)$$

Substituting $\emptyset$ for w_b in (5), we have

$$w_X(\theta) \subseteq w(\overline{\Gamma_1}) \cap w(\Gamma_2).$$

This statement is equivalent to $I_b'(w_X(\theta))$, and hence $I_b'(w_X(\theta))$ is valid.
Substituting $w_X(\theta)$ for w_b in (5), we get

$$I_b'(w_X(\theta)) \Rightarrow w_X(\theta) \cup w_X(\theta)^\circ \subseteq w(\overline{\Gamma_1}) \cap w(\Gamma_2),$$

and hence

$$w_X(\theta) \cup w_X(\theta)^\circ \subseteq w(\overline{\Gamma_1}) \cap w(\Gamma_2).$$

By the constructions of θ and ψ, $w_X(\theta(\psi)) \subseteq w_X(\theta) \cup w_X(\psi)^\circ$. On the other hand, since $w(\psi) \subseteq w^*(\theta)\backslash\{x_b^\circ\}$, we have $w(\psi)^\circ \subseteq w(\theta)^\circ\backslash\{x_b^\circ\}$, and hence $w_X(\psi)^\circ \subseteq w_X(\theta)^\circ$. Thus,

$$w_X(\theta(\psi)) \subseteq w_X(\theta) \cup w_X(\psi)^\circ \subseteq w_X(\theta) \cup w_X(\theta)^\circ \subseteq w(\overline{\Gamma_1}) \cap w(\Gamma_2). \tag{6}$$

From (4), we get

$$I'_b \Rightarrow \bigcup\{w_a^\circ \mid a \in BH(\pi)\} \cup \bigcup\{w_a \mid a \in H(\pi)\} \subseteq w(\overline{\Gamma_1}) \cap w(\Gamma_2).$$

Substituting $\emptyset$ for w_b, we obtain

$$\bigcup\{w_a^\circ \mid a \in BH(\pi)\} \cup \bigcup\{w_a \mid a \in H(\pi)\} \subseteq w(\overline{\Gamma_1}) \cap w(\Gamma_2).$$

From this and (6), we conclude that $\theta(\psi)$ satisfies the condition 4.

Case 5 Assume the last application of κ is $\blacksquare_n^l$ or $\blacksquare_n^r$, and $(\Gamma_1 \mid \Gamma_2)$ is connected with b by d.

$$\dfrac{\begin{array}{c}\pi_1\\ \vdots\\ [\theta_1]\ \Delta_1 \mid \Delta_2\end{array}}{[\theta]\ \Gamma_1 \mid \Gamma_2}$$

By the induction hypothesis,

$$\mathbf{wGL}_n \vdash \bigwedge\{[n]I_a : a \in BH(\pi_1)\} \wedge \bigwedge\{[n]^+ I_a : a \in H(\pi_1)\} \\ \rightarrow \left(\overline{\Delta_1^\#} \rightarrow \theta_1\right) \wedge \left(\theta_1 \rightarrow \Delta_2^\#\right).$$

By Proposition 2.2 and Lemma 9.6–7, we have

$$\mathbf{wGL}_n \vdash \bigwedge\{[n]I_a : a \in BH(\pi_1)\} \wedge \bigwedge\{[n]I_a : a \in H(\pi_1)\} \\ \rightarrow \left(\overline{\Gamma_1^\#} \rightarrow \theta\right) \wedge \left(\theta \rightarrow \Gamma_2^\#\right).$$

Note that $BH(\pi) = BH(\pi_1) \cup H(\pi_1)\backslash\{b\}$ and $H(\pi) = \emptyset$. We have

$$\mathbf{wGL}_n \vdash \bigwedge\{[n]I_a : a \in BH(\pi)\} \wedge [n]I_b \rightarrow \left(\overline{\Gamma_1^\#} \rightarrow \theta\right) \wedge \left(\theta \rightarrow \Gamma_2^\#\right).$$

By Theorem 5, we can construct the fixed-point ψ of $\theta(x_b)$. Applying a similar argument as in Case 4, we obtain

$$\mathbf{wGL}_n \vdash \bigwedge\{[n]I_a : a \in BH(\pi)\} \wedge [n]I_b(\psi) \rightarrow I_b(\psi).$$

By Proposition 2.3,

$$\mathbf{wGL}_n \vdash \bigwedge\{[n]I_a : a \in BH(\pi)\} \rightarrow I_b(\psi),$$

i.e.,

$$\mathbf{wGL}_n \vdash \bigwedge\{[n]I_a : a \in BH(\pi)\} \rightarrow \left(\overline{\Gamma_1^\#} \rightarrow \theta(\psi)\right) \wedge \left(\theta(\psi) \rightarrow \Gamma_2^\#\right).$$

From this and by the constructions of θ and ψ, $\theta(\psi)$ enjoys the conditions 1–3.

Condition 4 By the induction hypothesis,

$$T(\pi_1) \Rightarrow w_X(\theta_1) \cup \bigcup\{w_a^\circ \mid a \in BH(\pi_1)\} \cup \bigcup\{w_a \mid a \in H(\pi_1)\} \subseteq w(\overline{\Delta_1}) \cap w(\Delta_2).$$

Assume $T(\pi)$ is true. Since $T(\pi) \wedge I_b'$ implies $T(\pi_1)$, we have

$$I_b' \Rightarrow w_X(\theta_1) \cup \bigcup\{w_a^\circ \mid a \in BH(\pi_1)\} \cup \bigcup\{w_a \mid a \in H(\pi_1)\} \subseteq w(\overline{\Delta_1}) \cap w(\Delta_2).$$

As in Case 4, we obtain

$$I_b' \Longrightarrow w_X(\theta) \cup \bigcup\{w_a^\circ \mid a \in BH(\pi)\} \cup w_b^\circ \subseteq w(\overline{\Gamma_1}) \cap w(\Gamma_2).$$

Applying a similar argument as in Case 4, we can show that $w_X(\theta(\psi)) \subseteq w(\overline{\Gamma_1}) \cap w(\Gamma_2)$ and $\bigcup\{w_a^\circ \mid a \in BH(\pi)\} \subseteq w(\overline{\Gamma_1}) \cap w(\Gamma_2)$.

The proof of Lemma 10 is now completed. □

Lemma 11 *If* $^\circ\mathbf{wK4}_n^{\mathbf{Sp}} \vdash \Gamma_1 \mid \Gamma_2$, *then we can construct an interpolant* θ *of* $(\Gamma_1 \mid \Gamma_2)$ *in an effective way.*

Proof Let $\pi = (\kappa, d)$ be a split circular proof of $(\Gamma_1 \mid \Gamma_2)$. Then we can construct an interpolant of $(\Gamma_1 \mid \Gamma_2)$ by the following way. For each non-axiomatic leaf a, we put x_a as the provisional interpolant. From the leaf to the root, construct θ in accordance with rules of $\mathbf{wK4}_n^{\mathbf{Sp}}$. If we find a node n which is connected by a non-axiomatic leaf a, then apply Theorem 5, substitute the fixed-point for all occurrences of x_a, and continue to the next application. Since every non-axiomatic leaf a is connected to some interior node b by d, x_a must be eliminated. Therefore the resulting formula θ does not contain any literal of the form x_a, and satisfies that

$$\mathbf{wGL}_n \vdash \left(\overline{\Gamma_1^\#} \rightarrow \theta\right) \wedge \left(\theta \rightarrow \Gamma_2^\#\right) \text{ and}$$
$$w(\theta) \subseteq w(\overline{\Gamma_1}) \cap w(\Gamma_2),$$

by Lemma 10. Thus, θ is indeed an interpolant of $(\Gamma_1 \mid \Gamma_2)$. □

Theorem 4 follows from Lemma 11 immediately.

Proof of Theorem 4. Assume $\mathbf{wGL}_n \vdash \varphi \to \psi$. Then by Theorem 1 and Proposition 4, we can construct a proof π of $(\overline{\varphi}, \psi)$ in $\mathbf{wGL}_n^{\mathbf{G}}$. By Lemma 7 and Proposition 5, we obtain a split circular proof of $(\overline{\varphi} \mid \psi)$ in $^\circ\mathbf{wK4}_n^{\mathbf{Sp}}$. Let θ be an interpolant of $\overline{\varphi} \mid \psi$. Then θ is indeed a Lyndon interpolant of $\varphi \to \psi$. □

7 Concluding Remarks

We proved the interpolation properties for Sacchetti's logics $\mathbf{wGL}_n$, and the effectiveness of interpolants. However, we used Kripke semantics in the proof of the (cut)-admissibility for $\mathbf{wGL}_n^{\mathbf{G}}$. Therefore our argument is not purely syntactical at this time.

Question 1 Can we prove the (cut)-admissibility for $\mathbf{wGL}_n^{\mathbf{G}}$ syntactically? In particular, for a given proof π of Γ in $\mathbf{wGL}_n^{\mathbf{G}} + (cut)$, is there an effective way to obtain a proof π' of Γ without the rule (cut)?

Shamkanov [9] used the multi-set based sequent calculus which satisfies the admissibility of weakening and contraction. It is under consideration whether our proof works even in a multi-set setting. The proof transforming procedure described in Sect. 5 does not take care of degrees of formulas. Furthermore, the existence of the rule $(weak)$ is essential in proving Lemma 7. For these reasons, our proof transforming procedure needs to a little change in a multi-set setting.

Acknowledgements The author would like to thank Taishi Kurahashi, Makoto Kikuchi and an anonymous referee for the discussion and giving many valuable comments.

References

1. Boolos, G.: The Logic of Provability. Cambridge University Press, Cambridge (1993)
2. Iemhoff, R.: Reasoning in circles. In: van Eijck, J., Iemhoff, R., Joosten, J.J. (eds.) Liber amicorum Alberti/A Tribute to Albert Visser. Tributes, vol. 30, pp. 165–176. College Publications, London (2016)
3. Kurahashi, T.: Arithmetical soundness and completeness for Σ_2 numerations. Stud. Logica **106**, 1181–1196 (2018). https://doi.org/10.1007/s11225-017-9782-4
4. Kurahashi, T., Okawa, Y.: Effectively constructible fixed points in Sacchetti's modal logics of provability. https://arxiv.org/abs/1811.12827
5. Lindström, P.: Provability logic-a short introduction. Theoria **62**, 19-61 (1996). https://doi.org/10.1111/j.1755-2567.1996.tb00529.x
6. Sambin, G., Valentini, S.: The modal logic of provability. The sequential approach. J. Philos. Log. **11**(3), 311-342 (1982). https://doi.org/10.1007/BF00293433
7. Sacchetti, L.: The fixed point property in modal logic. Notre Dame J. Form. Log. **42**(2), 65–86 (2001). https://doi.org/10.1305/ndjfl/1054837934

8. Shamkanov, D.S.: Interpolation properties for provability logics GL and GLP. Proc. Steklov Inst. Math. **274**, 303–316 (2011). https://doi.org/10.1134/S0081543811060198
9. Shamkanov, D.S.: Circular proofs for the Gödel-Löb provability logic. Math. Notes **96**(4), 575–585 (2014). https://doi.org/10.1134/S0001434614090326
10. Smoryński, C.: Beth's theorem and self-referential sentences. In: Macintyre, A., Pacholski, L., Paris, J. (eds.) Logic Colloquium '77. Stud. Logic Foundations Math. vol. 96, pp. 253–261. North-Holland, Amsterdam (1978). https://doi.org/10.1016/S0049-237X(08)72008-1

Rosser Provability and the Second Incompleteness Theorem

Taishi Kurahashi

Abstract This paper is a continuation of Arai's paper on derivability conditions for Rosser provability predicates. We investigate the limitations of the second incompleteness theorem by constructing three different Rosser provability predicates satisfying several derivability conditions.

Keywords The second incompleteness theorem · Rosser provability predicates · Derivability conditions

MSC2010 03F40 · 03F30

1 Introduction

Gödel's second incompleteness theorem states that if T is a recursively axiomatized consistent extension of Peano Arithmetic PA, then T cannot prove the consistency of T. This statement of the theorem is somewhat ambiguous, and it should be stated more precisely. In 1939, the first detailed proof of the second incompleteness theorem appeared in their book [8] by Hilbert and Bernays. They introduced the following conditions **HB1**, **HB2** and **HB3** which are called the Hilbert–Bernays derivability conditions, and essentially proved that if T is as above and a Σ_1 provability predicate $\mathrm{Pr}_T(x)$ of T satisfies the Hilbert–Bernays derivability conditions, then the consistency statement $\forall x(\mathrm{Pr}_T(x) \to \neg \mathrm{Pr}_T(\dot{\neg} x))$ of T cannot be proved in T.

HB1 : If $T \vdash \varphi \to \psi$, then $T \vdash \mathrm{Pr}_T(\ulcorner \varphi \urcorner) \to \mathrm{Pr}_T(\ulcorner \psi \urcorner)$ for any formulas φ, ψ.
HB2 : $T \vdash \mathrm{Pr}_T(\ulcorner \neg\varphi(x) \urcorner) \to \mathrm{Pr}_T(\ulcorner \neg\varphi(\dot{x}) \urcorner)$ for any formula $\neg\varphi(x)$.
HB3 : $T \vdash f(x) = 0 \to \mathrm{Pr}_T(\ulcorner f(\dot{x}) = 0 \urcorner)$ for any primitive recursive term $f(x)$.

Moreover, Hilbert and Bernays proved that Gödel's provability predicate $\mathrm{Pr}_T(x)$ satisfies these conditions. In 1955, Löb [20] introduced the following conditions **D1**,

---

T. Kurahashi (✉)
Graduate School of System Informatics, Kobe University, 1-1 Rokkodai,
Nada, Kobe 657-8501, Japan
e-mail: kurahashi@people.kobe-u.ac.jp

T. Arai et al. (eds.), *Advances in Mathematical Logic*, Springer Proceedings
in Mathematics & Statistics 369, https://doi.org/10.1007/978-981-16-4173-2_4

D2 and **D3** which are called the Hilbert–Bernays–Löb derivability conditions, and proved that if $\mathrm{Pr}_T(x)$ satisfies these conditions, then Löb's theorem holds.

D1 : If $T \vdash \varphi$, then $T \vdash \mathrm{Pr}_T(\ulcorner \varphi \urcorner)$ for any formula φ.
D2 : $T \vdash \mathrm{Pr}_T(\ulcorner \varphi \to \psi \urcorner) \to (\mathrm{Pr}_T(\ulcorner \varphi \urcorner) \to \mathrm{Pr}_T(\ulcorner \psi \urcorner))$ for any formulas φ, ψ.
D3 : $T \vdash \mathrm{Pr}_T(\ulcorner \varphi \urcorner) \to \mathrm{Pr}_T(\ulcorner \mathrm{Pr}_T(\ulcorner \varphi \urcorner) \urcorner)$ for any formula φ.

Löb's theorem is known as a generalization of the second incompleteness theorem. Hence we obtain that if $\mathrm{Pr}_T(x)$ satisfies the Hilbert–Bernays–Löb derivability conditions, then T cannot prove the consistency statement $\neg \mathrm{Pr}_T(\ulcorner 0 \neq 0 \urcorner)$ of T. This seems to be the most well-known form of the second incompleteness theorem stated accurately.

Other sets of derivability conditions which are sufficient for the second incompleteness theorem have been proposed by Jeroslow [10], Montagna [21] and Buchholz [3] (see also [18]). On the other hand, the second incompleteness theorem does not hold for some provability predicates. Feferman [4] found a Π_1 formula $\tau(v)$ representing the set of all axioms of T in T such that the consistency statement $\forall x(\mathrm{Pr}_\tau(x) \to \neg \mathrm{Pr}_\tau(\dot{\neg} x))$ is provable in PA where $\mathrm{Pr}_\tau(x)$ is the provability predicate of T constructed from $\tau(v)$. Notice that Feferman's predicate satisfies **D1** and **D2**, but does not satisfy **D3** because it is not Σ_1.

An example of a Σ_1 provability predicate for which the second incompleteness theorem does not hold was given by Mostowski [22]. Let $\mathrm{Pr}_T^M(x)$ be the Σ_1 formula $\exists y(\mathrm{Prf}_T(x, y) \land \neg \mathrm{Prf}_T(\ulcorner 0 \neq 0 \urcorner, y))$ where $\mathrm{Prf}_T(x, y)$ is a $\Delta_1(\mathsf{PA})$ formula saying that "y is a T-proof of x". Then $\neg \mathrm{Pr}_T^M(\ulcorner 0 \neq 0 \urcorner)$ is trivially provable in PA. Since the formula $\mathrm{Pr}_T^M(x)$ satisfies **D1** and **D3**, it does not satisfy **D2**. Mostowski's example shows that for Σ_1 provability predicates $\mathrm{Pr}_T(x)$, the set $\{\mathbf{D1}, \mathbf{D3}\}$ of derivability conditions is not sufficient for the unprovability of $\neg \mathrm{Pr}_T(\ulcorner 0 \neq 0 \urcorner)$.

Rosser provability predicates were introduced by Rosser [24] to improve Gödel's first incompleteness theorem, and they are also examples of Σ_1 provability predicates for which the second incompleteness theorem does not hold ([12–14]). That is, $\mathsf{PA} \vdash \neg \mathrm{Pr}_T^R(\ulcorner 0 \neq 0 \urcorner)$ for any Rosser provability predicate $\mathrm{Pr}_T^R(x)$ of T. It follows that each Rosser provability predicate does not satisfy at least one of **D2** and **D3**. It is known that whether each Rosser provability predicate satisfies **D2** (and **D3**) or not depends on the choice of a Rosser predicate. Indeed, by using Kripke model theoretic method by Guaspari and Solovay [6], we obtain a Rosser provability predicate satisfying neither **D2** nor **D3**. Also Bernardi and Montagna [2] and Arai [1] proved the existence of Rosser predicates satisfying **D2**, and Arai proved the the existence of Rosser predicates satisfying **D3**.

Moreover, it can be shown that the consistency statement $\forall x(\mathrm{Pr}_T^R(x) \to \neg \mathrm{Pr}_T^R(\dot{\neg} x))$ is provable for each Arai's Rosser provability predicate $\mathrm{Pr}_T^R(x)$. Then Arai's results indicate that for Σ_1 provability predicates $\mathrm{Pr}_T(x)$, each of $\{\mathbf{D1}, \mathbf{D2}\}$ and $\{\mathbf{D1}, \mathbf{D3}\}$ is not sufficient for the unprovability of $\forall x(\mathrm{Pr}_T^R(x) \to \neg \mathrm{Pr}_T^R(\dot{\neg} x))$. Also these existence results show that $\{\mathbf{D1}, \mathbf{D2}\}$ and $\{\mathbf{D1}, \mathbf{D3}\}$ do not imply **D3** and **D2**, respectively.

The constructions of Rosser provability predicates are somewhat flexible, and thus actually, Rosser provability predicates satisfying several derivability conditions have also been investigated ([11, 16, 17]). In this paper, we construct three Rosser provability predicates satisfying several additional derivability conditions. As a consequence of these constructions, we obtain that some sets of conditions of provability predicates are not sufficient for some versions of the second incompleteness theorem. In particular, our second and third Rosser provability predicates satisfy the Hilbert–Bernays derivability conditions. Therefore in contrast to the Hilbert–Bernays–Löb derivability conditions, the Hilbert–Bernays derivability condition does not imply the unprovability of the consistency statement $\neg \mathrm{Pr}_T(\ulcorner 0 \neq 0 \urcorner)$ in general.

In Sect. 2, we introduce versions of derivability conditions, and also introduce some basic results from the paper [18] concerning derivability conditions. In Sect. 3, we introduce Rosser provability predicates, and describe background of the present paper. In the last section, we give constructions of our Rosser provability predicates.

2 Provability Predicates and Derivability Conditions

Throughout this paper, T denotes a recursively axiomatized consistent extension of Peano arithmetic PA in the language of first-order arithmetic $\mathcal{L}_A$. The numeral for each natural number n is denoted by $\overline{n}$. We fix some natural Gödel numbering, and let $\ulcorner \varphi \urcorner$ be the numeral for the Gödel number of a formula φ. We assume that 0 is not a Gödel number of any object. Let $\{\xi_k\}_{k \in \omega}$ be the effective reputation-free sequence of all $\mathcal{L}_A$-formulas arranged in ascending order of Gödel numbers. We assume that if ξ_k is a proper subformula of ξ_l, then $k < l$.

We say a Σ_1 formula $\mathrm{Pr}_T(x)$ is a *provability predicate* of T if it weakly represents the set of all theorems of T in T, that is, for any natural number n, $T \vdash \mathrm{Pr}_T(\overline{n})$ if and only if n is the Gödel number of some theorem of T. Provability predicates are expected to satisfy some natural conditions which are called derivability conditions. We introduce three versions of derivability conditions, that is, local version, uniform version and global version. See [18] for further details. In the following definitions, let Γ be either Δ_0 or Σ_1.

Definition 1 (*Local derivability conditions*)

D1 : If $T \vdash \varphi$, then $T \vdash \mathrm{Pr}_T(\ulcorner \varphi \urcorner)$ for any formula φ.
D2 : $T \vdash \mathrm{Pr}_T(\ulcorner \varphi \to \psi \urcorner) \to (\mathrm{Pr}_T(\ulcorner \varphi \urcorner) \to \mathrm{Pr}_T(\ulcorner \psi \urcorner))$ for any formulas φ, ψ.
D3 : $T \vdash \mathrm{Pr}_T(\ulcorner \varphi \urcorner) \to \mathrm{Pr}_T(\ulcorner \mathrm{Pr}_T(\ulcorner \varphi \urcorner) \urcorner)$ for any formula φ.
$\boldsymbol{\Gamma}$C : $T \vdash \varphi \to \mathrm{Pr}_T(\ulcorner \varphi \urcorner)$ for any Γ sentence φ.
B$_2$: If $T \vdash \varphi \to \psi$, then $T \vdash \mathrm{Pr}_T(\ulcorner \varphi \urcorner) \to \mathrm{Pr}_T(\ulcorner \psi \urcorner)$ for any formulas φ, ψ.

Definition 2 (*Uniform derivability conditions*)

D1$^{\mathbf{U}}$: If $T \vdash \forall \vec{x} \varphi(\vec{x})$, then $T \vdash \forall \vec{x} \mathrm{Pr}_T(\ulcorner \varphi(\dot{\vec{x}}) \urcorner)$ for any formula $\varphi(\vec{x})$.
D2$^{\mathbf{U}}$: $T \vdash \forall \vec{x}(\mathrm{Pr}_T(\ulcorner \varphi(\dot{\vec{x}}) \to \psi(\dot{\vec{x}}) \urcorner) \to (\mathrm{Pr}_T(\ulcorner \varphi(\dot{\vec{x}}) \urcorner) \to \mathrm{Pr}_T(\ulcorner \psi(\dot{\vec{x}}) \urcorner)))$ for any formulas $\varphi(\vec{x})$, $\psi(\vec{x})$.
D3$^{\mathbf{U}}$: $T \vdash \forall \vec{x}(\mathrm{Pr}_T(\ulcorner \varphi(\dot{\vec{x}}) \urcorner) \to \mathrm{Pr}_T(\ulcorner \mathrm{Pr}_T(\ulcorner \varphi(\dot{\vec{x}}) \urcorner) \urcorner))$ for any formula $\varphi(\vec{x})$.

ΓC$^{\mathbf{U}}$: $T \vdash \forall \vec{x}(\varphi(\vec{x}) \to \mathrm{Pr}_T(\ulcorner \varphi(\dot{\vec{x}}) \urcorner))$ for any Γ formula $\varphi(\vec{x})$.
B$_2^{\mathbf{U}}$: If $T \vdash \forall \vec{x}(\varphi(\vec{x}) \to \psi(\vec{x}))$, then $T \vdash \forall \vec{x}(\mathrm{Pr}_T(\ulcorner \varphi(\dot{\vec{x}}) \urcorner) \to \mathrm{Pr}_T(\ulcorner \psi(\dot{\vec{x}}) \urcorner))$ for any formulas $\varphi(\vec{x})$, $\psi(\vec{x})$.
CB : $T \vdash \mathrm{Pr}_T(\ulcorner \forall \vec{x} \varphi(\vec{x}) \urcorner) \to \forall \vec{x} \mathrm{Pr}_T(\ulcorner \varphi(\dot{\vec{x}}) \urcorner)$ for any formula $\varphi(\vec{x})$.

Definition 3 (*Global derivability conditions*)

D2$^{\mathbf{G}}$: $T \vdash \forall x \forall y(\mathrm{Pr}_T(x \dot{\to} y) \to (\mathrm{Pr}_T(x) \to \mathrm{Pr}_T(y)))$.
D3$^{\mathbf{G}}$: $T \vdash \forall x(\mathrm{Pr}_T(x) \to \mathrm{Pr}_T(\ulcorner \mathrm{Pr}_T(\dot{x}) \urcorner))$.
ΓC$^{\mathbf{G}}$: $T \vdash \forall x(\mathsf{True}_\Gamma(x) \to \mathrm{Pr}_T(x))$.

Here $\ulcorner \varphi(\dot{\vec{x}}) \urcorner$ is an abbreviation for $\ulcorner \varphi(\dot{x}_0, \ldots, \dot{x}_{k-1}) \urcorner$ which is a primitive recursive term corresponding to a primitive recursive function calculating the Gödel number of $\varphi(\overline{n_0}, \ldots, \overline{n_{k-1}})$ from $n_0, \ldots, n_{k-1}$. Also $x \dot{\to} y$ is a primitive recursive term such that $\mathsf{PA} \vdash \ulcorner \varphi \urcorner \dot{\to} \ulcorner \psi \urcorner = \ulcorner \varphi \to \psi \urcorner$ for all formulas φ and ψ. Furthermore $\mathsf{True}_\Gamma(x)$ is a natural formula defining the truth of Γ sentences (cf. Hájek and Pudlák [7]).

Notice that every provability predicate automatically satisfies **D1**. Since our provability predicates are Σ_1, **D3** is a particular case of **Σ_1C**. Also **D3$^{\mathbf{U}}$** and **D3$^{\mathbf{G}}$** are particular cases of **Σ_1C$^{\mathbf{U}}$**. The condition **CB** claims the provability of sentences corresponding to the Converse Barcan Formula (see [9]). It is easy to prove the following implications (see [18]).

Proposition 1

1. **Δ_0C** *and* **B$_2$** $\Rightarrow$ **D1**.
2. **Δ_0C$^{\mathbf{U}}$** *and* **B$_2$** $\Rightarrow$ **D1$^{\mathbf{U}}$**.
3. **D1** *and* **D2** $\Rightarrow$ **B$_2$**.
4. **D1$^{\mathbf{U}}$** *and* **D2$^{\mathbf{U}}$** $\Rightarrow$ **B$_2^{\mathbf{U}}$**.
5. **D1** *and* **CB** $\Rightarrow$ **D1$^{\mathbf{U}}$**.
6. **B$_2^{\mathbf{U}}$** $\Rightarrow$ **CB**.

Here the first clause of Proposition 1 means that for any Σ_1 formula $\mathrm{Pr}_T(x)$, if $\mathrm{Pr}_T(x)$ satisfies both **Δ_0C** and **B$_2$**, then it also satisfies **D1**.

Moreover, the following nontrivial implication holds.

Theorem 1 (Kurahashi [18]) **D1** *and* **B$_2^{\mathbf{U}}$** $\Rightarrow$ **Σ_1C$^{\mathbf{U}}$**.

By Proposition 1.4, we immediately obtain the following corollary which is due to Buchholz (see also [23]).

Corollary 1 (Buchholz [3]) **D1$^{\mathbf{U}}$** *and* **D2$^{\mathbf{U}}$** $\Rightarrow$ **Σ_1C$^{\mathbf{U}}$**.

We introduce several different consistency statements based on the provability predicate $\mathrm{Pr}_T(x)$.

Definition 4

- $\mathsf{Con}^H_{\mathrm{Pr}_T} :\equiv \forall x(\mathrm{Pr}_T(x) \to \neg\mathrm{Pr}_T(\dot{\neg} x))$.
- $\mathsf{Con}^L_{\mathrm{Pr}_T} :\equiv \neg\mathrm{Pr}_T(\ulcorner 0 \neq 0 \urcorner)$.
- For each formula φ, $\mathsf{Con}_{\mathrm{Pr}_T}(\varphi) :\equiv (\mathrm{Pr}_T(\ulcorner \varphi \urcorner) \to \neg\mathrm{Pr}_T(\ulcorner \neg\varphi \urcorner))$.
- $\mathsf{Con}^S_{\mathrm{Pr}_T} := \{\mathsf{Con}_{\mathrm{Pr}_T}(\varphi) : \varphi \text{ is a formula}\}$.[1]

Here $\dot{\neg} x$ is a primitive recursive term satisfying $\mathsf{PA} \vdash \dot{\neg}\ulcorner\varphi\urcorner = \ulcorner\neg\varphi\urcorner$ for any formula φ. Then for every formula φ, $\mathsf{Con}_{\mathrm{Pr}_T}(\varphi)$ follows from $\mathsf{Con}^H_{\mathrm{Pr}_T}$. Also $\mathsf{Con}^L_{\mathrm{Pr}_T}$ follows from $\mathsf{Con}_{\mathrm{Pr}_T}(0 = 0)$ because $\mathrm{Pr}_T(x)$ satisfies **D1**. Therefore we have that $\mathsf{Con}^H_{\mathrm{Pr}_T}$ is stronger than $\mathsf{Con}^S_{\mathrm{Pr}_T}$, and $\mathsf{Con}^S_{\mathrm{Pr}_T}$ is stronger than $\mathsf{Con}^L_{\mathrm{Pr}_T}$. In general, the converse implications do not hold.

Hilbert and Bernays introduced the conditions **HB1**, **HB2** and **HB3** described in the introduction which are sufficient for unprovability of $\mathsf{Con}^H_{\mathrm{Pr}_T}$. In our context, each of their conditions corresponds to $\mathbf{B_2}$, **CB** and $\mathbf{\Delta_0 C^U}$, respectively. Then we call the conditions $\mathbf{B_2}$, **CB** and $\mathbf{\Delta_0 C^U}$ the *Hilbert–Bernays derivability conditions*. Their result can be stated as follows (see [18]).

Theorem 2 (Hilbert and Bernays [8]) *If* $\mathrm{Pr}_T(x)$ *satisfies* $\mathbf{B_2}$, **CB** *and* $\mathbf{\Delta_0 C^U}$, *then* $T \nvdash \mathsf{Con}^H_{\mathrm{Pr}_T}$.

The following theorem is a well-known form of unprovability of consistency which is essentially due to Löb. The conditions **D1**, **D2** and **D3** are called the *Hilbert–Bernays–Löb derivability conditions*.

Theorem 3 (Löb [20]) *If* $\mathrm{Pr}_T(x)$ *satisfies* **D1**, **D2** *and* **D3**, *then* $T \nvdash \mathsf{Con}^L_{\mathrm{Pr}_T}$.

By Proposition 1.3, $\{\mathbf{D1}, \mathbf{D2}\}$ implies $\{\mathbf{D1}, \mathbf{B_2}\}$. In [18], it is proved that if $\mathrm{Pr}_T(x)$ satisfies **D1**, $\mathbf{B_2}$ and **D3**, then $T \nvdash \mathsf{Con}^H_{\mathrm{Pr}_T}$. This statement can be strengthened as follows.

Theorem 4 *If* $\mathrm{Pr}_T(x)$ *satisfies* **D1**, $\mathbf{B_2}$ *and* **D3**, *then* $T \nvdash \mathsf{Con}^S_{\mathrm{Pr}_T}$.[2]

Proof Suppose $\mathrm{Pr}_T(x)$ satisfies **D1**, $\mathbf{B_2}$ and **D3**. Let φ be any sentence satisfying $T \vdash \varphi \leftrightarrow \neg\mathrm{Pr}_T(\ulcorner\varphi\urcorner)$. Then we have $T \vdash \mathsf{Con}_{\mathrm{Pr}_T}(\varphi) \to \varphi$ as in [18]. Since $T \nvdash \varphi$, $T \nvdash \mathsf{Con}_{\mathrm{Pr}_T}(\varphi)$. Thus $T \nvdash \mathsf{Con}^S_{\mathrm{Pr}_T}$. □

Notice that if $\mathrm{Pr}_T(x)$ satisfies **D2**, then $\mathsf{Con}^S_{\mathrm{Pr}_T}$ follows from $\mathsf{Con}^L_{\mathrm{Pr}_T}$. Hence Theorem 3 is also a consequence of Theorem 4.

In his proof of the incompleteness theorems, Gödel constructed a $\Delta_1(\mathsf{PA})$ formula $\mathsf{Proof}_T(x, y)$ saying that y is the Gödel number of a T-proof of a formula with the Gödel number x. Gödel's provability predicate $\mathsf{Prov}_T(x)$ is defined as $\exists y \mathsf{Proof}_T(x, y)$. Then the formula $\mathsf{Prov}_T(x)$ is a Σ_1 provability predicate satisfying full derivability conditions $\mathbf{D1^U}$, $\mathbf{D2^G}$ and $\mathbf{\Sigma_1 C^G}$. Thus the sentence $\mathsf{Con}^L_{\mathsf{Prov}_T}$ is not provable in T by Theorem 3. This is Gödel's second incompleteness theorem. It is known that $\mathsf{Con}^H_{\mathsf{Prov}_T}$ and $\mathsf{Con}^L_{\mathsf{Prov}_T}$ are provably equivalent in PA (see [18]), and so let Con_T denote one of these consistency statements.

---

[1] Introducing this schematic consistency statement $\mathsf{Con}^S_{\mathrm{Pr}_T}$ was proposed by the referee.

[2] This means that $T \nvdash \mathsf{Con}_{\mathrm{Pr}_T}(\varphi)$ for some formula φ.

Theorem 5 (The second incompleteness theorem (Gödel [5])) *The consistency statement* Con_T *of* T *cannot be proved in* T.

3 Rosser Provability Predicates

In this section, we introduce Rosser provability predicates and survey on derivability conditions for Rosser provability predicates. We say a formula $\mathrm{Prf}_T(x, y)$ is a *proof predicate* of T if it satisfies the following conditions:

1. $\mathrm{Prf}(x, y)$ is $\Delta_1(\mathsf{PA})$;
2. $\mathsf{PA} \vdash \forall x(\mathsf{Prov}_T(x) \leftrightarrow \exists y \mathrm{Prf}_T(x, y))$;
3. for any natural number n and formula φ, $\mathbb{N} \models \mathsf{Proof}_T(\ulcorner \varphi \urcorner, \overline{n}) \leftrightarrow \mathrm{Prf}_T(\ulcorner \varphi \urcorner, \overline{n})$;
4. $\mathsf{PA} \vdash \forall x \forall x' \forall y(\mathrm{Prf}_T(x, y) \land \mathrm{Prf}_T(x', y) \to x = x')$.

The last clause means that our proof predicates are single-conclusion ones.

For each proof predicate $\mathrm{Prf}_T(x, y)$ of T, we can associate the Σ_1 formula

$$\exists y(\mathrm{Prf}_T(x, y) \land \forall z \leq y \neg \mathrm{Prf}_T(\dot{\neg} x, z))$$

which is said to be the *Rosser provability predicate* of $\mathrm{Prf}_T(x, y)$ or a Rosser provability predicate of T. Notice that every Rosser provability predicate of T is a Σ_1 provability predicate of T.

Rosser provability predicates were essentially introduced by Rosser [24] to improve Gödel's first incompleteness theorem. The following proposition is an important feature of Rosser provability predicates.

Proposition 2 *Let* $\mathrm{Pr}_T^R(x)$ *be any Rosser provability predicate of* T *and* φ *be any formula. If* $T \vdash \neg\varphi$, *then* $\mathsf{PA} \vdash \neg \mathrm{Pr}_T^R(\ulcorner \varphi \urcorner)$.

Since T proves $\neg 0 \neq 0$, $\mathsf{PA} \vdash \neg \mathrm{Pr}_T^R(\ulcorner 0 \neq 0 \urcorner)$ by Proposition 2. Thus we obtain the following proposition.

Proposition 3 *For any Rosser provability predicate* $\mathrm{Pr}_T^R(x)$ *of* T, $\mathsf{PA} \vdash \mathsf{Con}^L_{\mathrm{Pr}_T^R}$.

There are limitations of the existence of Rosser provability predicates satisfying certain derivability conditions. From Theorem 3 and Proposition 3, we obtain the following corollary.

Corollary 2 *There exists no Rosser provability predicate of* T *satisfying both* **D2** *and* **D3**.

The following proposition is implicitly stated in Jeroslow [10] without a proof.

Proposition 4 *There exists no Rosser provability predicate of* T *satisfying* $\mathbf{\Sigma_1 C}$.

Proof Let $\mathrm{Pr}_T^R(x)$ be the Rosser provability predicate of a proof predicate $\mathrm{Prf}_T(x, y)$. Let σ be a Σ_1 sentence satisfying the following equivalence:

$$\mathsf{PA} \vdash \sigma \leftrightarrow \exists x(\mathrm{Prf}_T(\ulcorner \neg\sigma \urcorner, x) \land \forall y < x \neg \mathrm{Prf}_T(\ulcorner \sigma \urcorner, y)).$$

Then $\mathsf{PA} \vdash \mathrm{Pr}_T^R(\ulcorner \sigma \urcorner) \to \neg\sigma$. If $T \vdash \sigma \to \mathrm{Pr}_T^R(\ulcorner \sigma \urcorner)$, then $T \vdash \sigma \to \neg\sigma$, and hence $T \vdash \neg\sigma$. Then $\mathbb{N} \models \sigma$ by the choice of σ. By Σ_1-completeness, $T \vdash \sigma$. This is a contradiction. We conclude $T \nvdash \sigma \to \mathrm{Pr}_T^R(\ulcorner \sigma \urcorner)$. □

Then by Proposition 4, Theorem 1 and Corollary 1, we obtain the following corollary.

Corollary 3

1. *There exists no Rosser provability predicate of T satisfying* $\mathbf{B}_2^{\mathbf{U}}$.
2. *There exists no Rosser provability predicate of T satisfying both* $\mathbf{D1^U}$ *and* $\mathbf{D2^U}$.

Kreisel and Takeuti [15] asked whether Rosser provability predicates satisfy **D2** or not. Guaspari and Solovay established a modal logical method of obtaining Rosser provability predicates without some certain conditions. From their method, we have:

Theorem 6 (Guaspari and Solovay [6]) *There exist Rosser provability predicates satisfying neither* **D2** *nor* **D3**.

Notice that Guaspari and Solovay's Rosser provability predicates are based on multi-conclusion proof predicates, and Shavrukov [25] proved the same result for Rosser provability predicates based on single-conclusion proof predicates.

The Rosser provability predicate of Gödel's proof predicate $\mathsf{Proof}_T(x, y)$ is denoted by $\mathsf{Prov}_T^R(x)$. Montagna [21] proved that the global version of **D2** does not hold for $\mathsf{Prov}_T^R(x)$.

Proposition 5 (Montagna [21]). *The Rosser provability predicate* $\mathsf{Prov}_T^R(x)$ *does not satisfy* $\mathbf{D2^G}$.

On the other hand, there are Rosser provability predicates satisfying some derivability conditions.

Theorem 7 (Bernardi and Montagna [2]; Arai [1]) *There exist Rosser provability predicates satisfying* $\mathbf{D2^G}$.

The existence of Rosser provability predicates satisfying **D2** was also mentioned in Shavrukov [25]. Kikuchi and Kurahashi [11] investigated Rosser provability predicates satisfying **D2** and an additional certain condition. Kurahashi [16] investigated provability logics of Rosser provability predicates satisfying **D2**.

The existence of Rosser provability predicates satisfying **D3** was proved by Arai.

Theorem 8 (Arai [1]) *There exist Rosser provability predicates satisfying* $\mathbf{D3^G}$.

Arai proved Theorems 7 and 8 under the assumption that formulas are formulated in negation normal form. Let $\mathsf{nnf}(\varphi)$ be one of negation normal forms of a formula φ such that $\mathsf{nnf}(\neg\neg\varphi) \equiv \mathsf{nnf}(\varphi)$. Then we can understand that Arai's Rosser provability predicates are defined as $\mathrm{Pr}^R_T(x) \equiv \exists y(\mathrm{Prf}_T(\mathsf{nnf}(x), y) \land \forall z \leq y \neg \mathrm{Prf}_T(\mathsf{nnf}(\dot{\neg} x), z))$. Then it is easy to see that $\mathsf{PA} \vdash \mathsf{Con}^H_{\mathrm{Pr}^R_T}$. Thus strictly speaking, from this point of view, Arai's Rosser provability predicates are different from ours. Notice that his proofs are applicable to our Rosser provability predicates with some modifications. In fact, our proof of Theorem 9 (resp. Theorem 11) improves Theorem 7 (resp. Theorem 8), and our proofs of these theorems are based on Arai's proofs.

Moreover, Arai's results indicate that neither $\{\mathbf{D1}, \mathbf{D2^G}\}$ nor $\{\mathbf{D1}, \mathbf{D3^G}\}$ is sufficient for the unprovability of the consistency statement $\mathsf{Con}^H_{\mathrm{Pr}_T}$. By Corollaries 2 and 3, $\{\mathbf{D1}, \mathbf{D2^G}\}$ implies neither $\mathbf{D3}$ nor $\mathbf{D1^U}$. A similar observation can be done for $\{\mathbf{D1}, \mathbf{D3^G}\}$, and then we obtain the following corollary on non-implications concerning derivability conditions and the second incompleteness theorem.

Corollary 4

1. $\{\mathbf{D1}, \mathbf{D2^G}\}$ *does not imply any of* $\mathbf{D1^U}$, $\mathbf{D3}$ *or* $T \nvdash \mathsf{Con}^H_{\mathrm{Pr}_T}$.
2. $\{\mathbf{D1}, \mathbf{D3^G}\}$ *does not imply any of* $\mathbf{B_2}$, $\mathbf{\Sigma_1 C}$ *or* $T \nvdash \mathsf{Con}^H_{\mathrm{Pr}_T}$.

For example, the first clause of this corollary means that there exists a Σ_1 provability predicate $\mathrm{Pr}_T(x)$ satisfying $\mathbf{D1}$ and $\mathbf{D2^G}$, and not enjoying any of $\mathbf{D1^U}$, $\mathbf{D3}$ or $T \nvdash \mathsf{Con}^H_{\mathrm{Pr}_T}$.

In this paper, we improve Theorems 7 and 8 by showing the existence of Rosser provability predicates satisfying more additional conditions. As a corollary to our results, we obtain several more non-implications.

4 Main Theorems

In this section, we prove three theorems which are main theorems of this paper. The first theorem is an improvement of Theorem 7.

Theorem 9 *There exists a Rosser provability predicate of T satisfying* $\mathbf{D2^G}$, $\mathbf{\Delta_0 C^G}$ *and* $\mathsf{PA} \vdash \mathsf{Con}^H_{\mathrm{Pr}^R_1}$. *That is, there exists a Rosser provability predicate* $\mathrm{Pr}^R_1(x)$ *of T satisfying the following conditions:*

1. $\mathsf{PA} \vdash \forall x \forall y(\mathrm{Pr}^R_1(x \dot{\to} y) \to (\mathrm{Pr}^R_1(x) \to \mathrm{Pr}^R_1(y)))$.
2. $\mathsf{PA} \vdash \forall x(\mathsf{True}_{\Delta_0}(x) \to \mathrm{Pr}^R_1(x))$.
3. $\mathsf{PA} \vdash \forall x(\mathrm{Pr}^R_1(x) \to \neg \mathrm{Pr}^R_1(\dot{\neg} x))$.

The second theorem shows that in the statement of Corollary 1, the condition $\mathbf{D2^U}$ cannot be replaced by $\mathbf{D2}$.

Theorem 10 *There exists a Rosser provability predicate of T satisfying* **CB**, **D2** *and* $\mathbf{\Delta_0 C^G}$. *That is, there exists a Rosser provability predicate* $\mathrm{Pr}_2^R(x)$ *of T satisfying the following conditions:*

1. $T \vdash \mathrm{Pr}_2^R(\ulcorner \forall \vec{x}\, \varphi(\vec{x}) \urcorner) \to \forall \vec{x}\, \mathrm{Pr}_2^R(\ulcorner \varphi(\dot{\vec{x}}) \urcorner)$ *for any formula* $\varphi(\vec{x})$.
2. $T \vdash \mathrm{Pr}_2^R(\ulcorner \varphi \to \psi \urcorner) \to (\mathrm{Pr}_2^R(\ulcorner \varphi \urcorner) \to \mathrm{Pr}_2^R(\ulcorner \psi \urcorner))$ *for any formulas* φ *and* ψ.
3. $\mathsf{PA} \vdash \forall x(\mathsf{True}_{\Delta_0}(x) \to \mathrm{Pr}_2^R(x))$.

The last theorem is an improvement of Theorem 8.

Theorem 11 *There exists a Rosser provability predicate of T satisfying* **CB**, $\mathbf{B_2}$, $\mathbf{D3^G}$ *and* $\mathbf{\Delta_0 C^G}$. *That is, there exists a Rosser provability predicate* $\mathrm{Pr}_3^R(x)$ *of T satisfying the following conditions:*

1. $T \vdash \mathrm{Pr}_3^R(\ulcorner \forall \vec{x}\, \varphi(\vec{x}) \urcorner) \to \forall \vec{x}\, \mathrm{Pr}_3^R(\ulcorner \varphi(\dot{\vec{x}}) \urcorner)$ *for any formula* $\varphi(\vec{x})$.
2. *For any formulas* φ *and* ψ, *if* $T \vdash \varphi \to \psi$, *then* $T \vdash \mathrm{Pr}_3^R(\ulcorner \varphi \urcorner) \to \mathrm{Pr}_3^R(\ulcorner \psi \urcorner)$.
3. $\mathsf{PA} \vdash \forall x(\mathrm{Pr}_3^R(x) \to \mathrm{Pr}_3^R(\ulcorner \mathrm{Pr}_3^R(\dot{x}) \urcorner))$.
4. $\mathsf{PA} \vdash \forall x(\mathsf{True}_{\Delta_0}(x) \to \mathrm{Pr}_3^R(x))$.

Remark 1 Notice that both of $\mathrm{Pr}_2^R(x)$ and $\mathrm{Pr}_3^R(x)$ satisfy the Hilbert–Bernays derivability conditions $\mathbf{B_2}$, **CB** and $\mathbf{\Delta_0 C^U}$. Thus by Theorem 2, T can prove neither $\mathsf{Con}^H_{\mathrm{Pr}_2^R}$ nor $\mathsf{Con}^H_{\mathrm{Pr}_3^R}$. Therefore our Theorems 10 and 11 cannot be proved under Arai's assumption that formulas are formulated in negation normal form. Moreover, since $\mathrm{Pr}_3^R(x)$ satisfies $\mathbf{B_2}$ and **D3**, $T \nvdash \mathsf{Con}^S_{\mathrm{Pr}_3^R}$ by Theorem 4.

By Proposition 3, T proves $\mathsf{Con}^L_{\mathrm{Pr}_2^R}$ and $\mathsf{Con}^L_{\mathrm{Pr}_3^R}$. Since $\mathrm{Pr}_2^R(x)$ satisfies **D2**, T can also prove $\mathsf{Con}^S_{\mathrm{Pr}_2^R}$. Therefore our Rosser provability predicates $\mathrm{Pr}_2^R(x)$ and $\mathrm{Pr}_3^R(x)$ indicate the difference of the unprovability of three consistency statements $\mathsf{Con}^H_{\mathrm{Pr}_T}$, $\mathsf{Con}^S_{\mathrm{Pr}_T}$ and $\mathsf{Con}^L_{\mathrm{Pr}_T}$. The following table summarizes the situation of the unprovability of consistency statements.

	$T \vdash \mathsf{Con}^H_{\mathrm{Pr}_T}$	$T \vdash \mathsf{Con}^S_{\mathrm{Pr}_T}$	$T \vdash \mathsf{Con}^L_{\mathrm{Pr}_T}$
$\mathrm{Pr}_1^R(x)$	Yes	Yes	Yes
$\mathrm{Pr}_2^R(x)$	No	Yes	Yes
$\mathrm{Pr}_3^R(x)$	No	No	Yes

We obtain the following corollary for provability predicates.

Corollary 5

1. $\{\mathbf{D1}, \mathbf{D2^G}, \mathbf{\Delta_0 C^G}\}$ *does not imply any of* $\mathbf{D1^U}$, **D3** *or* $T \nvdash \mathsf{Con}^H_{\mathrm{Pr}_T}$.
2. $\{\mathbf{D1}, \mathbf{CB}, \mathbf{D2}, \mathbf{\Delta_0 C^G}\}$ *does not imply any of* $\mathbf{D2^U}$, **D3** *or* $T \nvdash \mathsf{Con}^S_{\mathrm{Pr}_T}$.
3. $\{\mathbf{D1}, \mathbf{CB}, \mathbf{B_2}, \mathbf{D3^G}, \mathbf{\Delta_0 C^G}\}$ *does not imply any of* **D2**, $\mathbf{\Sigma_1 C}$ *or* $T \nvdash \mathsf{Con}^L_{\mathrm{Pr}_T}$.

Corollary 5 shows that $\mathsf{Con}^H_{\mathrm{Pr}_T}$ and $\mathsf{Con}^S_{\mathrm{Pr}_T}$ cannot be replaced by $\mathsf{Con}^S_{\mathrm{Pr}_T}$ and $\mathsf{Con}^L_{\mathrm{Pr}_T}$ in the statements of Theorems 2 and 4, respectively.

Remark 2 Since the set of all Δ_0 sentences are closed under negation, $\mathsf{PA} \vdash \forall x(\Delta_0(x) \land \neg\mathsf{True}_{\Delta_0}(x) \to \mathsf{True}_{\Delta_0}(\dot\neg x))$ where $\Delta_0(x)$ is a formula naturally representing the set of all Δ_0 sentences. By $\mathbf{\Delta_0 C^G}$ for $\mathrm{Pr}_1^R(x)$, we have that $\mathsf{PA} \vdash \forall x(\Delta_0(x) \land \neg\mathsf{True}_{\Delta_0}(x) \to \mathrm{Pr}_1^R(\dot\neg x))$. Since $\mathsf{Con}^H_{\mathrm{Pr}_1^R}$ is provable in PA, we obtain $\mathsf{PA} \vdash \forall x(\Delta_0(x) \land \neg\mathsf{True}_{\Delta_0}(x) \to \neg\mathrm{Pr}_1^R(x))$. Therefore the uniform Δ_0 reflection principle $\forall x(\Delta_0(x) \land \mathrm{Pr}_1^R(x) \to \mathsf{True}_{\Delta_0}(x))$ for $\mathrm{Pr}_1^R(x)$ is provable in PA.[3]

Before proving our results, we introduce some terminology and prove a lemma. We assume that our logical symbols are only $\land$, $\neg$ and $\forall$, and other logical symbols such as $\to$ and $\exists$ are introduced as abbreviations. We say a formula φ' is an *instance* of a formula φ if for some numbers $k, i_0, \ldots, i_{k-1}$, some variables $x_0, \ldots, x_{k-1}$ and some formula ψ, the formulas φ and φ' are of the forms $\forall x_0 \cdots \forall x_{k-1}\psi(x_0, \ldots, x_{k-1})$ and $\psi(\overline{i_0}, \ldots, \overline{i_{k-1}})$, respectively. For each natural number m, let F_m be the set of all formulas whose Gödel numbers are less than or equal to m. We say that a finite mapping $V : F_m \to \{0, 1\}$ is a *truth assignment on* F_m if V satisfies the usual conditions of truth assignments for propositional logic such as $V(\varphi \land \psi) = V(\varphi) \cdot V(\psi)$, $V(\neg\varphi) = 1 - V(\varphi)$, and so on. Let X be any finite set of formulas. Let $d(X) = \min\{n' : X \subseteq F_{n'}\}$. A truth assignment V on F_m is said to be a *model of* X if $d(X) \leq m$ and $V(\varphi) = 1$ for all $\varphi \in X$. Let $P_{T,m}$ be the finite set $\{\varphi : \mathbb{N} \models \exists y \leq \overline{m}\,\mathsf{Proof}_T(\ulcorner\varphi\urcorner, y)\}$ of formulas.

We introduce two conditions (A) and (B) about truth assignments V on F_m.

- (A): For any formulas $\varphi, \varphi' \in F_m$, if $V(\varphi) = 1$ and φ' is an instance of φ, then $V(\varphi') = 1$.
- (B): If $\varphi \in F_m$ is a Δ_0 sentence, then φ is true if and only if $V(\varphi) = 1$.

The above terminology and definitions are formalizable in PA. Then we can define a $\Delta_1(\mathsf{PA})$ formula $\mathsf{Sat}(u)$ saying that "there exists a model of $P_{T,u}$ satisfying the conditions (A) and (B)" by using the $\Delta_1(\mathsf{PA})$ formula $\mathsf{True}_{\Delta_0}(x)$. We prove the following lemma.

Lemma 1 $\mathsf{PA} \vdash \mathsf{Con}_T \leftrightarrow \forall u \mathsf{Sat}(u)$.

Proof We reason in PA.

($\leftarrow$): Suppose $\neg\mathsf{Con}_T$, then there exists a number m such that $P_{T,m}$ contains both $0 = 0$ and $0 \neq 0$. Obviously $P_{T,m}$ has no model. Therefore $\neg\mathsf{Sat}(m)$ holds for some m.

($\to$): Suppose Con_T. Then there is a definable complete consistent extension T' of T by the arithmetized completeness theorem (cf. Hájek and Pudlák [7]). Let m be any number. We define a finite mapping $V : F_m \to \{0, 1\}$ as follows: for every $\varphi \in F_m$, $V(\varphi) = 1$ if and only if $\varphi \in T'$. Since T' is complete and consistent, V is a truth assignment on F_m.

We prove that V satisfies the condition (A). Let φ and φ' be any formulas in F_m with $V(\varphi) = 1$ and φ' is an instance of φ. Then $\varphi \in T'$. Since $\varphi \to \varphi'$ is logically valid, $\varphi' \in T'$. Therefore $V(\varphi') = 1$.

[3] This is pointed out by the referee.

We prove that V satisfies the condition (B). Let $\varphi \in F_m$ be any Δ_0 sentence. If φ is true, then φ is provable in T by $\mathbf{\Delta_0 C^G}$ for $\mathsf{Prov}_T(x)$. Thus $\varphi \in T'$ and hence $V(\varphi) = 1$. If φ is false, then $\neg\varphi$ is a true Δ_0 sentence and is provable in T. By the consistency of T', $\varphi \notin T'$. Therefore $V(\varphi) = 0$.

Also since T' is an extension of T, V is a model of $P_{T,m}$. We conclude that $\mathsf{Sat}(m)$ holds. □

Notice that PA is essentially reflexive, that is, every consistent extension of PA can prove the consistency of every finite subtheory of itself (cf. [19]). Thus in a similar way as in our proof of Lemma 1, we obtain the following lemma.

Lemma 2 *For every natural number m, $T \vdash \mathsf{Sat}(\overline{m})$.*

4.1 Proof of Theorem 9

In this subsection, we prove Theorem 9. For each formula φ, we define the formula $-\varphi$ as follows:

$$-\varphi :\equiv \begin{cases} \varphi & \text{if } \varphi \text{ is not of the form } \neg\psi, \\ \psi & \text{if } \varphi \text{ is of the form } \neg\psi. \end{cases}$$

Let $\{\xi_k\}_{k\in\omega}$ be the effective enumeration of all formulas introduced in Sect. 2. Notice that if ξ_k is $-\xi_l$, then ξ_k is a subformula of ξ_l, and hence $k \leq l$. Therefore ξ_k is none of $-\xi_l$ for all $l < k$. This property will be used in our proofs.

We define a primitive recursive function $e(\varphi, V, n)$ as follows:

- If V is not a (code of) truth assignment on F_n, $e(\varphi, V, n) = 0$.
- If V is a truth assignment on F_n, then the value of $e(\varphi, V, n)$ is defined as follows by recursion on the construction of φ:

1. If φ is an atomic formula or a universal formula,

$$e(\varphi, V, n) := \begin{cases} V(\varphi) & \text{if } \varphi \in F_n, \\ 1 & \text{if } \varphi \notin F_n \ \&\ \varphi \text{ is a true } \Delta_0 \text{ sentence}, \\ 0 & \text{if } \varphi \notin F_n \ \&\ \varphi \text{ is a false } \Delta_0 \text{ sentence}, \\ 1 & \text{otherwise}. \end{cases}$$

2. If φ is $\neg\xi_0$, then $e(\varphi, V, n) := 1 - e(\xi_0, V, n)$.
3. If φ is $\xi_0 \wedge \xi_1$, then $e(\varphi, V, n) := e(\xi_0, V, n) \cdot e(\xi_1, V, n)$.

Then it can be proved that if V is a truth assignment on F_n, then for any formula $\varphi \in F_n$, $V(\varphi) = e(\varphi, V, n)$.

Here we state Theorem 9 again.

Theorem 9 *There exists a Rosser provability predicate* $\mathrm{Pr}_1^R(x)$ *of* T *satisfying the following conditions:*

1. $\mathsf{PA} \vdash \forall x \forall y(\mathrm{Pr}_1^R(x \dot{\to} y) \to (\mathrm{Pr}_1^R(x) \to \mathrm{Pr}_1^R(y)))$.
2. $\mathsf{PA} \vdash \forall x(\mathsf{True}_{\Delta_0}(x) \to \mathrm{Pr}_1^R(x))$.
3. $\mathsf{PA} \vdash \forall x(\mathrm{Pr}_1^R(x) \to \neg \mathrm{Pr}_1^R(\dot{\neg} x))$.

Proof We define a PA-provably recursive function $g_1(x)$ in stages. Let $\mathrm{Prf}_1(x, y)$ be the $\Delta_1(\mathsf{PA})$ formula $x = g_1(y) \land \mathsf{Fml}(x)$, where $\mathsf{Fml}(x)$ is a natural $\Delta_1(\mathsf{PA})$ formula saying that x is a formula. Let $\mathrm{Pr}_1^R(x)$ be the Rosser provability predicate of $\mathrm{Prf}_1(x, y)$. The definition of g_1 consists of Procedures 1 and 2. The definition of g_1 begins with Procedure 1, and enumerates theorems of T until appearing a number m such that $\mathsf{Sat}(m)$ does not hold. After appearing such a number m, the definition of g_1 shifts to Procedure 2. In Procedure 2, g_1 outputs all formulas in stages. In the definition of the function g_1, we identify each formula with its Gödel number.

Procedure 1.
Stage $1.m$:

- If $\mathsf{Sat}(m)$, then

$$g_1(m) = \begin{cases} \varphi & \text{if } m \text{ is a proof of } \varphi \text{ in } T, \text{ that is, } \mathsf{Proof}_T(\varphi, m) \text{ holds,} \\ 0 & m \text{ is not a proof of any formula in } T. \end{cases}$$

  Go to Stage $1.(m + 1)$.
- If $\neg\mathsf{Sat}(m)$, then go to Procedure 2.

Procedure 2.
Let m be the least number such that $\mathsf{Sat}(m)$ does not hold. Since $\mathsf{Sat}(m - 1)$ holds, there exists a model of $P_{T,m-1}$ satisfying the condition (B). Let V be the least such model. Let $n = d(P_{T,m-1})$, then V is defined on F_n.

We define the value $g_1(m + k)$ for $k \geq 0$ as follows:

$$g_1(m + k) = \begin{cases} \xi_k & \text{if } e(\xi_k, V, n) = 1, \\ \neg\xi_k & \text{if } e(\xi_k, V, n) = 0. \end{cases}$$

The definition of the function g_1 is completed.

First, we show that the formula $\mathrm{Prf}_1(x, y)$ is a proof predicate of T. Since $\forall u \mathsf{Sat}(u)$ is true in the standard model $\mathbb{N}$ by Lemma 1, $\mathbb{N} \models \mathsf{Proof}_T(\ulcorner \varphi \urcorner, \overline{n}) \leftrightarrow \mathrm{Prf}_1(\ulcorner \varphi \urcorner, \overline{n})$ for all φ and n by the definition of g_1. Also $\mathsf{PA} \vdash \forall x \forall x' \forall y(\mathrm{Prf}_1(x, y) \land \mathrm{Prf}_1(x', y) \to x = x')$ trivially holds. We show that the sentence $\forall x(\mathsf{Prov}_T(x) \leftrightarrow \exists y \mathrm{Prf}_1(x, y))$ is provable in PA. By the definition of g_1, this sentence is obviously proved in $\mathsf{PA} + \forall u \mathsf{Sat}(u)$. It suffices to show that this sentence is provable in $\mathsf{PA} + \exists u \neg \mathsf{Sat}(u)$.

We proceed in $\mathsf{PA} + \exists u \neg \mathsf{Sat}(u)$: Let m be the least number such that $\mathsf{Sat}(m)$ does not hold, and let $n = d(P_{T,m-1})$. Let V be the least model of $P_{T,m-1}$ satisfying (B). We show that for any k, g_1 eventually outputs the formula ξ_k. We distinguish the following three cases.

- $e(\xi_k, V, n) = 1$ and ξ_k is not of the form $\neg\psi$: Then $g_1(m + k) = -\xi_k = \xi_k$.
- $e(\xi_k, V, n) = 1$ and ξ_k is of the form $\neg\xi_l$: Then $e(\xi_l, V, n) = 0$ and $g_1(m + l) = \neg\xi_l = \xi_k$.
- $e(\xi_k, V, n) = 0$: Then for p with $\neg\xi_k = \xi_p$, $e(\xi_p, V, n) = 1$ and hence $g_1(m + p) = -\xi_p = \xi_k$.

We have shown that $\exists y \mathrm{Prf}_1(x, y)$ holds if and only if x is a formula. Since $\neg\mathsf{Con}_T$ holds by Lemma 1, x is provable in T if and only if x is a formula. Therefore $\forall x(\mathsf{Prov}_T(x) \leftrightarrow \exists y \mathrm{Prf}_1(x, y))$ holds.

Next, we show that our formula $\mathrm{Pr}_1^R(x)$ satisfies the required conditions. The following claim is a key property of our construction of $\mathrm{Pr}_1^R(x)$.

Claim 1 The following sentence is provable in PA: "Let m be the least number such that $\mathsf{Sat}(m)$ does not hold, let $n = d(P_{T,m-1})$ and let V be the least model of $P_{T,m-1}$ satisfying (B). Then for any formula φ,

$$e(\varphi, V, n) = 1 \text{ if and only if } \mathrm{Pr}_1^R(\ulcorner\varphi\urcorner) \text{ holds".}$$

Proof We proceed in PA. Let m, n and V be as indicated in the statement. Let φ be any formula.

($\Rightarrow$): Suppose $e(\varphi, V, n) = 1$. If $\neg\varphi \in F_n$, then $V(\neg\varphi) = 1 - V(\varphi) = 1 - e(\varphi, V, n) = 0$. Therefore $\neg\varphi$ is not in $P_{T,m-1}$ because V is a model of $P_{T,m-1}$. If $\neg\varphi \notin F_n$, then $\neg\varphi$ is not in $P_{T,m-1}$ because $n = d(P_{T,m-1})$. In either case, $\neg\varphi$ is not in $P_{T,m-1}$. Hence $\neg\varphi$ is not in $\{g_1(0), \ldots, g_1(m-1)\}$ because the construction of g_1 executes Procedure 1 before Stage 1.m.

Let $\varphi = \xi_k$. Then $\neg\varphi$ is not in $\{g_1(m), \ldots, g_1(m+k)\}$ because $\neg\varphi$ is neither $-\xi_{k'}$ nor $\neg\xi_{k''}$ for all $k' \leq k$ and $k'' < k$. If φ is not of the form $\neg\psi$, then $g_1(m + k) = -\varphi = \varphi$ because $e(\xi_k, V, n) = 1$. If φ is of the form $\neg\xi_l$, then $l < k$ and $g_1(m + l) = \neg\xi_l = \varphi$ because $e(\xi_l, V, n) = 0$. Therefore $\mathrm{Pr}_1^R(\ulcorner\varphi\urcorner)$ holds in either case.

($\Leftarrow$): Suppose $e(\varphi, V, n) = 0$. Then $\varphi \notin \{g_1(0), \ldots, g_1(m-1)\}$ because $n = d(P_{T,m-1})$ and V is a model of $P_{T,m-1}$. Let $\varphi = \xi_k$, then $g_1(m + k) = \neg\xi_k = \neg\varphi$. If φ is not of the form $\neg\psi$, then $\varphi \notin \{g_1(m), \ldots, g_1(m + k - 1)\}$ because φ is neither $-\xi_{k'}$ nor $\neg\xi_{k'}$ for all $k' < k$. If φ is of the form $\neg\xi_l$ for some $l < k$, then φ is neither $-\xi_{l'}$ nor $\neg\xi_{l'}$ for all $l' < l$. Since $e(\xi_l, V, n) = 1 - e(\varphi, V, n) = 1$, $g_1(m + l) = -\xi_l \neq \varphi$. Hence we have $\varphi \notin \{g_1(m), \ldots, g_1(m + k - 1)\}$. In either case, $\mathrm{Pr}_1^R(\ulcorner\varphi\urcorner)$ does not hold. $\square$

Claim 2 $\mathsf{PA} \vdash \forall x \forall y(\mathrm{Pr}_1^R(x \dot{\rightarrow} y) \rightarrow (\mathrm{Pr}_1^R(x) \rightarrow \mathrm{Pr}_1^R(y)))$.

Proof Since $\mathsf{PA} + \forall u \mathsf{Sat}(u) \vdash \mathsf{Con}_T$ by Lemma 1, $\mathsf{PA} + \forall u \mathsf{Sat}(u) \vdash \forall x(\mathsf{Prov}_T(x) \leftrightarrow \mathrm{Pr}_1^R(x))$ easily follows from the definition of g_1. Then $\mathbf{D2^G}$ for $\mathsf{Prov}_T(x)$ implies $\mathsf{PA} + \forall u \mathsf{Sat}(u) \vdash \forall x \forall y(\mathrm{Pr}_1^R(x \dot{\rightarrow} y) \rightarrow (\mathrm{Pr}_1^R(x) \rightarrow \mathrm{Pr}_1^R(y)))$.

We proceed in $\mathsf{PA} + \exists u \neg \mathsf{Sat}(u)$: Let n and V be as above. Let φ and ψ be any formulas with $\mathrm{Pr}_1^R(\ulcorner \varphi \to \psi \urcorner)$ and $\mathrm{Pr}_1^R(\ulcorner \varphi \urcorner)$ hold. Then $e(\varphi \to \psi, V, n) = e(\varphi, V, n) = 1$ by Claim 1. We have $e(\psi, V, n) = 1$ and hence $\mathrm{Pr}_1^R(\ulcorner \psi \urcorner)$ holds by Claim 1 again. We have shown that the theory $\mathsf{PA} + \exists u \neg \mathsf{Sat}(u)$ also proves $\forall x \forall y (\mathrm{Pr}_1^R(x \dot{\to} y) \to (\mathrm{Pr}_1^R(x) \to \mathrm{Pr}_1^R(y)))$. □

Claim 3 $\mathsf{PA} \vdash \forall x (\mathsf{True}_{\Delta_0}(x) \to \mathrm{Pr}_1^R(x))$.

Proof As in the proof of Claim 2, $\mathbf{\Delta_0 C^G}$ for $\mathsf{Prov}_T(x)$ implies $\mathsf{PA} + \forall u \mathsf{Sat}(u) \vdash \forall x (\mathsf{True}_{\Delta_0}(x) \to \mathrm{Pr}_1^R(x))$.

We work in $\mathsf{PA} + \exists u \neg \mathsf{Sat}(u)$: Let n and V be as above. First, we prove by induction on the construction of $\varphi \in \Delta_0$ that for all Δ_0 sentences φ, φ is true if and only if $e(\varphi, V, n) = 1$.

- (Base Case): φ is an atomic sentence or a universal sentence:
  ($\Rightarrow$): Suppose that φ is true. If $\varphi \notin F_n$, then $e(\varphi, V, n) = 1$ by the definition of e. If $\varphi \in F_n$, then $V(\varphi) = 1$ by the condition (B). Hence $e(\varphi, V, n) = 1$.
  The proof for ($\Leftarrow$) is similar.
- Induction cases are straightforward by the definition of e.

Let φ be any true Δ_0 sentence. Then $e(\varphi, V, n) = 1$ as shown above. By Claim 1, $\mathrm{Pr}_1^R(\ulcorner \varphi \urcorner)$ holds. We have shown $\mathsf{PA} + \exists u \neg \mathsf{Sat}(u) \vdash \forall x (\mathsf{True}_{\Delta_0}(x) \to \mathrm{Pr}_1^R(x))$. □

Claim 4 $\mathsf{PA} \vdash \forall x (\mathrm{Pr}_1^R(x) \to \neg \mathrm{Pr}_1^R(\dot{\neg} x))$.

Proof Since Con_T is equivalent to $\forall x (\mathrm{Pr}_T(x) \to \neg \mathrm{Pr}_T(\dot{\neg} x))$, $\mathsf{PA} + \forall u \mathsf{Sat}(u) \vdash \forall x (\mathrm{Pr}_1^R(x) \to \neg \mathrm{Pr}_1^R(\dot{\neg} x))$ by Lemma 1.

We reason in $\mathsf{PA} + \exists u \neg \mathsf{Sat}(u)$: Let n and V be as above. Suppose $\mathrm{Pr}_1^R(\ulcorner \varphi \urcorner)$ holds for a formula φ. Then $e(\varphi, V, n) = 1$ by Claim 1. Since $e(\neg \varphi, V, n) = 1 - e(\varphi, V, n) = 0$, $\mathrm{Pr}_1^R(\ulcorner \neg \varphi \urcorner)$ does not hold by Claim 1 again. □

This completes our proof of Theorem 9. □

4.2 *Proof of Theorem 10*

In this subsection, we prove Theorem 10.

Theorem 10 *There exists a Rosser provability predicate* $\mathrm{Pr}_2^R(x)$ *of* T *satisfying the following conditions:*

1. $T \vdash \mathrm{Pr}_2^R(\ulcorner \forall \vec{x}\, \varphi(\vec{x}) \urcorner) \to \forall \vec{x}\, \mathrm{Pr}_2^R(\ulcorner \varphi(\dot{\vec{x}}) \urcorner)$ *for any formula* $\varphi(\vec{x})$.
2. $T \vdash \mathrm{Pr}_2^R(\ulcorner \varphi \to \psi \urcorner) \to (\mathrm{Pr}_2^R(\ulcorner \varphi \urcorner) \to \mathrm{Pr}_2^R(\ulcorner \psi \urcorner))$ *for any formulas* φ *and* ψ.
3. $\mathsf{PA} \vdash \forall x (\mathsf{True}_{\Delta_0}(x) \to \mathrm{Pr}_2^R(x))$.

Proof The definition of our PA-provably recursive function g_2 corresponding to this theorem is different from the definition of the function g_1 in our proof of Theorem 9 only for Procedure 2. We describe only Procedure 2 of the definition of the function g_2.

Procedure 2. $\mathsf{Sat}(m-1)$ holds but $\mathsf{Sat}(m)$ does not hold.
Let $n = d(P_{T,m-1})$ and let V be the least model of $P_{T,m-1}$ satisfying the conditions (A) and (B).

We say a formula φ is *critical* if φ satisfies one of the following conditions:

1. $\varphi \in F_n$ and $V(\varphi) = 1$;
2. $\varphi \notin F_n$ and φ is a true Δ_0 sentence;
3. $\varphi \notin F_n$ and there exists a formula $\psi \in F_n$ such that φ is an instance of ψ and $V(\psi) = 1$.

Notice that if $\varphi \in F_n$, then φ is critical if and only if $V(\varphi) = 1$.

Let $\{\xi_k\}_{k\in\omega}$ be the effective enumeration of all formulas as above. We simultaneously define the values $g_2(m+k)$ for $k \geq 0$ and a sequence $\{i_k\}_{k\in\omega}$ of numbers as follows: Let $i_0 = 0$.

1. If ξ_k is not critical and $\neg\xi_k$ is critical, then let $g_2(m+i_k) = \neg\xi_k$, $g_2(m+i_k+1) = \xi_k$ and $i_{k+1} = i_k + 2$.
2. Otherwise, let $g_2(m+i_k) = \xi_k$ and $i_{k+1} = i_k + 1$.

The definition of g_2 is finished. As in the proof of Theorem 9, it can be shown that $\mathrm{Prf}_2(x, y)$ is a proof predicate of T, and we omit the proof.

Claim 1 The following sentence is provable in PA: "Let m be the least number such that $\mathsf{Sat}(m)$ does not hold, let n be $d(P_{T,m-1})$ and let V be the least model of $P_{T,m-1}$ satisfying the conditions (A) and (B). Then for any formula φ,

1. if φ is critical, then $\mathrm{Pr}_2^R(\ulcorner\varphi\urcorner)$ holds,
2. and if $\neg\varphi \in F_n$ and $\neg\varphi$ is critical, then $\mathrm{Pr}_2^R(\ulcorner\varphi\urcorner)$ does not hold".

Proof We proceed in PA. Let m, n and V be as indicated in the statement. Let φ be any formula. Then for some k, φ is ξ_k.

1. Suppose ξ_k is critical. First, we prove that $\neg\xi_k$ is not in $\{g_2(0), \ldots, g_2(m-1)\}$. If $\xi_k \in F_n$, then $V(\xi_k) = 1$, and thus $\neg\xi_k$ is not in $P_{T,m-1}$ because V is a model of $P_{T,m-1}$. If $\xi_k \notin F_n$, then $\neg\xi_k \notin P_{T,m-1}$ because $n = d(P_{T,m-1})$. In either case, $\neg\xi_k \notin P_{T,m-1}$. Therefore $\neg\xi_k \notin \{g_2(0), \ldots, g_2(m-1)\}$. By the definition of g_2, $\neg\xi_k$ is also not in $\{g_2(m), \ldots, g_2(m+i_k)\}$. Since ξ_k is critical, $g_2(m+i_k) = \xi_k$. Hence $\mathrm{Pr}_2^R(\ulcorner\xi_k\urcorner)$ holds.

2. Suppose $\neg\xi_k \in F_n$ and $\neg\xi_k$ is critical. Then $V(\neg\xi_k) = 1$. Thus $\xi_k \in F_n$ and $V(\xi_k) = 0$. It follows that ξ_k is not critical. Also ξ_k is not in $P_{T,m-1}$ nor $\{g_2(0), \ldots, g_2(m-1)\}$. Even if ξ_k is of the form $\neg\psi$, ξ_k is not in $\{g_2(m), \ldots, g_2(m+i_k-1)\}$ because ξ_k is not critical. Since $g_2(m+i_k) = \neg\xi_k$, $\mathrm{Pr}_2^R(\ulcorner\xi_k\urcorner)$ does not hold. □

Claim 2 $T \vdash \mathrm{Pr}_2^R(\ulcorner\forall\vec{x}\,\varphi(\vec{x})\urcorner) \to \forall\vec{x}\,\mathrm{Pr}_2^R(\ulcorner\varphi(\dot{\vec{x}})\urcorner)$ for any formula $\varphi(\vec{x})$.

Proof Let $\varphi(\vec{x})$ be any formula. As in our proof of Theorem 9, it suffices to show that the sentence is provable in $T + \exists u \neg \mathsf{Sat}(u)$. We reason in $T + \exists u \neg \mathsf{Sat}(u)$. Let n and V be as above. Notice that n is larger than the Gödel number of the formula $\neg \forall \vec{x} \varphi(\vec{x})$ by Lemma 2. Suppose $\mathrm{Pr}_2^R(\ulcorner \forall \vec{x} \varphi(\vec{x}) \urcorner)$ holds. Since $\neg \forall \vec{x} \varphi(\vec{x}) \in F_n$, $\neg \forall \vec{x} \varphi(\vec{x})$ is not critical by Claim 1. Then $V(\neg \forall \vec{x} \varphi(\vec{x})) \neq 1$, and hence $V(\forall \vec{x} \varphi(\vec{x})) = 1$. Let $\varphi(\vec{a})$ be any instance of $\forall \vec{x} \varphi(\vec{x})$. If $\varphi(\vec{a}) \in F_n$, then $V(\varphi(\vec{a})) = 1$ by the condition (A). Thus $\varphi(\vec{a})$ is critical. If $\varphi(\vec{a}) \notin F_n$, then $\varphi(\vec{a})$ is also critical because $V(\forall \vec{x} \varphi(\vec{x})) = 1$. In either case, $\varphi(\vec{a})$ is critical. Then $\mathrm{Pr}_2^R(\ulcorner \varphi(\vec{a}) \urcorner)$ holds by Claim 1. □

Claim 3 $T \vdash \mathrm{Pr}_2^R(\ulcorner \varphi \to \psi \urcorner) \to (\mathrm{Pr}_2^R(\ulcorner \varphi \urcorner) \to \mathrm{Pr}_2^R(\ulcorner \psi \urcorner))$ for any formulas φ and ψ.

Proof Let φ and ψ be any formulas. We work in $T + \exists u \neg \mathsf{Sat}(u)$. Let n and V be as above. Notice that n is larger than the Gödel numbers of the formulas $\neg \varphi$ and $\neg(\varphi \to \psi)$ by Lemma 2. Suppose $\mathrm{Pr}_2^R(\ulcorner \varphi \to \psi \urcorner)$ and $\mathrm{Pr}_2^R(\ulcorner \varphi \urcorner)$ hold. Since $\neg(\varphi \to \psi)$ and $\neg \varphi$ are in F_n, these sentences are not critical by Claim 1. Then $V(\neg(\varphi \to \psi)) \neq 1$ and $V(\neg \varphi) \neq 1$. Thus $V(\varphi \to \psi) = V(\varphi) = 1$. Hence $V(\psi) = 1$ and ψ is critical. Therefore $\mathrm{Pr}_2^R(\ulcorner \psi \urcorner)$ holds by Claim 1. □

Claim 4 $\mathsf{PA} \vdash \forall x(\mathsf{True}_{\Delta_0}(x) \to \mathrm{Pr}_2^R(x))$.

Proof We proceed in $\mathsf{PA} + \exists u \neg \mathsf{Sat}(u)$. Let n and V be as above. Let φ be any true Δ_0 sentence. If $\varphi \in F_n$, then $V(\varphi) = 1$ by the condition (B). Thus φ is critical. If $\varphi \notin F_n$, then φ is critical because φ is a true Δ_0 sentence. In either case, φ is critical. Therefore we have $\mathrm{Pr}_2^R(\ulcorner \varphi \urcorner)$ by Claim 1. □

This completes our proof of Theorem 10. □

4.3 *Proof of Theorem 11*

In this subsection, we prove Theorem 11. Before proving the theorem, for each natural number m, we recursively define the sequence $\{X_{m,n}\}_{n \in \omega}$ of finite sets of negated formulas as follows:

1. $X_{m,0} := \{\neg \varphi : \neg \varphi \in P_{T,m}\}$.
2. $\neg \varphi \in X_{m,n+1}$ if and only if at least one of the following conditions holds:
   - $\neg \varphi \in F_m$ and for some instance φ' of φ, $\neg \varphi' \in X_{m,n}$.
   - There is a formula ψ such that $\neg \psi \in X_{m,n}$ and $\neg \psi \to \neg \varphi \in P_{T,m}$.

Let $X_m := \bigcup_{n \in \omega} X_{m,n}$.

Lemma 3 *Let φ be any formula and m be any natural number.*

1. *If $\varphi \in X_m$, then φ is provable in T.*
2. $X_m \subseteq X_{m+1}$.

3. *$X_m \subseteq F_m$. As a consequence, $X_m = \bigcup_{n \leq |F_m|} X_{m,n}$ where $|F_m|$ is the number of elements of the finite set F_m.*

Proof 1. We prove by induction on n that for any $n \in \omega$, if $\neg\varphi \in X_{m,n}$, then $\neg\varphi$ is provable in T.

- If $\neg\varphi \in X_{m,0}$, then $\neg\varphi \in P_{T,m}$ and hence $\neg\varphi$ is provable in T.
- Assume that the statement is true for n. Suppose $\neg\varphi \in X_{m,n+1}$. If $\neg\varphi' \in X_{m,n}$ for some instance φ' of φ, then $\neg\varphi'$ is T-provable by induction hypothesis. Since $\varphi \to \varphi'$ is logically valid, $\neg\varphi$ is also provable in T.

  If there is a formula ψ such that $\neg\psi \in X_{m,n}$ and $\neg\psi \to \neg\varphi \in P_{T,m}$, then $\neg\psi$ is provable in T by induction hypothesis. Since $\neg\psi \to \neg\varphi$ is T-provable, $\neg\varphi$ is also T-provable.

2. We prove $X_{m,n} \subseteq X_{m+1,n}$ for all $n \in \omega$ by induction on n. The $n = 0$ case is immediate from $P_{T,m} \subseteq P_{T,m+1}$. Assume $X_{m,n} \subseteq X_{m+1,n}$. Suppose $\neg\varphi \in X_{m,n+1}$. If $\neg\varphi \in F_m$ and $\neg\varphi' \in X_{m,n}$ for some instance φ' of φ, then $\neg\varphi \in F_{m+1}$ and $\neg\varphi' \in X_{m+1,n}$, and hence $\neg\varphi \in X_{m+1,n+1}$.

If there is a formula ψ such that $\neg\psi \in X_{m,n}$ and $\neg\psi \to \neg\varphi \in P_{T,m}$, then $\neg\varphi \in X_{m+1,n+1}$ because $\neg\psi \in X_{m+1,n}$ and $\neg\psi \to \neg\varphi \in P_{T,m+1}$.

3. This is proved by induction and by using the fact $P_{T,m} \subseteq F_m$. □

Notice that Lemma 3 is formalizable in PA. Also notice that there is a PA-provably recursive computation calculating X_m from m.

Here we give our proof of Theorem 11.

Theorem 11 *There exists a Rosser provability predicate $\mathrm{Pr}_3^R(x)$ of T satisfying the following conditions:*

1. *$T \vdash \mathrm{Pr}_3^R(\ulcorner \forall\vec{x}\,\varphi(\vec{x}) \urcorner) \to \forall\vec{x}\,\mathrm{Pr}_3^R(\ulcorner \varphi(\dot{\vec{x}}) \urcorner)$ for any formula $\varphi(\vec{x})$.*
2. *For any formulas φ and ψ, if $T \vdash \varphi \to \psi$, then $T \vdash \mathrm{Pr}_3^R(\ulcorner \varphi \urcorner) \to \mathrm{Pr}_3^R(\ulcorner \psi \urcorner)$.*
3. *$\mathsf{PA} \vdash \forall x(\mathrm{Pr}_3^R(x) \to \mathrm{Pr}_3^R(\ulcorner \mathrm{Pr}_3^R(\dot{x}) \urcorner))$.*
4. *$\mathsf{PA} \vdash \forall x(\mathsf{True}_{\Delta_0}(x) \to \mathrm{Pr}_3^R(x))$.*

Proof We define a PA-provably recursive function g_3 corresponding to this theorem in stages. In the definition, as in Guaspari and Solovay [6], the bell which plays a role of a flag is prepared. As in our proofs of Theorems 9 and 10, the construction of g_3 consists of Procedures 1 and 2, and the bell may ring during the execution of Procedure 1. When the bell rings, the construction switches to Procedure 2. Also in the definition of the function g_3, we can use the formula $\mathrm{Pr}_3^R(x)$ by the recursion theorem.

Procedure 1.
Stage 1.m:

- If there exists some formula φ satisfying at least one of the following conditions, then ring the bell and go to Procedure 2:

1. $X_m \cup P_{T,m}$ contains both φ and $\neg\varphi$;
2. $\neg\varphi \notin X_m$ and $\neg\mathrm{Pr}_3^R(\ulcorner\varphi\urcorner) \in X_m$;
3. φ is a true Δ_0 sentence and $\neg\varphi \in X_m$.

– Otherwise,

$$g_3(m) = \begin{cases} \varphi & \text{if } m \text{ is a proof of } \varphi \text{ in } T, \\ 0 & m \text{ is not a proof of any formula in } T. \end{cases}$$

Go to Stage $1.(m+1)$.

Procedure 2.

The bell rings at Stage $1.m$. Let $\chi_0, \dots, \chi_{k-1}$ be a list of all elements of the finite set X_{m-1}. For $i < k$, let

$$g_3(m+i) = \chi_i.$$

After that, let $\{\xi_i\}_{i\in\omega}$ be the effective enumeration of all formulas introduced in Sect. 2. For $i \geq 0$, let

$$g_3(m+k+i) = \xi_i.$$

Our definition of the function g_3 has just been completed. The following claim shows up an important feature of the construction of the function g_3.

Claim 1 The following statement is provable in PA: "If the bell rings at Stage $1.m$, then for any formula φ,

$$\neg\varphi \in X_{m-1} \text{ if and only if } \neg\mathrm{Pr}_3^R(\ulcorner\varphi\urcorner) \text{ holds". \quad (1)$$

Proof We work in PA: Suppose that the bell rings at Stage $1.m$.

($\Rightarrow$): Suppose $\neg\varphi \in X_{m-1}$. Then $\neg\varphi$ is χ_i for some $i < k$ where $k = |X_{m-1}|$. Also $g_3(m+i) = \neg\varphi$ by the definition of g_3. If φ were in $X_{m-1} \cup P_{T,m-1}$, then the bell would ring before Stage $1.m$, and this contradicts the choice of m. Thus $\varphi \notin X_{m-1} \cup P_{T,m-1}$. Then φ is not in the list $g_3(0), \dots, g_3(m-1), g_3(m), \dots, g_3(m+k-1)$. Therefore $\neg\mathrm{Pr}_3^R(\ulcorner\varphi\urcorner)$ holds.

($\Leftarrow$): We prove the contrapositive. Suppose $\neg\varphi \notin X_{m-1}$. In particular, $\neg\varphi \notin X_{m-1,0}$ and hence $\neg\varphi \notin P_{T,m-1}$. Let i and j be such that ξ_i is φ and ξ_j is $\neg\varphi$. Then $g_3(m+k+i) = \varphi$ and $g_3(m+k+j) = \neg\varphi$. Since $\neg\varphi \notin X_{m-1} \cup P_{T,m-1}$, $\neg\varphi$ does not appear in $g_3(0), \dots, g_3(m+k+j-1)$. Since ξ_i is a proper subformula of ξ_j, we have $i < j$. It follows that $\mathrm{Pr}_3^R(\ulcorner\varphi\urcorner)$ holds. □

Claim 2 PA $\vdash$"the bell rings" $\leftrightarrow \neg\mathsf{Con}_T$.

Proof We reason in PA:

($\rightarrow$): Suppose that the bell rings at Stage $1.m$. We distinguish the following three cases.

- $X_m \cup P_{T,m}$ contains both φ and $\neg\varphi$: By Lemma 3, both φ and $\neg\varphi$ are provable in T. Then T is inconsistent.
- $\neg\varphi \notin X_m$ and $\neg\mathrm{Pr}_3^R(\ulcorner\varphi\urcorner) \in X_m$: By Lemma 3, $\neg\mathrm{Pr}_3^R(\ulcorner\varphi\urcorner)$ is provable in T. On the other hand, since $\neg\varphi \notin X_{m-1}$ by Lemma 3, $\mathrm{Pr}_3^R(\ulcorner\varphi\urcorner)$ holds by Claim 1. Then $\mathrm{Pr}_3^R(\ulcorner\varphi\urcorner)$ is provable because it is a true $\boldsymbol{\Sigma}_1$ sentence. Therefore T is inconsistent.
- φ is a true Δ_0 sentence and $\neg\varphi \in X_m$: Then φ is provable in T by $\mathbf{\Delta_0C^G}$ for $\mathsf{Prov}_T(x)$. Also by Lemma 3, $\neg\varphi$ is provable in T. Hence T is inconsistent.

($\leftarrow$): If T is inconsistent, then for some m and φ, $P_{T,m}$ contains both φ and $\neg\varphi$. Then the bell rings at some stage. □

Claim 3 For any natural number n, T proves "If the bell rings at Stage $1.m$, then m is larger than $\overline{n}$".

Proof Let n be any natural number. We discuss in T. Suppose that the bell rings at Stage $1.m$ for some $m \leq n$. Then from our proof of Claim 2, there are formulas φ, $\neg\varphi \in F_n$ such that both φ and $\neg\varphi$ are T-provable. Thus $P_{T,n}$ is inconsistent. This contradicts the reflexiveness of T. □

Our formula $\mathrm{Prf}_3(x, y)$ is a proof predicate of T.

Claim 4

1. $\mathsf{PA} \vdash \forall x(\mathsf{Prov}_T(x) \leftrightarrow \exists y\mathrm{Prf}_3(x, y))$.
2. For any $n \in \omega$ and formula φ, $\mathbb{N} \models \mathsf{Proof}_T(\ulcorner\varphi\urcorner, \overline{n}) \leftrightarrow \mathrm{Prf}_3(\ulcorner\varphi\urcorner, \overline{n})$.

Proof 1. In the theory $\mathsf{PA} + \neg$"the bell rings", $\forall x(\mathsf{Prov}_T(x) \leftrightarrow \exists y\mathrm{Prf}_3(x, y))$ holds by the definition of g_3.

In $\mathsf{PA}+$"the bell rings", g_3 outputs all formulas, and T proves all formulas by Claim 2. Thus $\forall x(\mathsf{Prov}_T(x) \leftrightarrow \exists y\mathrm{Prf}_3(x, y))$ also holds.

2. Since $\mathbb{N} \models \neg$"the bell rings" by Claim 2, we obtain $\mathbb{N} \models \mathsf{Proof}_T(\ulcorner\varphi\urcorner, \overline{n}) \leftrightarrow \mathrm{Prf}_3(\ulcorner\varphi\urcorner, \overline{n})$ holds for any $n \in \omega$ and formula φ by the definition of g_3. □

Claim 5 $\mathsf{PA} \vdash \forall x(\mathrm{Pr}_3^R(x) \to \mathrm{Pr}_3^R(\ulcorner\mathrm{Pr}_3^R(\dot{x})\urcorner))$.

Proof We work in PA: First, suppose that the bell never rings. Then T is consistent by Claim 2. Assume that $\mathrm{Pr}_3^R(\ulcorner\varphi\urcorner)$ holds. Since $\mathrm{Pr}_3^R(\ulcorner\varphi\urcorner)$ is a $\boldsymbol{\Sigma}_1$ sentence, $\mathrm{Pr}_3^R(\ulcorner\varphi\urcorner)$ is provable in T, and hence $\mathrm{Pr}_3^R(\ulcorner\varphi\urcorner) \in P_{T,m}$ for some m. By the consistency of T, we have $\neg\mathrm{Pr}_3^R(\ulcorner\varphi\urcorner) \notin P_{T,m}$. Therefore $\mathrm{Pr}_3^R(\ulcorner\mathrm{Pr}_3^R(\ulcorner\varphi\urcorner)\urcorner)$ holds by the definition of g_3. We have proved that $\neg$"the bell rings" implies $\forall x(\mathrm{Pr}_3^R(x) \to \mathrm{Pr}_3^R(\ulcorner\mathrm{Pr}_3^R(\dot{x})\urcorner))$.

Secondly, we assume that the bell rings at Stage $1.m$. Suppose $\neg\mathrm{Pr}_3^R(\ulcorner\mathrm{Pr}_3^R(\ulcorner\varphi\urcorner)\urcorner)$ holds. By Claim 1, $\neg\mathrm{Pr}_3^R(\ulcorner\varphi\urcorner) \in X_{m-1}$. If $\neg\varphi \notin X_{m-1}$, then the bell rings before Stage $1.m$. This is a contradiction. Thus $\neg\varphi \in X_{m-1}$. By Claim 1 again, $\neg\mathrm{Pr}_3^R(\ulcorner\varphi\urcorner)$ holds.

We have proved that $\forall x(\mathrm{Pr}_3^R(x) \to \mathrm{Pr}_3^R(\ulcorner\mathrm{Pr}_3^R(\dot{x})\urcorner))$ is also implied by the assumption "the bell rings". Thus we conclude that $\forall x(\mathrm{Pr}_3^R(x) \to \mathrm{Pr}_3^R(\ulcorner\mathrm{Pr}_3^R(\dot{x})\urcorner))$ holds. □

Claim 6 $T \vdash \mathrm{Pr}_3^R(\ulcorner \forall \vec{x}\varphi(\vec{x})\urcorner) \to \forall \vec{x}\mathrm{Pr}_3^R(\ulcorner \varphi(\dot{\vec{x}})\urcorner)$ for any formula $\varphi(\vec{x})$.

Proof We reason in T: As in our proof of Claim 5, it suffices to prove the sentence under the assumption "the bell rings". We assume that the bell rings at Stage $1.m$. Suppose $\neg\mathrm{Pr}_3^R(\ulcorner \varphi(\dot{\vec{a}})\urcorner)$ for some $\vec{a}$. Then by Claim 1, $\neg\varphi(\dot{\vec{a}}) \in X_{m-1}$, and hence $\neg\varphi(\dot{\vec{a}}) \in X_{m-1,n}$ for some $n \leq |F_{m-1}|$ by Lemma 3. Since $\varphi(\dot{\vec{a}})$ is an instance of $\forall \vec{x}\varphi(\vec{x})$ and $\neg\forall \vec{x}\varphi(\vec{x}) \in F_{m-1}$ by Claim 3, we have $\neg\forall \vec{x}\varphi(\vec{x}) \in X_{m-1,n+1} \subseteq X_{m-1}$. Therefore $\neg\mathrm{Pr}_3^R(\ulcorner \forall \vec{x}\varphi(\vec{x})\urcorner)$ holds by Claim 1 again. □

Claim 7 If $T \vdash \varphi \to \psi$, then $T \vdash \mathrm{Pr}_3^R(\ulcorner \varphi \urcorner) \to \mathrm{Pr}_3^R(\ulcorner \psi \urcorner)$ for any formulas φ and ψ.

Proof Suppose $T \vdash \varphi \to \psi$. Then $T \vdash \neg\psi \to \neg\varphi$. It suffices to show that the sentence $\mathrm{Pr}_3^R(\ulcorner \varphi \urcorner) \to \mathrm{Pr}_3^R(\ulcorner \psi \urcorner)$ is provable in T+"the bell rings".

We reason in T+"the bell rings": Suppose that the bell rings at Stage $1.m$ and that $\neg\mathrm{Pr}_3^R(\ulcorner \psi \urcorner)$ holds. Then by Claim 1, $\neg\psi \in X_{m-1}$. Thus $\neg\psi \in X_{m-1,n}$ for some $n \leq |F_{m-1}|$ by Lemma 3. Let k be the least proof of $\neg\psi \to \neg\varphi$ in T. Then $k \leq m-1$ by Claim 3 (because k is standard), and hence $\neg\psi \to \neg\varphi \in P_{T,m-1}$. We obtain $\neg\varphi \in X_{m-1,n+1} \subseteq X_{m-1}$. Therefore $\neg\mathrm{Pr}_3^R(\ulcorner \varphi \urcorner)$ holds by Claim 1. □

Claim 8 $\mathsf{PA} \vdash \forall x(\mathsf{True}_{\Delta_0}(x) \to \mathrm{Pr}_3^R(x))$.

Proof We proceed in PA+ "the bell rings": Assume that the bell rings at Stage $1.m$. Let φ be any true Δ_0 sentence. If $\neg\varphi \in X_{m-1}$, then the bell rings before Stage $1.m$. Thus $\neg\varphi \notin X_{m-1}$. By Claim 1, $\neg\mathrm{Pr}_3^R(\ulcorner \varphi \urcorner)$ does not hold. This means $\mathrm{Pr}_3^R(\ulcorner \varphi \urcorner)$ holds. □

Our proof of Theorem 11 is completed. □

Acknowledgements We would like to thank the referee for carefully reading our manuscript and for giving valuable suggestions and comments. This work was supported by JSPS KAKENHI Grant Numbers 16K17653 and 19K14586.

References

1. Arai, T.: Derivability conditions on Rosser's provability predicates. Notre Dame J. Formal Logic **31**(4), 487–497 (1990)
2. Bernardi, C., Montagna, F.: Equivalence relations induced by extensional formulae: classification by means of a new fixed point property. Fundam. Math. **124**(3), 221–233 (1984)
3. Buchholz, W.: Mathematische Logik II. http://www.mathematik.uni-muenchen.de/~buchholz/articles/LogikII.ps (1993)
4. Feferman, S.: Arithmetization of metamathematics in a general setting. Fundam. Math. **49**, 35–92 (1960)
5. Gödel, K.: Über formal unentscheidbare Sätze der Principia Mathematica und verwandter Systeme I. (in German). Monatshefte für Mathematik und Physik **38**(1), 173–198 (1931). English translation in Kurt Gödel, Collected Works, Vol. 1 (pp. 145–195)

6. Guaspari, D., Solovay, R.M.: Rosser sentences. Ann. Math. Logic **16**(1), 81–99 (1979)
7. Hájek, P., Pudlák, P.: Metamathematics of First-Order Arithmetic. Perspectives in Mathematical Logic, Springer, Berlin (1993)
8. Hilbert, D., Bernays, P.: Grundlagen der Mathematik, vol. II. Springer, Berlin (1939)
9. Hughes, G.E., Cresswell, M.J.: A New Introduction to Modal Logic. Routledge, London (1996)
10. Jeroslow, R.G.: Redundancies in the Hilbert-Bernays derivability conditions for Gödel's second incompleteness theorem. J. Symb. Logic **38**(3), 359–367 (1973)
11. Kikuchi, M., Kurahashi, T.: Universal Rosser predicates. J. Symb. Logic **82**(1), 292–302 (2017)
12. Kreisel, G.: Ordinal logics and the characterization of informal concepts of proof. In: Proceedings of International Congress of Mathematicians 1958, pp. 289–299. Cambridge University Press, New York (1960)
13. Kreisel, G.: Mathematical logic. In: Lectures in Modern Mathematics, vol. 3, pp. 95–195. Wiley, New York (1965)
14. Kreisel, G.: A survey of proof theory II. In: J.E. Fenstad, (ed.), Proceedings of the Second Scandinavian Logic Symposium. Studies in Logic and the Foundations of Mathematics, vol. 63, pp. 109–170. North Holland (1971)
15. Kreisel, G., Takeuti, G.: Formally self-referential propositions for cut free classical analysis and related systems. Dissertationes Mathematicae (Rozprawy Matematyczne) **118** (1974)
16. Kurahashi, T.: Rosser provability and normal modal logics. Studia Logica. **108**(3), 597–617 (2020)
17. Kurahashi, T.: Henkin sentences and local reflection principles for Rosser provability. Ann. Pure Appl. Logic **167**(2), 73–94 (2016)
18. Kurahashi, T.: A note on derivability conditions. In: Preparation. J. Symb. Logic **85**(3), 1224–1253 (2020)
19. Lindström, P.: Aspects of Incompleteness. Lecture Notes in Logic, vol. 10, 2nd edn. AK Peters (2003)
20. Martin Hugo Löb: Solution of a problem of Leon Henkin. J. Symb. Logic **20**(2), 115–118 (1955)
21. Montagna, F.: On the formulas of Peano arithmetic which are provably closed under modus ponens. Bollettino dell'Unione Matematica Italiana **16**(B5), 196–211 (1979)
22. Mostowski, A.: Thirty years of foundational studies: lectures on the development of mathematical logic and the study of the foundations of mathematics in 1930–1964. Acta Philosophica Fennica **17**, 1–180 (1965)
23. Rautenberg, W.: A Concise Introduction to Mathematical Logic, 3rd edn. Universitext. Springer, New York (2010)
24. John Barkley Rosser: Extensions of some theorems of Gödel and Church. J. Symb. Logic **1**(3), 87–91 (1936)
25. Vladimir Yurievich Shavrukov: On Rosser's provability predicate. Zeitschrift für Mathematische Logik und Grundlagen der Mathematik **37**(4), 317–330 (1991)

On Takeuti's Early View of the Concept of Set

Hidenori Kurokawa

Abstract Gaisi Takeuti occasionally talked about the concept of set. Based on these remarks, we can easily see that Takeuti had an original view of the concept of set. However, Takeuti gave no systematic development of his own view of the concept of set in its entirety. In this paper, we try both to put together Takeuti's remarks on the concept of set, most of which are currently available only in Japanese, and to reconstruct Takeuti's view of the concept of set in a manner as systematic as possible. In particular, we will take a look at a relatively early period of his career, and we will focus first on Takeuti's distinction between two concepts of set and second on how a Hilbert-Gentzen finitistic proof-theoretic analysis can clarify one of these concepts of set.

Keywords Gaisi Takeuti · Concept of set · Takeuti's fundamental conjecture · Higher-order logic · Proof theory · History of logic

1 Introduction

Gaisi Takeuti (1926–2017) is well-known primarily as a proof theorist. His first major accomplishments in proof theory are (i) to formulate Takeuti's fundamental conjecture, which states that the cut-elimination theorem holds for a logical system called 'generalized logical calculus (GLC),' which is a Gentzen-style sequent calculus for finite-order predicate logic formulated by himself (or for its special case G^1LC, a sequent calculus for second-order logic), and (ii) to prove that consistency of second-order arithmetic finitistically follows from the conjecture [29][1]. Later in [41] he also proved the consistency of Π^1_1-analysis by using a constructive method called 'ordinal diagrams.'

[1]The results were both stated and "proved" in a Japanese journal already in 1949 [28]. The proof of (ii) is correct and hence the credit for this result should be attributed to [28], but the proof of (i) in [28] had a problem. About the history related to this, see Sect. 3.2.3.

H. Kurokawa (✉)
Kanazawa University, Kakuma, Kanazawa Ishikawa 920-1192, Japan

T. Arai et al. (eds.), *Advances in Mathematical Logic*, Springer Proceedings in Mathematics & Statistics 369, https://doi.org/10.1007/978-981-16-4173-2_5

As a proof theorist, Takeuti was mainly interested in the finisitic foundations of mathematics. However, for a couple of reasons, Takeuti was also particularly interested in some foundational issues related to the concept of set. First, his major works in proof theory including Takeuti's fundamental conjecture focus on finite-order predicate calculus, which necessarily involves the concept of set. Hence, for Takeuti, since the time of his early career, sets have been mathematical objects to be studied by using proof-theoretic methods for handling finite-order predicate calculus. Naturally, he needed to consider the foundational status of the concept of set even from a proof-theoretic point of view. Second, perhaps independently of his proof-theoretic works, Takeuti had broad interests in the foundations of mathematics and did some work in set theory. Having interests both in proof theory and set theory, Takeuti was one of few proof-theorists whose view of the concept of set can significantly contribute to a proof-theoretic analysis of the concept of set. In this sense, his view of the concept of set may be of particular interest.

However, Takeuti was a mathematical logician and he seems to have wanted to express his philosophical view mainly by proving mathematical results themselves rather than by writing prose. Hence, he did not leave any systematic development of his own philosophical view of the foundations of mathematics. Even compared with the case of his view of finitism, his remarks on the concept of set look less systematic. Nonetheless, a careful reading of his prose has made it clear that he had a coherent view of the concept of set, or so we claim.

The aim of this historical paper is both to put together Takeuti's remarks on the concept of set, most of which are currently available only in Japanese and to present his view of the concept of set as systematically as possible. In particular, we mostly focus on his remarks in the relatively early period of his career before a significant change of his view around 1970. Around that time, Takeuti's foundational view of the concept of set underwent a significant change, which was caused by his adoption of the concept of 'infinite mind' in [44]. However, here we confine ourselves to discuss his view of the concept of set before this. Only on the occasions in which we need to discuss developments in the later period do we mention some later remarks of his. Also, let us note that, although we sometimes use the word 'finitism,' we do not go into the details of this concept. As is well-known, Takeuti advocates a variant of Hilbert-Gentzen's finitism, whose precise content has not yet been fully examined. We simply assume that the reader has some familiarity with the rough idea.

In this paper we proceed as follows. In Sect. 2, we briefly take up the issues to be discussed in this paper by giving an overview of Takeuti's view of the concept of set. In Sect. 3, we present his view of the concept of set in detail. In Sect. 3.1, we take a look at his critical discussions on traditional views of the concept of set. In Sect. 3.1, we will discuss Takeuti's positive account of the concept of set (Sects. 3.2.1, 3.2.2), in particular his view of how the concept of set should be analyzed by a proof-theoretic method (Sect. 3.2.3).

2 An Overview of Takeuti's View of the Concept of Set

Takeuti's remarks on the concept of set spread over many years. It starts in the 1950s and the last remarks are around 2000. Among all the documents that Takeuti left, we will mainly focus on Takeuti's Japanese articles, [33–35, 40, 44, 45, 47, 48, 53, 54]. Besides, we will use other sources in English, such as [43, 49, 51].

Takeuti's investigations into proof theory of finite-order predicate calculus already started in 1940s. In fact, his first publication of the subject [28] in Japanese came up in 1949. Around the time, Takeuti must have already had a strong interest in the concept of set related to his proof-theoretic work, but he did not publish his philosophical view of it. For instance, [28] contains no philosophical remark but contains only technical results. Takeuti started publishing his view of the concept of set in a series of papers [33–35]. The view is neither philosophically sophisticated nor systematically developed, but we take the series of the papers to be particularly important since some themes in these papers suggest that already at this stage Takeuti had a core of his view of the concept of set. The themes can be summarized as follows.

1. The nature of the concept of set. In particular, a distinction between the concept of set based on the notion of proposition and the concept of set in set theory
2. The role of formalistic, finitisitic proof theoretic analysis of the concept of set
3. Discontent with the standard axiomatic set theory (ZF or NBG)

Takeuti [40] contains a little more systematic development of the item 1 and 3. The idea of making a distinction between the concept of set of individuals and the concept of set in set theory appeared to be further deepened into a kind of methodological distinction in foundations of mathematics in general as stated in [49]. The latter concept of set was pursued separately from the former in the introduction of [42], which is a paper on Takeuti's transfinite type theory, and in a survey paper of set theory [44]. In particular, in [44] Takeuti presents his own view of the core idea of the concept of set as it appears in set theory.

Then, in [45], Takeuti introduced the terms 'a finite mind' and 'an infinite mind,' by which Takeuti modified his framework of discussing the foundations of mathematics. All of the forgoing themes including the issues related to the concept of 'infinite mind' are discussed in [51].

In addition, only in the first edition of [51] published in 1975 did Takeuti start discussing the following important issue by using a new terminology, although essentially the same issue is a motivating background for the system presented in [42].

4. Is the universe of set 'a completed totality' or 'in the process of creation,' i.e. 'a closed universe' or 'a growing universe'?

Takeuti's discussion on this issue is not much developed but he kept thinking about this issue until his (almost) last publication.

This roughly explains Takeuti's view on the concept of set over the years. Taking a look at the whole text, one can easily observe that the foregoing themes have been persistently kept throughout his career. Due to such a persistence of his view, in our study of Takeuti's view of the concept of set (including this paper and the subsequent one), in general we adopt a *methodological* principle in exploring Takeuti's text. We minimize discussions on changes over time, except when they are really needed to present Takeuti's view. In particular, in this paper we treat the introduction of the concept of 'infinite mind' as the only major change in his view of the concept of set. This is because some structural transformation indeed seems to have occurred to Takeuti's thought at the occasion of his adoption of the concept, although there is a significant amount of continuation and deepening of ideas. However, due to the limitation of space, we leave our treatment of Takeuti's later view of the concept of set after the introduction of 'an infinite mind' (hence the theme 4) to the extended version of this paper. As a result, minor changes of his view are more or less treated as clarifications or improvements of essentially the same point in this paper.

3 Themes in Takeuti's View of the Concept of Set

In this section, we start talking about those themes which Takeuti discussed in more details. In doing this, we will roughly follow the foregoing list of themes.

3.1 Critical Remarks on the Concept of Set

This topic is discussed in [33], i.e. the first one in the series of papers [33–35] on formalism. This series of papers appears to have been commissioned by the Japan Association of Philosophy of Science and is one of his earliest ones in Takeuti's career, but most topics that we want to talk about in this paper already appear in this series of papers. Contrary to an impression that one may have by taking a look at the title 'From a standpoint of formalism,' they do not contain any detailed discussion of Hilbert's program but the following points concerning the concept of set. (a) The 'definition' of set given in the standard reference is by no means satisfactory. (b) We can handle the concept of set only by the idea of 'formalism.' (c) The point of formalism consists in an analysis of the concept of set. (d) The concept of set can be understood as a kind of reification of a propositional function. (e) The concept of set so understood is different from the abstract one used in set theory. (f) The adoption of axiomatic set theories based on mere habits cannot be justified. The points a, f correspond to the theme 3, b, c, d do to 2, and e does to 1, respectively. We now discuss Takeuti's critiques of the traditional 'definition' of set.

3.1.1 Critiques of the Traditional Definition of the Concept of Set

In [33], instead of talking about Hilbert's formalism, Takeuti argues that in order to understand what Hilbertian formalism is, one needs to understand what contemporary mathematics is, what the concept of set is, and how essential the concept of set is in contemporary mathematics. Then Takeuti complains that the 'definition' of set given in the standard Japanese dictionary of mathematics is by no means satisfactory. The 'definition' that Takeuti quotes can be translated as: 'Consider those objects [taishō][2] of our intuition [chokkan] or thought [shikō] which are in a certain definite realm [han-i][3] as a unified totality. Let us call it a *set* [shūgō] (of those objects) and an individual object in the realm an element or a member of the set ([33], p. 26 (268), left, l. 31).' Takeuti's complaint against this goes as follows.

> Is this a mathematical definition at all, which is supposed to be clear? The definition looks like an item in a dictionary of philosophy. Isn't it so unclear that even mathematicians who have already grasped the concept [gainen] of set may be lost (provided that they try to understand the literary meaning of the definition)? First of all, the meaning of each of these phrases such as 'intuition,' 'objects of thought,' 'a certain definite realm,' 'a unified totality' is unclear. These sound really non-mathematical. Can't we interpret them in many ways? Moreover, the definition as it is would give rise to a contradiction. ([33], p. 26 (268), right, l. 2).[4]

However, Takeuti immediately adds a comment that this is not because the quality of the definition of the dictionary is poor but because this is inevitable. Indeed, the definition quoted above is not much different from the well-known 'definition' of set given by Cantor, whose English translation quoted by Takeuti in [43] goes: 'any collection into a whole M of definite and separate objects m of our intuition or our thought. ([43], p. 2)' This probably means that Takeuti admitted that it is inevitable here to start talking about sets from the foregoing definition but he did not believe that we would be able to go too far only by this 'definition.'

Although Takeuti does not explicitly discuss a true cause of such a confusion on the concept of set in [33], he later summarizes the point as follows: 'from the viewpoint of contemporary mathematics, the concept of set cannot be defined but can be formulated only via formalism [keishiki-shugi] ([34], p. 18 (298), left, l. 31).' Takeuti does not explicitly mention which version of formalism he means here, but we suspect that what he means is merely a sort of axiomatic treatment of the concept of set in the sense of the word 'formalism' closer to the one in Hilbert's *The Foundations of Geometry* than in the sense of Hilbert's program since Takeuti uses the word 'formulate' here[5] instead of using the word 'illuminate.' As we will see, Takeuti uses the latter word for discussing the clarification of the concept of set from

---

[2] In the following, we write the original Japanese words for important concepts in the forms transliterated to Roman alphabets in square brackets. Incidentally, we also use square brackets in a quote in order to supplement the quoted passage.

[3] We have deliberately avoided using a word such as 'domain' or 'range' which has been used as a technical term in logic.

[4] All the quotes from Takeuti's Japanese text are translated by ourselves, except [56], which has already been translated into English. We quote [56], unless we need to modify the translation. Also, for each quote from Japanese text, we put the page and the line number.

[5] Considering Takeuti's later statements, Takeuti could not have taken such an axiomatic treatment to be one of the standard axiomatic set theories such as ZF. See Sect. 3.2.2 for what kind of axiomatic approach to the concept of set Takeuti approves.

the formalist standpoint in the sense of Hilbert's program. Anyways, this discussion sets the initial stage of Takeuti's discussion of the concept of set. The traditional 'definition' does not work. Then what should we do?

3.1.2 Discontent with the Standard Axiomatic Set Theories (ZF or NBG)

One may imagine that a natural way of moving from the critique of the naive 'definition' to the next stage of discussion on the concept of set is just to approve the standard axiomatic set theories such as ZF or NBG. However, interestingly Takeuti does *not* take this way, although he of course does not deny the axiomatic method itself (as we have just seen). He is actually quite critical of such axiomatic set theories at least in his early career, and he is still so later but in a certain restricted manner.

In order to make clear Takeuti's general attitude towards the standard axiomatic set theories, we quote some passages from his mature period[6] before we get into more detailed discussions of his on the topic in [33–35], etc. In one of his expositions of set theory, after admitting that axiomatic set theories have been successful to some extent, Takeuti states the following advantages and disadvantages of Zermelo's set theory ([54], p. 105, l. 6)

> ⟨ Advantages ⟩: It does not ask an essential question such as 'what is a set?'
>
> Hence, it is pragmatic [jitsuyō-teki] and functional [kinō-teki], and it is convenient to work within.
>
> ⟨ Disadvantages ⟩: It does not ask an essential question such as 'what is a set?' Hence, it lacks thought [shisō][7] behind it, and it does not reach a profound issue.

Although the expressions are less radical than those in the earlier period that we will see soon, Takeuti is still somewhat sarcastic. Anyways, these appraisals have substantial connections with Takeuti's view of the concept of set. The latter of the foregoing two points that axiomatic set theories lack 'thought' implies that Takeuti must have been sure that there is something missing from the current axiomatization of set theory. Indeed, Takeuti expresses even more radical discontent with the standard axiomatic set theories in [34, 35].

> Then what are symbols such as $a \in A$, $\{x \mid \mathfrak{A}(x)\}$? ... These are really mistaken ways [go-yō] of using symbols. That is, quite a few concepts, such as 'transfinite ordinals,' 'cardinals,' 'axiomatic set theories,' etc., widely rooted in contemporary mathematics, are indeed mistakes [ayamari] or a kind of illusions [gen-ei] that arise from misapprehension. ([34], p. 19 (298), left, l. 37)

---

[6] We use the material from a later period, since we try to emphasize that whereas Takeuti's view against axiomatic set theories in his mature period is milder than before, he is still critical of them.

[7] The original Japanese word for 'thought' has a meaning close to 'philosophy' in Japanese.

> It is true that our concept of set[8] does not give us a clear definition of it, but (if we are allowed to use such an expression) it satisfies our logical sense.[9] However, the Fraenkel-von Neumann[10] set theory is not pleasant set theories to work with that satisfy our logical sense. We cannot feel the senses of the systems except that they are sheer formal axiomatic systems .... These systems would not make any sense once a formal inconsistency is derived in the systems ([35], p. 27 (351), left, l. 18).

They indeed look radical critiques of axiomatic set theories.[11] However, before dismissing them by saying that they are too extreme, we should think about what were the specific reasons why Takeuti thought about axiomatic set theory this way. In fact, we have deliberately quoted these passages somewhat out of context, because our intention is to pose these as exegetical problems to be solved along the way of presenting Takeuti's view of the concept of set.

Then what does Takeuti mean by 'mistakes' or 'illusions'? How do these critiques cohere with Takeuti's remark that the concept of set can be formulated only via formalism? Takeuti's critiques are not mere complaints. Takeuti actually has specific reasons to think that the current axiomatizations of set theory are wanting. In the following, we will at least partially[12] show both how we can reasonably locate these remarks against the axiomatic set theories in his entire picture of the concept of set and that Takeuti's view is not so much insane after all as it may have initially looked. Roughly, an answer to the first question is initially an abuse of symbols, but Takeuti's discussions later becomes less radical and more specific about which particular points of set theory he criticizes. An answer to the second question is that what he thought was missing in axiomatic set theory is concerned with either logic or formalistic (finitistic) analysis of the concept in proof theory, as one can imagine by such a phrase as 'logical sense.'

---

[8] This is not meant to be the naive definition of set criticized above, but Takeuti's own view of the concept.

[9] A comment on the original Japanese phrase for 'logical sense' may be worthwhile, for this phrase may sound a little odd. Takeuti here essentially coined the phrase 'logical sense.' It is a genuine exegetical problem what he meant exactly by the phrase. Takeuti actually uses a Japanese word which is often misused for a Japanese word used as a translation of the word 'intuition.' Hence a Japanese reader with a philosophical background might think that Takeuti misuses the word. However, in other places Takeuti correctly uses the Japanese word for 'intuition.' Thus Takeuti must have purposefully chosen the phrase to express a different meaning from that of the word 'intuition.' The Japanese word for 'sense' here is the same one as used in, e.g., 'sense perception.'

[10] It is not clear which set theory Takeuti had in mind. But this does not affect the point.

[11] One of the reasons why Takeuti was so critical of set theory may have been that he was not really working in transfinite set theory in 1950s. See the following recollection of Takeuti. 'However, my own interest was in the foundations of mathematics in a narrow sense [proof theory or consistency proof], and my real encounter of set theory had been postponed until I met Gödel in 1959 in Princeton for the first time (p. 4, [46]).' In order to consider the foundations of analysis, the entire transfinite set theory may not be needed. Accordingly, Takeuti seems to have been unfamiliar with the iterative conception of set. Note also that Takeuti then developed his theory of ordinals [30] which takes ordinals as primitive concepts, presumably for the purpose of giving consistency proofs. Takeuti may have thought the concept of set and that of ordinals separately to some extent.

[12] We leave Takeuti's later discussions on the concept of set to the extended version.

3.2 The Concept of Set to Be Analyzed by Finitistic Methods

In this subsection, we start discussing some core ideas of Takeuti's view of the concept of set. Let us first describe how Takeuti understands the concept of set came about. In [34] Takeuti describes an explanation of the origin of the concept of set, and he makes a crucial distinction between two different concepts of set. Then we discuss his own view of how formalistic and finitistic analyses of the concept of set can contribute to clarifying one of them.

3.2.1 The Concept of Set as the Thought-Objectification of a Proposition

Takeuti argues that the Cantor-style 'definition' of the concept of set is unclear but he gives no positive account of the concept in [33]. However, in [34] Takeuti gives one by discussing both (i) the role of symbols in mathematics and (ii) the concept of set from the viewpoint of propositions expressed by the symbols.

According to Takeuti, the role of symbols in mathematics is often thought of as an auxiliary medium for expressing concepts or it is sometimes thought that inventions of symbols precede sophistications of concepts. However, he is sympathetic to the view that 'mathematical concepts and mathematical symbols are simultaneously generated and completed by the interactions between them' ([34], p. 295). This is illustrated by the use of symbols in elementary algebra, which Takeuti claims made us aware of the notion of arbitrariness. Besides, he claims both that the power of symbols goes beyond that and that such an awareness of arbitrariness paved the way to contemporary mathematics and abstract 'number concepts.'

> ... [W]hat is the significance of such symbolizations [kigō-ka]? ... [T]he significance would consists in that which is symbolized becoming an object of our thought, through symbolization.... Once a subordinate concept [jūzoku gainen] [an auxiliary concept coming with a main object of thought] becomes an object of thought, such a thought-objectification [shikōtaishō-ka][13] of what used to be incapable of being an object of thought creates a new conceptual world at a higher dimension [jigen][14] that was previously unimaginable ([34], p. 16 (296), left, l. 13).

In this quote, Takeuti essentially identifies the primary role of symbolization as creating an object of thought. He does not yet mention the concept of set here, but immediately after this he points out, 'the magic of symbols in mathematics culminates in the notation of "set" ([34], p. 17(297), left, l. 2) and starts specifically talking about the concept of set. He introduces the notation for a 'set' together with logical symbols and explains how to use it when we are given 'a proposition [meidai] $\mathfrak{A}(x)$ about

[13] Takeuti coined the original Japanese word for 'thought-objectification.' This means something like 'objectification in thought.'

[14] Obviously, this is not intended as a technical term.

x (ibid).' ('Proposition' here actually means something with free variables. So this should be called 'propositional function' or 'open formula,' but let us keep using this term, unless we have particular reason why we prefer 'propositional function,' since this phrase is already quite long.)

> When we have a proposition $\mathfrak{A}(x)$ about x, let us introduce the expression $\{x \mid \mathfrak{A}(x)\}$. We call such a thing 'the collection of x that satisfies $\mathfrak{A}(x)$' or 'the set of x that satisfies $\mathfrak{A}(x)$.' Moreover, we put it as A, and we simply call such an A a *set*. ... What is the meaning of 'set' here? The basis of the intuition of the set consists in both 'collection of things' or 'geometric figures,' but ... it consists in the thought-objectification of a proposition. ... In the actual history of mathematics, the thought-objectification was realized by the introduction of logical symbols such as $\forall$, $\exists$, $\neg$, $\wedge$, $\vee$ or the notation such as $\mathfrak{A}(x)$ independently of the emergence of the notion of set, but the thought-objectification can be accomplished only through the use of such notation ([34], p. 17 (297), left, l. 25).

This is a clear statement of Takeuti's fundamental idea of both the concept of set as a thought-objectification of a proposition and also the role that logical symbols play in its formulation. This 'thought-objectification' is what we mentioned in 1 of the three themes as 'reification.' The idea is to transform the content of proposition as a kind of abstract object, but he is not explicit about how it can be constructed or formed. This is to be explained in his further treatment of the issue shortly. Incidentally, he seems to have (correctly) realized that the development of set theory and the use of formal language in mathematical logic were done separately.

In [33], the concept of set is claimed to be essential in contemporary mathematics, but Takeuti gives no reason for that there. An explanation of why so is given in [34].

> But if you tell contemporary mathematicians to give up the set-theoretic notation and think only by means of the notation for propositions, then we would be almost in the same position, but in addition most contemporary mathematicians would not be able to think at all ([34], p. 17 (297), left, l. 39).

Then he claims, 'In fact, this is because the essence of [the concept of] set consists in the notation $\{x \mid \mathfrak{A}(x)\}$ and $a \in A$ ([34], p. 17 (297), right, l. 1).'

Based on these quotes, we can notice that already at this point Takeuti's view of the concept of set seems to contain two different ideas, i.e., i) the thought-objectification of a proposition and ii) that the essence of set consists in the set-theoretic notations. They are in a way inseparable since the notion of proposition itself depends on how symbols work, but the exact relationship between them is not clearly explained.

In fact, Takeuti's view of the relation between them seems to be even more complicated. He gives further discussions on the concept of set.

> A 'set' – the thought-objectification of a proposition – leads to considering a proposition concerning itself by means of its own objectification. That is, this requires us to consider a 'set of sets.' Moreover, as a case of a 'set of sets of sets,' etc., ... the higher-dimensional evolution of the concept and the notation of $a \in A$ further create the concept of set as a general concept vaguely based on the concept of 'belongs to' ([34], p. 17 (297), right, l. 14).

By this, Takeuti seems to imply that the inclination towards iterations of objectification is intrinsic in the thought-objectification of a proposition.

Moreover, Takeuti identifies both the cause of both Russell's paradox and the establishment of axiomatic set theories with an effect of symbolization.

> What is the cause of such a big historical event [Russell's paradox] that trembled contemporary mathematics? This is clearly due to the two notations $a \in A$, $\{x \mid \mathfrak{A}(x)\}$. That is by no means due to the concept of set as the thought-objectification of a 'proposition'! The cause is a separation of elements and sets that can be expressed by $a \in A$, $\{x \mid \mathfrak{A}(x)\}$. ... and also putting them *on an equal footing* through elements and sets being simultaneously objectified ([34], p. 18 (298), left, l. 14, emphasis is by the present author).

In addition, Takeuti claims that it has a crucial consequence that ∈ is meaningful only in set-theoretic notation.

> ... the concept 'belongs to' '∈' in $a \in A$ by no means consists in the original proposition or the thought-objectification of the proposition. That is, if one carefully considers the situation in this sense, then one may notice that the liberal notion of ∈ considered in classical set theory[15] or axiomatic set theory never exists anywhere. At least it never occurs in the thought-objectification of a proposition. We should say that this can have a *raison d'être* only in the notation of $a \in A$ ([34], p. 18 (298), left, l. 22).

In these quotes, Takeuti makes a distinction between the concept of set as the thought-objectification of a 'proposition' and a separation between elements and sets through set-theoretic notations. Also he attributes the cause of Russell's paradox exclusively to the separation between sets and elements. This is natural, since according to Takeuti the separation is due to the set-theoretic notations. Besides, the membership relation ∈ in set theory is claimed to be merely an artifact produced by the notation. (Accordingly, naive and axiomatic set theories are treated as in the same category.)

However, these statements put together, Takeuti's idea of the relation between the concept of set as the thought-objectification of a proposition and the concept of set based on the separation between elements and sets becomes somewhat unclear. This unclarity is directly connected with the problem concerning the relation between propositions and symbols that we pointed out above. The unclarity is raised by the following facts. Takeuti states that the concept of set as the thought-objectification of a proposition also leads to its own objectification and its iteration. And such iterations of set formation and the abstract concept of 'belong to,' which emerges with the iteration, are all claimed to be caused ultimately by symbolization.

The quote given in the Sect. 3.1.1 containing a claim that set-theoretic notions are mistaken and illusions is actually located after the previous quote. Takeuti essentially claims that set-theoretic notations go out of control at some point. The power of symbolization initially makes possible the thought-objectification of a proposition. Although this power is necessary for mathematicians to do mathematics, the power of symbolization makes other things possible. One is the separation between the notion of element and the notion of set, i.e., elements and sets are put on equal footings through set-theoretic notations (from $\mathfrak{A}(x)$ to $x \in A$), and then the concept of 'belongs to' ∈ is also entirely separated from the original concept of proposition $\mathfrak{A}(x)$, which is potentially harmful. Doesn't this imply that the concept of set as the thought-objectification is also problematic?

Before handling this issue, let us finish summarizing Takeuti's discussions on set in [34]. After discussing the abstract concept of set produced by the symbolization

[15] It is not clear what he means here. Presumably, this means naive set theory.

such as $a \in A$ or $\{x \mid \mathfrak{A}(x)\}$, Takeuti goes into a topic somewhat different from the one discussed above. He gives an account of how the magic of symbols is raised from a viewpoint of mathematicians' faculty of mind.

> Isn't this the function of the mind to intuit, through volition [ishi], that which (otherwise) cannot necessarily be seen or thought as if it were visible and conceivable? Then the symbols give a basis for such an intuition as the symbolization of it ([34], p. 19 (299), left, l. 6).

Then Takeuti asks the reader to re-read what he stated by using the word 'illusions' by replacing it with 'intuitions' and even coins a new term for that.

> I call the intuition which captures general functions and sets 'enactive, volitional intuition [kōi-teki ishi-teki chokkan].' ... It is like Don Juan's intuition which captures a woman's heart if I am allowed to use quite an inappropriate metaphor. Such an intuition is like a wicked[16] suspicion [jasui], at the same time penetrating the core of the matter ([34], p. 19 (299), left, l. 15).

Takeuti appears to think that the pertinent faculty of mind is partly responsible for the illusions he talked about, taking a kind of voluntarism, and he claims that the illusions should be understood as a sort of 'intuitions.'

This concludes our summary of Takeuti's discussions on the concept of set in [34], but we have an exegetical challenge here. Let us quickly reformulate our problem. The entire picture of the concept of set contains both the thought-objectification and the symbolization. On the one hand, Takeuti appears to claim that the problematic part, i.e. the cause of Russell's paradox, consists only in the symbolization. The mere thought-objectification of a proposition appears to be benign. On the other hand, he also seems to claim that, in order for the thought-objectification of a proposition to be carried out, we need the symbolization. Thus, let us pose again the question: how are the thought-objectification of a proposition and the symbolization related to each other? One might think that Takeuti's view of the thought-objectification and the symbolization is simply incoherent. Or, it might seem that Takeuti implies the following. The concept of set is based on the thought-objectification of a proposition, and the latter is due to symbolization. But symbolization inevitably leads to a separation of elements and sets, which gives rise to a paradox. Although some concept of set is necessary in mathematics, any concept of set may be ultimately problematic.

However, such interpretations may not be inevitable. Let us suggest a more charitable interpretation. First, the fundamental idea of set consists in a reification of the concept of proposition or propositional function. Second, the thought-objectification itself requires a certain degree of symbolization of a proposition $\mathfrak{A}(x)$. Third, however, the symbolization involving $a \in A$ or $\{x \mid \mathfrak{A}(x)\}$ leads to an extreme consequence such as Russell's paradox, hence a concept of set goes beyond the one in the first point. Fourth, as a result, symbolization can play both a good and a bad role in the rise of the concept of set. Under this interpretation, we have only to be careful not to abuse symbols when we handle the concept of set as the thought-objectification of a proposition. Takeuti does not explicitly mention where we can draw a distinction between a benign role and a harmful role of symbols, but such a distinction could be

[16] In the sense that the guess is done with an evil intention and without evidence.

made between (i) the iteration of set formation based on finite(second)-order predicate calculus, which Takeuti uses in GLC and endorses in [51])[17] and (ii) putting an element and a set on an equal footing. While (i) contains a kind of iteration of set formation, we take this kind of iteration to be unproblematic, since there are distinctions of levels among predicates. It is (ii) that matters. This is a subtle difference, but such a distinction can provide a coherent interpretation of what Takeuti states here.

Nevertheless, even this interpretation may not have everything that we need. In order for Takeuti's explanation of the concept of set to be completed, an account of the concept of proposition should be needed, but such a thing is still missing here. When one allows an object created by a proposition (a propositional function) to be contained in the range of quantifiers that occur in the proposition used for reification of the object itself, one may have to worry about impredicativity or circularity between objects and propositions, i.e., the kind of problem that Frege suffered from. This implies that not only the concept of set but the concept of proposition itself must be appropriately restricted. Also, apparently, unless the concept of proposition is explained, one cannot give an adequate ground for drawing a line between the concept of set or symbolization that is benign and the one that is harmful. However let us put off discussing propositions, for Takeuti raises the issue again in [40].

Before leaving this subsection, let us give some comments on Takeuti's concept of 'enative, volitional intuition.' First, we briefly mention the history of the term. The term is a modification of another term that comes originally from Kitaro Nishida, who is the founder of the Kyoto School of philosophy. We cannot go into the details of Nishida's philosophy here, but we quickly mention the background. Nishida's own term was 'enactive intuition' in [15, 16].[18] The term stands for a common place or topos, which plays an essential role in his explanation of the formation of a historical world, in which agents' actions and intuitions interact reciprocally and they constitute an inseparable unity.[19] Nishida was purposefuly against the Kantian tradition of taking the intuition to be a passive faculty of mind. The term 'enactive intuition' itself was adopted by a mathematician whose name is Zyoiti Suetuna in [24, 25] to explain constructions of natural numbers, etc. Suetuna was a professor of Tokyo Imperial University, when Takeuti was a student. Suetuna was a specialist of analytic number theory but he was also interested in the foundations of mathematics and somewhat familiar with both Hilbert's and Brouwer's works on foundations.[20] Takeuti apparently picked up the term from Suetsuna's work and modified it. Second, we give some comment on a conceptual issue. Takeuti states that this sort of intuition is behind abstract sets, but the analogy of Don Juan seems to suggest that this does

[17] In [40], a distinction is drawn between a set of individuals (second-order case) and pure set theory (with only one kind of variable). Although the former uses a little narrower concept of set, the difference between second- and higher-order is not important. See our discussion on [40] below.

[18] This is a key phrase in Nishida's late philosophy, occurring in various articles, e.g., [17].

[19] The word 'enactive' may sound obscure, but it seems to be commonly used in translations of Nishida's text. It is also occasionally translated into 'active,' or 'acting.' Our explanation here partly owes to [14].

[20] Suetuna not only was interested in Brouwer's works but also he himself had some idea of constructive mathematics based on Nishida's enactive intuition [25].

not really give conclusive evidence for mathematical truth. The role that the term plays is more like giving a name to the difficulty and connecting it with a mental faculty rather than resolving the difficulty.

We interpret the passage in this way because it is quite hard to imagine that Takeuti thought this intuition would get rid of the burden of proof for the consistency of the pertinent set theory. This would make a sharp contrast with the role that Brouwer's notion of intuition plays in his view of mathematics. From Brouwer's point of view, intuitionistic mathematics needs no consistency proof since intuitionistic mathematics has been actually constructed by intuitions of the idealized mathematician (or the creating subject). In contrast, from Takeuti's viewpoint, the pertinent set theory is not justifiable by the intuition at issue (without proof) and its consistency is hard to prove in spite of our having the intuition. Then certainly we are not exempt from proving its consistency. All in all, we take the epistemological status of the enative, volitional intuition to be unclear.

This intuition aside, the foregoing interpretation looks largely coherent with what Takeuti says in a later paper [40]. Hence, we can interpret the passages quoted here to be a rough sketch of the view that is refined later. Let us now move on to Takeuti's further discussion on the distinction between the two concepts of set.

3.2.2 a Set as a Thought-Objectifiction and a Pure Set in Set Theory

We continue to discuss the foregoing two concepts of set. Before going into Takeuti's discussions on the distinction, we start from his methodological remark given at the beginning of [40]. Takeuti states that sets did not belong to the world of things (individuals) that originally existed.

> ... [T]he concept of set did not originally exist, and it exists in a much higher dimensional and abstract world than the world of originally extant 'things [mono].' Compared with the world of originally extant things – they themselves can be abstract – they are entities that exist only as abstract imaginations or fictive[21] entities ([40], p. 1 (197), right, l. 1).

Takeuti seems to mean that no matter which concept of set we take, sets are different from non-sets, i.e., individuals, and hence fictive entities. From this premise, he argues not that existentially quantified statements over sets are literally false, but that the concept of set is not pre-determined and the concept of set is to some extent our own creation.

> The concept of 'set' is not a priori determined! There is no requirement that a set has to be such and such in an *a priori* manner. In fact, we have quite a few different 'concepts of set.' It is not uncommon that an axiom that is taken to be correct relative to one concept of set is taken to be wrong according to another concept of set. To consider whether a conjecture is correct or not is not to consider whether it is correct with respect to the initially pre-determined concept of set, but it is always essential to consider both what kind of concept of set we should think about and thus also how we should create the concept of set. ...

[21] Takeuti here somehow uses English word 'fictive' in the original text.

> [I]t seems that a problem concerning the concept of set not merely consists in solving the problem but also consists in creating a new thought [shisō] on the concept of set ([40], p. 2 (98), left, l. 5).[22]

Takeuti's point here is that it is a matter of one's decision which concept she takes. Although this is indeed a crucial methodological decision, one may wonder whether this is a methodological comment given to the abstract concept in set theory discussed in the previous section or he merely talks about a broader concept of set. We take this remark to be literally applied to all the concepts of set that he discusses in the paper. That is, we take the point presented in the second paragraph to be self-applied to his view of set. The view in [33–35] presented above and a clearer version to be presented below are both his own creations of the concept of set. Indeed, immediately after the foregoing methodological remark, Takeuti talks about the concept of set as the thought-objectification of a proposition. In presenting the concept of set, his expository plan here is to present things in a step-by-step manner, namely first to start from individuals before talking about a collection of them, to talk about a property of such individuals, to make a connection between the property and a proposition, and ultimately to make a contrast between the concept of set based on the notion of proposition and the abstract concept in set theory. So, in the next step, Takeuti talks about individuals and a collection thereof.

> Now let us consider a collection of things. We assume that we do not know anything about the nature of things and their arbitrary collection. In such a case, how should we rationalize [gōri-ka][23] this state of affairs? As things stand, the only method of rationalizing it seems to be the one based on an axiomatic method [kōri-shugi].[24] In an axiomatic method, we have to rationalize a domain of 'things' before we rationalize 'a collection of things.' In order to make our story a little more concrete, let us now take the domain of 'natural numbers' as a domain of 'things.' I.e., we consider the theory of natural numbers before we rationalize the concept of set of natural numbers ([40], p. 2 (98), left, l. 20).

Let us note in passing that, by reading this passage, one can reconfirm that Takeuti's aforementioned complaint on axiomatic set theories cannot be directed to the axiomatic method in general. His critique is not a global one but a local one.

To make the discussion based on a concrete example, Takeuti takes natural numbers as a case of individuals. He first formulates a theory of natural numbers, i.e. (first-order) Peano arithmetic and discusses what kind of concept of set can be introduced in the theory of natural numbers. Then he formulates predicative and impredicative subsystems of second-order arithmetic, i.e. extensions of PA, as theories of collections of individuals based on a property and a propositional function (or an open formula).[25] He does this both by introducing set variables and $\in$-relation and

---

[22] This passage may remind the reader of Gödel's view presented in [10] to some extent. That is no wonder since this passage seems to have been written after Takeuti met Gödel.

[23] It is unclear what he means here even in Japanese. Our interpretation is that this means 'a rational way of considering the state of affairs,' or 'making sense of the state of affairs.'

[24] The Japanese word used here means more like 'axiomatism,' but this is not commonly used in English. So we have chosen this phrase. See the comment on 'formalism' in Sect. 3.1.

[25] The terms 'predicative/impredicative' was, historically, introduced by Poincaré, e.g., [19], and extensively discussed by Russell [23]. There are different non-equivalent formulations (cf. e.g., [6,

by postulating the axiom of induction and the comprehension principle of a set of natural numbers. Here $\mathfrak{A}(x)$ is a predicate in the language of second-order arithmetic, and X is a variable for a set of natural numbers.

(1) $\forall X \forall x(0 \in X \wedge \forall y(y \in X \rightarrow y+1 \in X) \rightarrow x \in X)$
(2) $\forall \mathfrak{A} \exists Y \forall x(x \in Y \leftrightarrow \mathfrak{A}(x))$.

In both predicative and impredicative cases, we require Y not to occur in $\mathfrak{A}$ to avoid a circularity. Besides, in a predicative case, we require quantified set variables such as $\forall Z$ or $\exists X$ not to occur in $\mathfrak{A}(x)$, but in an impredicative case bound set variables are allowed to occur in $\mathfrak{A}(x)$. Then Takeuti remarks that the former is a conservative extension of PA and the latter is a non-conservative one.[26]

These theories are given to illustrate the following ideas of how the concept of set is used in such theories of collections of individuals based on a property and a propositional function or an open formula. (Takeuti uses expressions 'property' and 'proposition,' but the context suggests that he talks about formal systems.)

> ... [T]he central idea of the concept of set consists in the idea that when we have a property $\mathfrak{A}(x)$, we denote the totality of x that satisfies the property $\mathfrak{A}(x)$ as $\{x \mid \mathfrak{A}(x)\}$ and call it a 'set,' and we consider a satisfying the condition $\mathfrak{A}(a)$ to be an element thereof. What do we think by this? This is nothing but to consider a proposition $\mathfrak{A}$ about x [i.e. a propositional function $\mathfrak{A}(x)$]. In other words, this is nothing but both to start considering anew $\mathfrak{A}(x)$ itself to be a subject [shugo] [in a grammatical sense] in the world in which so far only [words or phrase standing for] natural numbers are subjects and $\mathfrak{A}(x)$ is a predicate, in a way similar to the one in which we consider 0, 1, 2 ,..., to be a subject and also to introduce new variables ranging over such sets of natural numbers. This is very similar to a leap from the standpoint in which one considers only individual numbers 0, 1, 2, ... to the standpoint in which one considers arbitrary natural numbers and all natural numbers in general ([40], p. 2 (98), right, l. 31).

The discussions given here make clearer Takeuti's points summarized earlier. First, the comprehension principle seems to work as a principle of objectification of a proposition, but Takeuti explains this phenomenon in grammatical terms, i.e., a nominalization of a predicate. Second, however, in the case of the basic concept of set which has a close connection with the concept of proposition, we only consider a set

9]). However, roughly, a definition is called 'impredicative' provided that, when we define a set (of, say, natural numbers), we quantify over the set to be defined in the proposed definition itself (i.e., implicitly considered to be in the domain of second-order quantification). Otherwise, we call the definition 'predicative.' Takeuti applies the terminology when we talks about the comprehension principle (2) below. This may not be a 'definition' but these days it is a common usage of the term..

[26] Due to the combination of (1) and the predicative (2), what Takeuti calls a predicative extension must be essentially equivalent to a subsystem of second-order arithmetic ACA_0 (since the induction principle (1) does not allow any set parameter), and an impredicative extension is essentially equivalent to Π^1_1-CA_0 or stronger systems. Takeuti's comment on conservativeness is correct, since the predicative extension is conservative and the impredicative extension is not conservative, but there is a predicative system in his criterion but not a conservative extension for arithmetic. To obtain such a system, we can make the induction principle (1) slightly stronger, a schematic formulation of induction in which set-parameters are allowed. So Takeuti's comments do not seem to intend to make a sharp distinction between conservativeness and non-conservativeness.

of individuals, i.e., natural numbers. In this case, although we have both a predicative and an impredicative concept of set, a set is always a set of natural numbers, and we never go beyond this. A set variable X, Y, etc. are even syntactically distinguished from an individual variable x, y, etc. and the former can occur only at the right side of $\in$ and the latter only at its left side. Hence, there is no iteration of construction of sets going beyond $\mathfrak{A}(x)$, although the comprehension principle allows us to have a notation such as $a \in A$ or $\{x \mid \mathfrak{A}(x)\}$.

Then Takeuti more explicitly states the important distinction alluded to in [34].

> We have thus far talked about the concept of set quite a while, but what is the bridge between the (aforementioned) concept of set and set theory? In my view, set theory is in a way independent of the concept of set, and it contains a much higher-level [kōdo-no] concept. Wasn't the transition from the concept of set to set theory a big leap? The motivation of set theory must have consisted in the iterated use of the concept of set, such as a set of things, a set of sets of things, and a set thereof, etc. This required the purification of the concept of set. Another cause of the transition must have consisted in the beauty of the fact that we can consider the set of all natural numbers by considering 0 to be the empty set, {0} [to be 1], {0, {0}}, etc. ([40], p. 3 (99), right, l. 40)

Takeuti claims that there is a big leap between the basic concept of set of individuals without its purification and the concept of set in set theory created through iterations and purification. Although he does not use the terminology of the thought-objectification anymore in [40], the former concept of set is practically identical with this concept of set. The purification of the latter concept of set occurs by considering only pure sets without ever considering the realm of individuals. In the pure concept of set, even natural numbers are given not as individuals but as sets as discussed in the above quote. Takeuti claims that this makes it possible to use both (i) only one kind of variable since everything is a set and (ii) $\in$ as the only relational symbol in the language. The comprehension principle, unlike (2), should accordingly be modified with only one kind of variable as follows.

(3) $\forall \mathfrak{A} \exists y \forall x (x \in y \leftrightarrow \mathfrak{A}(x))$.

As mentioned before (see the fifth quote in Sect. 3.2.1), this indeed looks like putting an element and a set *on an equal footing*. In these quotes, Takeuti raised the same topic as discussed in [34], but the discussion here is based on concrete cases. Also, when we use set variables in both sides of $\in$, it is not only very natural to consider iterations of constructions of set via the comprehension principle, since a set itself can occur at the place of element in $x \in y$, but also to consider 'only pure sets.' It must be due to these differences that Takeuti treats a set as a thought-objectification and a pure set in set theory differently.

However, as Takeuti points out in [40], (3) immediately leads to Russell's paradox, and we need to modify the axiomatization of the concept of pure set. Then Takeuti both presents Zermelo's separation axiom: (4) $\forall \mathfrak{A} \forall x \exists y \forall z (z \in y \leftrightarrow z \in x \wedge \mathfrak{A}(z))$, whose idea is to partly restore of the idea of 'a collection of things,' since an 'already formed set x' plays the role of it, and gives a brief exposition of axiomatic theories. Takeuti here makes a distinction within the pure concept of set between the one given by naive set theory and the one given by axiomatic set theories and discusses some philosophical issues of the latter, focusing on the development of set theory for its own sake. However, here let us confine our discussions to the difference between between a set of individuals and a pure set for a reason to be given shortly.

One consequence of the distinction at issue between the concepts of set is the following. The foregoing worry about the necessity of restricting the concept of proposition and its thought-objectification can be handled if the concept of set as the thought-objectification of a proposition is naturally restricted to that which can be constructed out of either a proposition (a propositional function) or a property of individuals which are given before we start talking about sets of them. In this case, we do not have to worry about the kind of problem mentioned before (see the tenth paragraph of Sect. 3.2.1). This corresponds to the predicative concept of set. However, it may still be problematic how we can handle a highly impredicative set of individuals. We will come back to Takeuti's solution to this problem later.

Although in [40] Takeuti keeps the fundamental distinction between the two concepts of set that he made in the series of papers [33–35], there is a difference between his discussions in the series of papers and [40]. First, in the series of papers, Takeuti did not explicitly make the distinction, and hence we needed to have a charitable interpretation to avoid a collapse of the two concepts of set. Here the distinction is made between a set of individuals and a pure set.[27] Second, Takeuti used to be more dismissive to the latter concept of set than in [40], but in [40] the emphasis is put on the distinction itself. This is partly because, unlike the previous series of papers, in [40] Takeuti discusses not only naive set theory (which he quickly dismisses) but also axiomatic set theories. Takeuti apparently started thinking that pursuing the concept of set in the context of second-order arithmetic and doing so in axiomatic theories are essentially different projects, the latter having its own goal.

By drawing the conceptual distinction discussed here (and dismissing naive set theory due to its inconsistency), Takeuti seems to draw a consequence related to the nature of foundational studies of the concept of set, i.e., a distinction between the concept of set that is suitable for a finitistic proof-theoretic analysis and the concept of set used in set theory which realizes Cantor's original idea of set-generation principle. We will see shortly his view of how the proof-theoretic analysis should be given to the former concept of set. Concerning the latter concept, Takeuti identifies the abstract concept of set in set theory to be the idea of generating as many sets as possible. His thought on this abstract concept of set will be significantly more sophisticated later partly due to an influence of Gödel's thought. The quote from [40] shows the basic idea.

[27] About this issue, we have suggested above that the distinction could be drawn between the concept of set in finite- or second-order predicate calculus and the concept of set putting elements and sets on equal footing. But here Takeuti talks about only second-order arithmetic and a set of individuals. We do not take this to be a serious difference. Takeuti probably wanted to make the contrast sharp here, but the same point could have been made for higher-order arithmetic or logic. We claim this because over the years Takeuti consistently regards finite-order predicate calculus as benign in its use of the concept of set. Indeed, GLC is a Gentzen-style higher-order predicate calculus which Takeuti formulates and, in [51], Takeuti draws a similar distinction between finite-order predicate calculus and ZF set theory on a new ground adopted after 1970 [51], p. 161.

> Once we make a transition from the concept of set [of things] to sets in set theory, we can focus only on how the idea of 'as many as possible' can be expressed ... ([40], p. 4 (100), left, l. 10).

Actually, one of Takeuti's reasons why he regards standard axiomatic set theories as wanting is that they may not be adequate from the viewpoint of this principle. However, we put off going further into the latter concept of set to the extended version of this paper, since its development made by Takeuti is mainly done after 1970.

3.2.3 Logical or Formalistic Analysis of the Concept of Set

Takeuti's view of how proof-theorists should analyze the concept of set is already discussed in [33] and other papers. In this subsection, we first discuss Takeuti's general view of formalistic analyses (based on Hilbert-Gentzen's finitism) of the concept of set. Briefly put, his general view is that 'logic' should play a more important role in analyzing the concept than usually thought, although the logic which Takeuti has in mind is second-order or finite-order predicate calculus. Second, we focus on a concrete proposal which Takeuti claims can provide a finitistic analysis of the concept of set.

In the part of [33] where Takeuti claims 'the concept of set can be formulated only by formalism (see p. 5 above),' Takeuti states as follows.

> What we currently call 'logic' – something which we use in an almost instinctive manner – is also a creation of human beings. Logic must have been created by the long history of human beings and completed by mathematicians ... This seems to be determined in such an instinctive manner that there is no other way of thinking probably because logic is 'the most excellent' [mottomo suguretaru] way of thinking. ... Isn't contemporary mathematics now in the process of creating *the second logic* [daini no ronri] through the concept of set? ... In spite of its vague character, the concept of set makes it possible for many mathematicians to make a clear, intuitive and instinctive inference. ... Isn't this because the 'second logic' has been uniquely determined as 'the most excellent'? ([33], p. 28 (268), left, l. 13. The emphasis is due to the present author)

What Takeuti has in mind 'logic' and 'the second logic' here is not entirely clear. Still let us first give some comments on Takeuti's terminology. His original phrase in Japanese corresponding to 'the most excellent' is by no means standard in this context. The phrase expresses a general idea of 'the best,' 'superb,' "extremely good,' or something like these. Takeuti seems to have wanted to avoid any philosophical or mathematical cliché. As a result, we understand that he wanted to say something special, but he does not seem to be able to convey his message very well, since the expression is too general. But we suspect that by 'logic' he meant the standard first-order logic, and 'the second logic' is finite (second)-order predicate calculus or (Hilbert's and Gentzen's) proof-theoretic methods for analyzing the structure of a proof, or both. The ground for claiming this is based on the structure of Takeuti's text itself. Immediately after the foregoing quote, Takeuti states as follows.

> I am exclusively working on formalism [keishiki-shugi] because I feel 'what is the most excellent' also in an illumination [kaimei] of the nature of 'this second logic,' 'the excellent.' I never pursue the problem of the consistency of mathematics per se. I have an opinion that *an illumination of 'the concept of set' is all that matters in formalism.* Lots of people would have a cynical feeling concerning the phrase such as 'an illumination of the nature of the concept of set,' but I believe that if 'the fundamental conjecture for GLC,' which I am always talking about, is solved affirmatively, then we would be able to say that it means that we have obtained an illumination of the nature of the concept of set to some extent. ([33], p. 28 (268), left, l. 35, the emphasis is by the present author).

In addition, Takeuti appears to think that the next stage of the thought-objectification of a proposition is a theorization of the notion of proof via a formalistic, proof-theoretic analysis. Indeed, Takeuti states towards the end of [34],

> [T]he thought-objectification through symbolization in mathematics has developed, through that of propositions (sets) and that of the concept of logic, up to the thought-objectification of the notion of proofs in mathematics itself in formalism ([34], p. 19 (299), right, l. 11).

In the former passage, Takeuti clearly talks about finite(second)-order predicate calculus, and such a logic explicitly involves the concept of set, so it is natural to interpret that 'the second logic' which he talked about in the previous passage is really such a logic. If so, the logic used before that must be first-order logic.

Reading these passages under this interpretation, one can observe the following. First, Takeuti was convinced already at this moment (as of 1956) that what matters in Hilbertian formalism is not only proving the consistency of mathematics but a clarification of the concept of set. Second, he was convinced, in particular, that solving Takeuti's conjecture would contribute to clarifying the concept of set. In other words, proving cut-elimination for GLC would be helpful in illuminating the concept of set. This view may have a partial realization in some technical results of Takeuti's such as the combination of cut-elimination in a subsystem of GLC and ordinal diagrams in [38, 39] (see further discussions on this below). Third (in the latter quote), Takeuti considers a formalistic analysis of proofs to be the next step to the thought-objectification of a proposition, i.e., the thought-objectification of a proof. This can be naturally interpreted as implying that Takeuti has in mind treating "proofs" as mathematical objects.[28] At the time when this passage was written, only few people shared the view expressed by the third point, namely understanding the concept of proposition, set and proofs from a uniform perspective and taking proofs as mathematical objects. Curry-Howard isomorphism (propositions-as-types) may be one of few such cases, but Takeuti's paper precedes the emergence of propositions-as-types and the underlying view of proofs in the propositions-as-types is a 'constructivist' one, not a formalist one. These points suggest that, when he criticizes axiomatic set theories, Takeuti means that one of things that are missing from axiomatic set theories is their connection with finitistic logical analysis à la Gentzen.

---

[28] After stating this, Takeuti poses a question of what the significance of this is and says, 'it seems to me that the problem of consistency is a touch stone of formalism and proving that would give the foundations of formalism rather than giving the foundations of mathematics ([34], p. 19 (299), right, l. 23).' This is intriguing, but what he meant by this needs to be further investigated.

However, these are merely expressions of his opinions and contain no concrete proposal. Of course, as mentioned above, Takeuti has some technical bases for these philosophical claims. It is probably worthwhile to recall the technical works that Takeuti published around the same time as these 'philosophical' papers. Takeuti formulated a sequent calculus for second-order predicate calculus (called ELK) already in a paper published in 1949 [28], which was actually accepted on November 1, 1948.[29] Note that this is only ten years after Gentzen's best-known consistency proof of Peano Arithmetic, based on transfinite induction up to ϵ_0, in [7]. Due to the problem contained in [28], Takeuti published his idea of proving cut-elimination of GLC as a conjecture in [29] and then his partial results concerning the conjecture in [31, 32, 36, 37]. The series of these results culminated in [39], combined with the idea of ordinal diagrams in [38]. Hence, around 1958, Takeuti already had sufficient technical results which look sufficient to give (at least partial) support for the foregoing general philosophical view. However, Takeuti gave practically no philosophical justification of these works in words. Perhaps, Takeuti believed at this moment that the mathematical results must be sufficiently convincing to show why these results give illuminations of the concept of set. However, these are the kind of mathematical results such that after having mathematically understood these results, one might still ask how such a formalistic analysis illuminates the concept of set. Takeuti did not give an answer to this question in [33–35], but, in [51], Takeuti had a more concrete proposal concerning the point. (Note that, in [51], Takeuti starts discussions on the issue by stating, '[w]e think it important that consistency problems be formulated in an appropriate philosophical framework ([51], p. 297).')

3.2.3.1 The Concept of Set and Consistency Proofs In the following let us take a look at some of Takeuti's philosophical discussions by using sources including some text from later periods and his own later recollections on his works before 1970. This may sound anachronistic, considering the scope of this paper. However, our point of discussing the text here is to show that the programmatic idea of Takeuti in the 50's and 60's would be substantially developed later into a concrete proposal. Also, since the idea that Takeuti uses here has nothing to do with the philosophical idea that he adopts after 1970s and the first edition of [51] is published in 1975, we can reasonably guess that Takeuti already had this idea before 1970s.

Let us first quote a passage from [51]. After claiming that a consistency proof of number theory is not particularly important since we have a good intuition of natural numbers, he states as follows.

> In contrast, we can conceive of the world of sets only through our imagination and our mathematical experience. Consequently, the problem of the consistency of the comprehension axiom is a serious and important foundational question. ... [T]he fact that no contradiction arises does not explain *what it means to say that a theorem is provable from the comprehension axiom. Nonconstructive proofs provide no insight into this important question.* On

[29] Takeuti must have started working on Gentzen's paper around 1946, according to his recollection mentioned in [55]. He gave a talk at the spring meeting of the Mathematical Society of Japan in 1947 before he graduated Tokyo Imperial University.

> the other hand a constructive proof strengthens our intuition and adds meaning to the theorem. In particular, *a constructive proof of the cut-elimination theorem would give us greater confidence in the comprehension axioms* and hence *strengthen our convictions about our imaginary world of sets*. ([51], p. 298, the emphasis is by the present author.)

This passage might look like merely a restatement of what Takeuti said in the passages quoted above. Still, the quoted passage already contains a few new points that are worth explaining. First, this is a clear statement that we *need* a consistency proof of a system containing the comprehension axiom and hence the concept of set, whereas by the statement quoted above Takeuti expressed his opinion that a proof theoretic analysis of finite-order predicate calculus *can* clarify the concept of set. Second, the passage states that it also matters which method of proving consistency we use. He claims that a nonconstructive proof of consistency is uninformative, since it does not clarify the significance of the fact that a theorem is proven by a certain form of comprehension axiom, i.e. the role played by the comprehension axiom in a proof. Here Takeuti must have had in mind the nonconstructive proof of Takeuti's conjecture [20, 26, 27].[30] Since the comprehension axiom can be formulated as the left rule for higher(second)-order universal quantifier in Takeuti's GLC (L∀), Takeuti essentially claims here that the rule L∀ plays a key role in consistency proof, and that we need to understand its role when we prove cut-elimination by proving this constructively. Let us recall the higher(second)-order L∀ rule.

$$\text{L}\forall \quad \frac{F(V), \Gamma \Rightarrow \Delta}{\forall \varphi F(\varphi), \Gamma \Rightarrow \Delta}$$

Here V is an abstract w.r.t. an arbitrary abstract in the language of higher(second)-order logic. In the same way as we can by some versions of the comprehension axiom, we can formulate certain restricted versions of L∀, and by using these we can formulate subsystems of higher(second)-order logic, which Takeuti used in his partial results leading to [39]. Third, the last sentence explicitly states that a constructive proof of cut-elimination of a system containing the comprehension axiom would increase the degree of our confidence on the concept of set.

[30] As far as we know, Takeuti did not give any substantial comment on Girard's seminal result [8] on the normalization theorem of higher-order polymorphic lambda-calculus, based on Curry-Howard isomorphism, which implies cut-elimination for the corresponding second-order logic. The proof method used here appears to be much more constructive than the aforementioned semantic proof, but Takeuti was quiet about this. Arai's comment on the semantic proof of Takeuti conjecture may be helpful for us to understand a reason of Takeuti's silence. Arai states, 'we can by no means say that the fundamental conjecture is positively solved. This is not because the proof is semantic but because the proof can be formalized only in a theory that properly contains the theory whose consistency is supposed to prove ([2], p. 2).' Girard's proof of normalization of F shares such a property. The proof is not formalizable in second-order arithmetic due to Gödel's second incompleteness theorem. To formalize the proof, we need a properly stronger theory. Takeuti may have been unsatisfied with the proof in this sense.

One might think that the first and third opinions are commonly shared by proof-theorists. But this is not quite the case.[31] Takeuti himself has a recollection on his motivation of his study of the consistency of analysis in [53], 'I can claim confidently that there was no general urge to study the consistency problem, especially to study [the impredicative concept of set] proof-theoretically. However, as my objective was a proof theoretic study of [the concept of set] from the beginning, I plunged into such a study without misgivings ([53], p. 131).'[32] Takeuti was isolated when he began the project of proving Takeuti's fundamental conjecture.[33]

The outstanding features of Takeuti's view can be made clearer in comparison with a view of other proof theorists. For instance, Kreisel, who was one of the most influential proof-theorists in the twentieth century and a friend of Takeuti, had quite a different view. Kreisel says, 'the widespread idea of connecting formalization and reliability is equally incoherent. When we really want to make sure that a mathematical result is correct we do not formalize its proof, ... but we try to make it intelligible. We rely on mathematical, not logical foundations! (p. 361. [11])' This comment is not directed to cut-elimination or consistency proofs, but if one cannot rely on formalizations or logical foundations, then there is no reason why a constructive proof of cut-elimination of a system containing the impredicative comprehension principle can strengthen our confidence of the principle. Hence, Kreisel's comment is almost directly opposed to Takeuti's view. Also, unlike Kreisel who apparently thought that the cumulative hierarchy of sets is 'clear' and hence sufficiently reliable, see e.g., [12], Takeuti (at least at the moment) appears to have thought that the concept of set is problematic and a finitistic analysis of the concept is helpful to make it secure.[34]

---

[31] Concerning the second point, there is indeed an almost uniform agreement among proof-theorists that a constructive proof is preferable.

[32] We modified the translation of the bracketed phrases to uniformize the translating terms.

[33] It is not easy to trace how Takeuti came up with the idea of proving the consistency of analysis via cut-elimination of finite-order predicate calculus. But Takeuti recalls a heuristic process of how he started working on GLC. After talking about an episode that Takeuti asked Suetuna how he can learn the foundations of mathematics but he got an answer that since there was no established material, one would need to think oneself, Takeuti states, 'Thus, the first thing that I started doing is to formulate the system of GLC by introducing second-order variables to LK and developed Dedekind's cut in GLC. Although the comprehension principle for higher-order cases is complicated, that for second-order cases is very straightforward. By doing this, I was able to roughly grasp how such a logical system works. Then I believed that cut can be eliminated from the system and a consistency proof of the theory of real numbers, and I then quickly did prove it, or more precisely I thought did prove it (p. 55, [55]).' He seems to have got directly into the heart of the matter.

[34] Feferman [5] summarizes Kreisel's view roughly as follows, (i) there is not a shadow of a doubt as the consistency of ZF set theory, (ii) the paradoxes stemmed from a confusion between sets and properties and Zermelo offered a perfect explanation of what his axioms are supposed to be about, namely the cumulative hierarchy of sets, (iii) proofs which use abstract principles are not only more intelligible but also more reliable, and the professed aim of proof theory to secure the foundations of mathematics is an ideological hangover, (iv) the point of proof-theoretic work consists in finding out further concepts or results and in short applications or by-products thereof, e.g. the 'proof-mining' program. Although Takeuti may not have any problem with agreeing to iv), the other points clearly have an opposing inclination against Takeuti's viewpoint.

3.2.3.2 Vicious Circularity in the Concept of Set and Cut-Elimination Now we move on to the question: what kind of particular contribution Takeuti thought can be made by a proof-theoretic analysis of the pertinent concept of set? Here is a little more specific reflection by him. He starts from describing what is a potential problem of the concept of set when one has a vicious circle in defining a set.

> Let us consider the concept of set not to be a given, already known concept. Instead, let us consider a set to be constructed in such a way that, for instance, a set A is defined by stating, 'let A be the totality of x which satisfies the property $P(x)$.' When we consider a set A, if the property $P(x)$ has the notion 'for all sets x' [in its defining conditions], then the aforementioned constructing process would not hold since it contains a vicious circle. That is, in order to define A, the notion of 'all sets' needs to be defined, and it must contains A itself, so it is a vicious circle. Russell and Whitehead tried to construct mathematics only by definitions that contain no vicious circles and ended up with a complete failure. ... The foregoing problem is not only a philosophical problem but also a problem that necessarily appears as an essential difficulty provided that we try to pursue and analyze the concept of set played in mathematics. How should we disentangle [tokihogusu] such a vicious circle? This is one of the most essential problems in foundations of mathematics ([50], p. 75, l. 1).

Then Takeuti proposes his own solution to this problem of vicious circularity (i.e., impredicativity explained in the footnote 25) by focusing the notion of cut-free proof.

> I thought that the focal point of the problem is not to disentangle a vicious circle in a concept but rather to disentangle a vicious circle that occurs when the concept of set is used in *proofs or inference* in mathematics. In particular, I came to understand my own variant of Hilbert's program as that of considering Gentzen's cut-elimination in a system that has the concept of set along with vicious circularity (although Gentzen treated a system without vicious circularity) and reveal how the vicious circularity can be disentangled (that is my fundamental conjecture) and further make lucid the process of such a disentanglement as if we saw this with eyes ([50], p. 75, l. 13, the emphasis is by the present author).

Here Takeuti explicitly states in what way he thought his version of Hilbert's program can contribute to a conceptual analysis of the concept of set which involves vicious circularity. He makes it clear that his aim in doing proof theory was to disentangle a vicious circle with respect to the impredicative concept of set via cut-elimination, i.e. not by a direct conceptual analysis but by a proof-theoretic analysis via the notion of cut-free proof. Focusing on the notion of proof in this manner seems highly original.

Feferman finds Kreisel's view persuasive 'at least to the following extent: proofs themselves should be the central objects of interest; proof theory is developing precise theories related to informal ideas about proofs and which illuminate our daily experience with them' ([5], p. 359). Takeuti cannot agree more to the idea that proofs themselves should be the central objects in proof theory.

However, Feferman also expresses a disagreement to Kreisel's acceptance of current systems of analysis and set theory. 'Not that the mental picture of the cumulative hierarchy is not convincing enough to warrant believing the consistency of set theory—indeed this picture provides quite an infallible guide when working within ZF; rather, it is that we really do not know what set-theoretical statements mean in the same way that we know what arithmetical statements mean. For, we have a complete mental picture of the totality N of natural numbers as generated from 0 by the successor operation, but no such picture of the supposed totality $\mathcal{P}(N)$ of subsets of N (ibid.).' This is a reason why Feferman took a predicativist view of analysis. Concerning predicativism and ZF, Takeuti has a view different both from Kreisel and Feferman. (See section "Disentangling Vicious Circularity via Modulations".).

In order to highlight this point, let us briefly review other well-known attitudes towards impredicativity. There are roughly three different views of impredicativity. The first is to prohibit impredicativity in mathematics, like Poincaré [19], Russell [23], Weyl [57], Feferman [4, 6],[35] since they take seriously the view that a set is an entity that we actually construct by using a comprehension principle. The second is to take a form of platonism about sets and to consider the impredicativity to be entirely harmless, since a set is something existing independent of constructions by mathematicians and we just pick them up by using a comprehension principle. People like Gödel [9] and Ramsey [22] take this view. The third, e.g., Carnap [3], Longo [13], take a view that impredicativity looks problematic under certain understanding of higher (second)-order quantifiers in such a way as to 'run through all the instances' of what are quantified over. This view understands the meaning of second-order quantifiers in such a way that a complete verification of a (second-order) universally quantified statement consists in 'nothing more than its logical (more exactly tautological) validity for an arbitrary property ([3], p. 51).' They think that the uniformity and the genericity of the second-order quantification make the impredicative quantification harmless. Takeuti's view can be identified with none of these. Takeuti takes the impredicativity to be problematic, but, instead of prohibiting such a quantification in mathematics, he takes it a kind of epistemological challenge and tries to understand better by using a proof-theoretic method.[36]

Let us now go back to the foregoing quote. The passage specifically expresses what was Takeuti's aim when he talked about a proof-theoretic analysis of the concept of set, but this may still look somewhat too general, and he does not specify how this can be done. However, not only did Takeuti have his general view of the concept of set that it should be analyzed proof-theoretically, he had a concrete proposal concerning how a constructive proof-theoretic analysis can be helpful to illuminate the concept of set. What is at issue is the fact that, in finite(second)-order predicate calculus, higher(second)-order left $\forall$ rule (equivalent to the comprehension axiom) may have a more complicated formula in the uppersequent than the principal formula, which severely spoils the subformula property. This is usually taken to be a negative evidence against the significance of cut-elimination for finite(second)-order predicate calculus. Takeuti himself succinctly summarizes the problem as follows.

---

[35] We do not talk about subdivisions of positions among these people here.

[36] Not surprisingly, if we consider the finite/infinite distinction, instead of the predicative/impredicative distinction, then we can observe that Takeuti's view is schematically parallel to Hilbert's finitism. But Hilbert's emphasis seems to have been put on a different point, i.e. distinction between contentual (= finite) and formal (= infinite). Takeuti added some new aspects to Hilbert's original idea, and Takeuti's version is a kind of generalization of the Hilbert-Gentzen finitistic method.

> [T]he fundamental conjecture on GLC has a feature that is entirely different from Gentzen's fundamental theorem. From Gentzen's fundamental theorem, an important result follows, that is, a sequent is derived in the system from simple sequents by accumulating more complex ones. This does not follow from the fundamental conjecture of GLC. The cause of this failure is in
>
> the rule of 'left second order ∀': $\dfrac{F(\{x\}A(x)), \Gamma \Rightarrow \Delta}{\forall\phi F(\phi), \Gamma \Rightarrow \Delta}$
>
> In this inference, if $A(x)$ is a complex formula, then $F(\{x\}A(x))$ is far more complex than $\forall\phi F(\phi)$, and hence the principle that the upper sequent is simpler than the lower sequent does not hold ([56], p. 132).

This is usually taken to be a vicious circularity or impredicativity that still persists even when cut is admissible.[37] However, it should be pointed out that the aforementioned partial results can be used to back up Takeuti's view. In [38, 39], Takeuti proved the consistency of Π^1_1-CA_0 via the cut-elimination of a fragment of $\mathrm{G}^1\mathrm{LC}$. This is one of the best technical achievements of Takeuti, and there is a hope that we can extract some evidence that Takeuti is indeed correct in claiming his disentanglement of impredicativity via cut-elimination. Despite its importance, we do not go into the detailed discussions of the philosophical significance of this result (whose reason will be explained shortly), but we confine ourselves to a related argument given by Takeuti.

3.2.3.3 Disentangling Vicious Circularity via Modulations In the following, we focus on Takeuti's concrete (but apparently neglected) proposal concerning how cut-elimination and the issue of predicative/impredicative distinction are connected. In particular, let us focus on an argument given in [51] (in the first ed.), in which Takeuti presents a theory of 'modulations,' which Takeuti claims can accommodate impredicativity. We do not have enough space to precisely define the concept of modulation, but roughly it goes as follows. In a concretely given cut-free proof, we can take a modulation of a formula on the antecedent containing a universal second (or higher)-order quantifier introduced by the Left rule for ∀. Then we obtain a formula by replacing the quantifier with abstracts occurring in an application of the rule by which the quantifier is introduced. The statement in [51] is the following. (The term *P-modulation* below suggests that a modulation of $\Gamma \Rightarrow \Delta$ is determined depending on a cut-free proof P, and the term *predicative proof* means a proof without any impredicative application of L∀.)

Proposition 1 ([51], p. 301) *Let P be a cut-free proof of $\Gamma \Rightarrow \Delta$, and let $\Gamma' \Rightarrow \Delta'$ be the P-modulation of $\Gamma \Rightarrow \Delta$. Then there exists a cut-free predicative proof of $\Gamma' \Rightarrow \Delta'$.*

Prawitz [21] summarizes the content of this proposition as follows: 'by proving a cut-elimination theorem for such a system constructively, we obtain a predicative

[37] See [8] concerning how circularity should be handled even in a successful case of cut-elimination. Concerning some nuances related to the circularity involved here, see [18].

interpretation of all the theorems of the system. In defense of this view Takeuti adduces the following immediate corollary of the cut-elimination theorem: given an impredicative proof of $\Gamma \Rightarrow \Delta$, there is a predicative proof of a modulation $\Gamma^* \Rightarrow \Delta^*$ obtained from $\Gamma \Rightarrow \Delta$ by replacing certain subformulas of the form $\forall X F(X)$, namely those subformula that in the given proof were introduced in the antecedent as the result of impredicative $\forall$-generalizations. They are to be replaced by $Q(F(V_1) \wedge F(V_2) \wedge \cdots \wedge F(V_n))$, where $V_1, V_2, \ldots, V_n$ are abstracts occurring in $\forall$-generalizations by which $\forall X F(X)$ is introduced, and Q is a string of $\forall$-quantifiers binding all free variables in any V_i. As is easily seen, the modulations give (classically) stronger theorems but have nevertheless elementary proofs' ([21], p. 1095).

Takeuti explains the foundational significance of this proposition as follows.

> Our theory of modulations, ... makes clear one important reason for our interest in cut-free proofs, namely, a cut-free proof of a theorem enables us to give to that theorem an interpretation that avoids Russell's vicious circle. ... If we had a uniform "method" that transformed every formal proof into a cut-free proof, then that "method" would itself determine an interpretation of mathematics that avoids Russell's vicious circle ([51], p. 301–2).

Takeuti's point seems to be that, intuitively, if we forget about the problematic introduction of $\forall$ on the left, the theory of modulations can at least make the abstract in the complicated premise explicit and then we can avoid the vicious circularity. Hence, if he is right about this, then through a modulation a cut-free proof can give a predicative interpretation of a theorem that originally contained impredicativity.[38]

Before moving on, let us note that one of Takeuti's reasons for presenting the theory of modulations is to give a critique of what he calls 'quasi foundations,' i.e., predicativism (most likely, Takeuti has in mind Feferman's view here). This is a view which develops a significant portion of classical analysis without using impredicative definitions or comprehension principles. His argument against predicativism is too cryptic, but Takeuti probably implies that predicativists' way of doing foundational studies is to distort the original meaning of theorems.

Takeuti gives two arguments against predicativism. First, after pointing out that predicativism does not identify analysis as 'a system of axiom' but 'a collection of results,' he complains, 'when one develops a collection of results that are "similar" to a given set of theorems of analysis it is not at all clear what metamathematical conclusions can be drawn or what kind of theory has been founded ([51], p. 302).' By this, he probably means that when we obtain a metamathematical result, e.g., consistency, the significance of it is hard to evaluate (may become incomparable with the original one) since the results are different from the original analysis.

However, Takeuti's second and main critical point is that, in predicativism, cut-elimination is taken to be unimportant.

---

[38] Since this idea of handling impredicativity requires a constructive cut-elimination, one might think that this for the full system may be just based on a promissory note. However, concerning the subsystem whose cut-elimination has been proven by [39], the theory gives more than a mere promissory note. But the ultimate appraisal of it requires us to examine on what the method of proving cut-elimination depends (see below).

> We may, therefore, regard the predicative comprehension axioms as what is really important. The theorems of classical mathematics are simply approximations to stronger theorems. Consequently, so long as our concern is with practical foundations we need not prove the cut-elimination theorem; we only need to justify the predicative comprehension axioms ([51], p. 302).

This is because 'If we wish a "quasi foundation" for a certain collection of results we simply start with predicative comprehension axioms and develop the modulations of the results that we wish ([51], p. 302).' Then cut-elimination would be become irrelevant, since we only handle predicative comprehension anyways. However, this is not the view to which Takeuti can agree. Takeuti gives a critique from his viewpoint.

> We ... advocate a different viewpoint with respect to which proving the cut-elimination theorem constructively is a matter of paramount importance. Indeed, only through a constructive proof of cut-elimination can the theory of modulations become truly a significant theory in a world of predicative mathematics. The proof of the cut-elimination theorem as presented in Chap. 3 is set-theoretical, and therefore useless for our purpose ([51], p. 302).

In this somewhat condensed argument, Takeuti highlights the importance of both cut-elimination itself and also of a 'constructive' proof of cut-elimination, claiming that a set-theoretical (or semantical) proof of the admissibility of cut for GLC is not sufficient.

Aside from the issue of whether Takeuti's critique of predicativism works or not, one might wonder whether this can get rid of vicious circularity or not. Indeed, after giving the explanation of the modulations, Prawitz [21] expresses his skepticism.

> Takeuti's discussion of philosophical issues is refreshingly open, and the reviewer finds it interesting but in some cases problematic. For instance, two provable theorems $A \Rightarrow B$ and $B \Rightarrow C$ will be interpreted by modulations as $A^* \Rightarrow B^*$ and $B^\# \Rightarrow C^\#$, respectively. Although B^* implies $B^\#$, this cannot in general be seen predicatively. It is true that $A^* \Rightarrow C^\#$ can be proved predicatively, but this cannot be seen from the theorems when reinterpreted, which hence may lose much of their interest and usefulness (p. 1095, [21]).

Interestingly, Takeuti actually gives a reply to this in an unpublished handout [52].

> What I discussed there is the predicative elementary mathematics is good enough for quasi-foundations in the following sense. Whenever the impredicative mathematician proves a Theorem A, then the predicative mathematician proves a stronger theorem at least in the eye of the impredicative mathematician. Of course if $A \Rightarrow B$ and $B \Rightarrow C$ are proved by the predicative mathematician, then $A \Rightarrow C$ is also proved by him. The predicative mathematician lives independent of the impredicative mathematical world. He does not have to watch the impredicative mathematician at all. Suppose that the impredicative mathematician asks him $A \Rightarrow B$ and $B \Rightarrow C$, and the predicative mathematician answers $A^* \Rightarrow B^*$ and $B^\# \Rightarrow C^\#$ and then the impredicative mathematician asks him about $A^* \Rightarrow C^\#$. The predicative mathematician may not answer $A^* \Rightarrow C^\#$ immediately. Nevertheless he will answer yes after a while. There are [sic] nothing wrong about it. I don't understand why it loses much of its interest and usefulness (p. 6, [52])

Both Prawitz's critique and Takeuti's reply seem to be somewhat unclear, but let us try to give at least a tentative interpretation of the discussion here.[39] First, Prawitz

[39] This is not a definitive or decisive interpretation but an attempt of interpretation.

critique is based on a conceptual distinction between 'predicatively provable' and 'predicatively seen.' Prawitz does not appear to give a precise definition of this concept 'predicatively seen.' However, probably, this means that one can immediately see that a modulation of a certain sequent is provable by taking a look at the form of the modulation of a sequent, i.e. by explicitly showing an abstract occurring in a sequent before the pertinent application of L∀. Then, Prawitz's point seems to be that since the abstract occurring in B^* and $B^\#$ can be very different, there is no reason why one can immediately see that this can be predicatively proved. As a result, even if $A^* \Rightarrow B^*$ and $B^\# \Rightarrow C^\#$ can be immediately, i.e. predicatively, seen, this property is not preserved when we consider $A^* \Rightarrow C^\#$ (not preserved under Cut with respect to certain modulations B^* and $B^\#$). Hence, $A^* \Rightarrow C^\#$ may not be predicatively seen. But Takeuti appears to mean in his reply that what matters to him is not Prawitz's version of 'predicatively seen,' but 'predicatively provable' via taking modulations. (Note that Prawtiz admits that B^* implies $B^\#$. Since these do not involve impredicative application of L∀, this may be sufficient for predicative provability of $A^* \Rightarrow C^\#$, although it may not be 'predicatively seen.')

We have discussed the following two issues related to the theory of modulations: whether the theory of modulations can really avoid Russell's vicious circle or not; whether Takeuti's critique of predicativism by the theory of modulations is sound or not. Such issues of interpretation of cut-free proofs aside, concerning the aforementioned achievement of the cut-elimination of a fragment of G^1LC in [39] and the consistency proof of Π^1_1-CA_0, Takeuti gives a recollection as follows.

> After several years of struggle, I was never able to understand why such a disentanglement can occur. I wanted to solve at least the first level of the problem of how to disentangle the vicious circle, and I managed to solve only the first level. However, I still cannot visualize the general case of the disentanglement as if I saw it with my eyes. In the first place, concerning the vicious circle occurring the actual mathematics, Friedman and Simpson showed that most of such cases of vicious circularity can be covered by the case whose consistency I proved. Hence, it is even hard to get an essentially new example from the ordinary mathematics ([50], p. 75, left, l. 29).

This quote suggests that, as far as his mathematical program of proving the consistency of finite(second)-order logic or arithmetic (i.e. analysis) by using a constructive method, Takeuti has been significantly successful. In particular, concerning the mathematical data to be handled, having shortage of an essentially new case of vicious circularity means that most cases have already been dealt with, since, as Takeuti mentions here, the program of reverse mathematics due to Friedman and Simpson has made it clear that the overwhelming majority of theorems in contemporary mathematics can be developed within the subsystem of second-order arithmetic Π^1_1-CA_0.

However, the philosophical or epistemological significance of this work has to be carefully examined in relation to the disentanglement of vicious circularity or impredicativity via cut-elimination. To do that, we need to fully analyze *exactly in what sense* these results give a partial realization of Takeuti's view. This would ultimately require us to discuss *how* not only cut-elimination but also his method of ordinal diagrams is particularly helpful for 'disentangling' the vicious circle involved here. For instance, we have to make sure that the proof-method that Takeuti uses does not smuggle in any sort of impredicativity or circularity. (Then the heart of the matter is what kind of assumptions we need in order to prove that ordinal diagrams are well-ordered, for Takeuti already comments: 'the only idea which is beyond the Hilbert's finite standpoint is the proof of the well-ordering of ordinal diagrams ([41], p. 300).'[40]) We are not ready to discuss this difficult issue in this historical paper, whose main aim is to handle a broader issue of Takeuti's view of the concept of set. Anyways, it is probably clear at this point that Takeuti's view that the concept of set should be analyzed via a finitistic method is not only a slogan but has a concrete basis.

4 Closing Remarks

Let us take stock of our discussions. Takeuti starts his discussions on the concept of set by criticizing the Cantorian definition of the concept of set (Sect. 3.1.1) and expresses his discontent with the concept of set in axiomatic set theories (Sect. 3.1.2). Then Takeuti gives his positive account of the concept of set. Pointing out the importance of the role of symbols in our mathematical thought, he claims that the concept of set is given as the thought-objectification of a proposition via effects of symbolization (Sect. 3.2.1).

However, he claims in addition both that symbolization occasionally goes out of control and that a more abstract concept of set in set theory is also produced. Accordingly, he introduces a distinction between two concepts of set. One of them is a benign concept of set as the thought-objectification of a proposition (a kind of reification of proposition), and the other is a potentially harmful concept of pure set in set theory (naive or axiomatic!), which 'put elements and sets on an equal footing.' It is somewhat surprising that such a distinction is drawn between the concept of set as the reification of proposition and that of naive or axiomatic set theories. This seems to be Takeuti's original view.[41] Takeuti criticizes the abstract concept of set in set theory (naive or axiomatized) because according to him it is like an 'illusion' produced by an abuse of symbols. About the concept of set, he provides an explanation of how

40 The context naturally suggests that Takeuti assumes that the contentual mathematics in Hilbert's finite standpoint can be identified with or not stronger than primitive recursive arithmetic, and he points out that other things needed to prove cut-elimination for the appropriate subsystem of GLC are all primitive recursive.

41 It is worthwhile, however, to note here that there seem to have been two different aspects in the traditional concept of set since Cantor, Dedekind (and probably Frege), one of which is the concept of set useful for 'concept formation' and the other is a combinatorial concept of set, which is suitable for developing a theory of transfinite ordinals and cardinals. The former concept (Dedekind's) is the one used by mathematicians as tools for developing mathematics, and the second one (Cantor's) is the concept used by professional set theorists. Although ultimately the distinction may not be so clear-cut, historically it seems clear that these two general inclinations have existed. The majority of mathematicians take ZF set theory to have both aspects, but Takeuti *as a proof-theorist* seems to have been interested only in the former aspect of the concept of set and to take a version of higher-order logic to be a paradigmatic theory for giving a (proof-theoretic) foundation of mathematics.

such an illusion arises by a faculty of mind called 'enactive, volitional intuition' (Sect. 3.2.2).

Takeuti then provides concrete examples of the concept of set as the thought-objectification of a proposition. In these cases, unlike pure sets, we consider individuals first and a collection of these individuals introduced via a property of these individuals is considered to be a set. Such a concept of set is adopted in second-order predicate calculus or arithmetic. The principle of comprehension plays a role in objectifying the notion of proposition and constructing a set (also in Sect. 3.2.2).

Takeuti had a strong belief that a logical method of analyzing proofs can provide a clarification of the concept of set as the thought-objectification of a proposition (Sect. 3.2.3). Probably, this belief is partly based on the fact that, in his own GLC, set comprehension principle is formulated as a higher(second)-order logical rule L∀. The concept of set in finite-order predicate calculus is intertwined with 'logic' but involves vicious circularity (impredicativity). He claims that the vicious circularity involved in the concept of set can be disentangled by cut-elimination. This claim does not make sense unless logic and the concept of set are closely related in this particular formulation. Accordingly, the specific ground for Takeuti's critiques of axiomatic set theories appears to include that, although consistent, logic and the construction of set are quite separated in these theories.

Concerning the issue of what kind of method can be used for clarification of the concept of set, Takeuti argues for a proof of cut-elimination. Cut-elimination for GLC is Takeuti's fundamental conjecture itself, but merely showing the existence of a cut-free proof by using a set-theoretic method is not sufficient. The method of the proof needs to be *constructive* in order for it to illuminate the concept of set through an analysis of the concept of proof. He argues that by focusing not on the definition of a concept but on a cut-free proof of a proposition containing a concept with a vicious circle, we can avoid the vicious circle. (In particular, he makes a concrete proposal that a theory of modulations can disentangle vicious circularity contained in applications of higher(second)-order L∀.)

Takeuti's view of the concept of set, which is a combination of Hilbert-Gentzen finitism and impredicativism, is different from platonists' view, from predicativists' view, and from the schematic (or generic) understanding of second-order quantifiers. Takeuti thinks that a vicious circle is inevitable in formulating classical analysis but a problem, and he tries to disentangle the vicious circularity by eliminating cuts. As far as we know, the view that the vicious circularity involved in the concept of set should be disentangled by cut-elimination is Takeuti's original one.

The significance of Takeuti's view may be made more understandable if we take into consideration the possibility that we may know less than we have thought we do concerning dichotomies such as finitistic/non-finitistic and predicative/impredicative. Let us quote a remark of an outstanding contemporary proof theorist.

> Takeuti's work is often discussed in terms of dichotomies such as finitary/ transcendent, predicative/impredicative, constructive/non-constructive. However, in my opinion, the meaning, the significance and the substance of these dichotomies have not been sufficiently examined in the literature and no theory as a decent mathematical one has been formed yet. At least, I do not understand where we should draw distinctions or boundaries or why some sort of

results are meaningful only when they are proven in a finitistic method ([1], p. 334, translated by the current author).

Actually Takeuti himself occasionally uses the traditional distinctions such as predicativity/impredicativity in his discussions of foundational matters, as we have mentioned. But it is true that Takeuti does not entirely rely on such traditional distinctions, as predicativism/impredicativism, axiomatic/naive set theories, or (Hilbert's version of) finitism/non-finitism, when he formulates his own foundational view. In view of the foregoing skeptical opinion on the dichotomies, we cannot but feel that the significance of Takeuti's view may not yet be sufficiently appreciated. Even if his view had only a historical significance, the view should be more closely explored.

Anyways, fully discussing the problem of whether Takeuti's view of the concept of set discussed above can be philosophically defended or not is out of scope of this historical paper whose aim is to present our understanding of Takeuti's view of the concept of set. We leave to another occasion both this problem and discussions on Tekeuti's later view (after 1970) of the concept of set.

Acknowledgements The author is grateful to Walter Dean for helping him edit English translations of quotations from Takeuti's text, Ryota Akiyoshi and the anonymous referee for their helpful comments, but the author is ultimately responsible for any error and unclarity.

References

1. Arai, T.: On Takeuti's fundamental conjecture (in Japanese). Sugaku (Mathematics) **40**(4), 322–337 (1988)
2. Arai, T.: What is and what should be Takeut's fundamental conjecture? (in Japanese). In: Proof Theory and Computational Theory. RIMS Kôkyorôku, vol. 1442, pp. 1–7 (2005)
3. Carnap, R.: Die logizistische Grundlegung der Mathematik. Erkenntnis **2**, 91–105, (1931). English translation in The Philosophy of Mathematics, eds. by P. Benacerraf and H. Putnam, pp. 41–52. Cambridge University Press (1964)
4. Feferman, S.: Systems of predicative analysis. J. Symb. Logic **29**(1), 1–30, 03 (1964)
5. Feferman, S.: Review: Gaisi Takeuti, Proof Theory. Bull. Am. Math. Soc. **83**(3), 351–361 05 (1977)
6. Feferman, S.: In the Light of Logic. Oxford University Press, Oxford (1998)
7. Gentzen, G.: Neue fassung des widespruchsfreiheitsbeweises für die reine Zahlentheorie, Forschungen zur Logik und zur Grundlegung der exacten Wissenschaften, Neue Folge, pp. 19–44 (1938)
8. Girard, J.-Y.: Une extension de l'interprétation de Gödel l'analyse, et son application à l'elimination des coupures dans l'analyse et la théorie des types. In: J.E. Fenstad, (ed.), Proceedings of the Second Scandinavian Logic Symposium, pp. 63–92. North Holland (1971)
9. Gödel, K.: Russell's mathematical logic. In: P. Benacerraf, H. Putnam (ed.), Philosophy of Mathematics, pp. 447–469. Prentice-Hall (1964)
10. Gödel, K.: What is Cantor's continuum problem (1964 version). In P. Benacerraf, H. Putnam, (ed.), Philosophy of Mathematics, pp. 470–485. Prentice-Hall (1964)
11. Kreisel, G.: A survey of proof theory. J. Symb. Log. **33**(3), 321–388 (1968)

12. Kreisel, G.: Two notes on the foundations of settheory. Dialectica **23**(2), 93–114 (1969)
13. Longo, G.: Prototype proofs in type theory. Math. Log. Q. **46**(2), 257–266 (2000)
14. Maraldo, J.C.: Nishida Kitaro. In: E.N. Zalta, (ed.), The Stanford Encyclopedia of Philosophy. Metaphysics Research Lab, Stanford University, Springer (2019)
15. Nishida, K.: The standpoint of enactive intuition (in Japanese) (Kōi-teki chokkan no tachiba). In: Collected Papers in Philosophy I (in Japanese). Iwanami Shoten, Tokyo (1935). English translation, The standpoint of active intuition, in Ontology of Production: 3 essays, trans. by W. Haver, Duke University Press (2012)
16. Nishida, K.: Enactive intuition (in Japanese) (kōi-teki chokkan). In: Collected Papers in Philosophy II (in Japanese). Iwanami Shoten, Tokyo (1937)
17. Nishida, K.: Logic and life (in Japanese) (Ronri to seimei). In: Collected Papers in Philosophy II (in Japanese). Iwanami Shoten, Tokyo (1937). English translation in Place and Dialectic: Two Essays by Nishida Kitaro, trans. by J.W.M. Kummel and S. Nagatomo, O. U. P (2012)
18. Pistone, P.: Polymorphism and the obstinate circularity of second order logic: a victim's tale. Bull. Symb. Logic (1) (2018)
19. Poincaré, H.: Science et Méthode. Flammarion (1908)
20. Prawitz, D.: Hauptsatz for higher order logic. J. Symb. Logic **33**(3), 452–457, 09 (1968)
21. Prawitz, D.: Review of Gaisi Takeuti "Proof Theory," second ed. J. Symb. Logic **56**(3), 1094–1096 (1991)
22. Ramsey, F.P.: The Foundations of Mathematics and Other Logical Essays. Paterson, N.J., Littlefield, Adams (1931)
23. Russell, B.: Mathematical logic as based on the theory of types. Am. J. Math. **30**(3), 222–262 (1908)
24. Suetsuna, J.: Mathematics and the History of Mathematics (in Japanese). Kobundo (1944)
25. Suetsuna, J.: The Foundations of Mathematics (in Japanese). Iwanami Shoten (1952)
26. Tait, W.W.: A nonconstructive proof of Gentzen's Hauptsatz for second order predicate logic. Bull. Am. Math. Soc. (1966)
27. Takahashi, M.: A proof of cut-elimination theorem in simple type-theory. J. Math. Soc. Japan **19**(4), 399–410, 10 (1967)
28. Takeuti, G.: On the foundations of mathematics (in Japanese). Sugaku (Mathematics) **2**(1), 16–32 (1949)
29. Takeuti, G.: On a generalized logical calculus. Japan. J. Math. **23**, 39–96 (1953)
30. Takeuti, G.: Construction of the set theory from the theory of ordinal numbers. J. Math. Soc. Japan **6**(2), 196–220, 09 (1954)
31. Takeuti, G.: On the fundamental conjecture of GLC I. J. Math. Soc. Japan **7**(3), 249–275, 06 (1955)
32. Takeuti, G.: On the fundamental conjecture of GLC II. J. Math. Soc. Japan **7**(4), 394–408 10 (1955)
33. Takeuti, G.: From the standpoint of formalism I (in Japanese). Kagaku-kisoron Kenkyu (Stud. Philos. Sci.) **2**(2), 25–29 (1956)
34. Takeuti, G.: From the standpoint of formalism II (in Japanese). Kagaku-kisoron Kenkyu (Stud. Philos. Sci.) **2**(2), 15–19 (1956)
35. Takeuti, G.: From the standpoint of formalism III (in Japanese). Kagaku-kisoron Kenkyu (Stud. Philos. Sci.) **2**(2), 24–27 (1956)
36. Takeuti, G.: On the fundamental conjecture of GLC III. J. Math. Soc. Japan **8**(1), 54–64 (1956)
37. Takeuti, G.: On the fundamental conjecture of GLC IV. J. Math. Soc. Japan **8**(2), 145–155 (1956)
38. Takeuti, G.: Ordinal diagrams. J. Math. Soc. Japan **9**(4), 386–394, 10 (1957)
39. Takeuti, G.: On the fundamental conjecture of GLC V. J. Math. Soc. Japan **10**(2), 121–134 (1958)
40. Takeuti, G.: On set theory (in Japanese). Kagaku-kisoron Kenkyu (Stud. Philos. Sci.) **5**(3), 1–7 (1963)
41. Takeuti, G.: Consistency proofs of subsystems of classical analysis. Ann. Math. **86**(2), 299–348 (1967)

42. Takeuti, G.: The universe of set theory. In Foundations of Mathematics (Symposium Commmorating Kurt Gödel, Columbus, Ohio, 1966, pp. 74–128 (1969)
43. Takeuti, G.: Introduction to Axiomatic Set Theory. Springer, New York (1971)
44. Takeuti, G.: Recent set theory (in Japanese). Sugaku (Mathematics) **23**(1), 18–26 (1971)
45. Takeuti, G.: On mathematics (in Japanese). Kagaku-kisoron Kenkyu (Stud. Philos. Sci.) 18–22 (1972)
46. Takeuti, G.: Set theory (in Japanese). Sūgaku Seminar (Susemi) – Math. Seminar 4–5 (1972)
47. Takeuti, G.: A Mathematical World-view (Sūgaku-teki Sekai-kan, in Japanese). Kinokuniya-Shoten (1984)
48. Takeuti, G.: On proof theory (in Japanese). Kagaku-kisoron Kenkyu (Stud. Philos. Sci.) **17**(1), 1–5 (1984)
49. Takeuti, G.: Proof theory and set theory. Synthese **62**(2), 255–263 (1985)
50. Takeuti, G.: Like a sunset (in Japanese). Sūgaku Seminar (Susemi) – Math. Seminar 75 (1987)
51. Takeuti, G.: Proof theory. In: Studies in Logic and the Foundations of Mathematics, 2nd edn. North-Holland (1987)
52. Takeuti, G.: My foundational point of view: old and new. In: A Handout Delivered at a Conference in Keio University (1996)
53. Takeuti, G.: Birth of second order proof theory by the fundamental conjecture on GLC (in Japanese). Sūri-kagaku (Mathematical Sciences) translated in Memoire of a proof theorist, special volume, 10–17 (1998)
54. Takeuti, G.: What is a Set? – a New Edition (Shūgō to-ha nani ka?, in Japanese). Kodansha (2001)
55. Takeuti, G., Iwamura, T., Shimauchi, T., Nanba, K., Hirose, K., Maehara, S.: A round table discussion on the foundations of mathematics – past, present, future (in Japanese). Sūgaku Seminar (Susemi) – Mathematics Seminar, pp. 53–64 (1979)
56. Takeuti, G.: Trans. by N. Passell and M. Yasugi, Memoirs of a Proof Theorist: Gödel and Other Logicians (2003)
57. Weyl, H.: Das Kontinuum. Veit & Co., Leipzig (1918)

On Countable Stationary Towers

Yo Matsubara and Toshimichi Usuba

Abstract In this paper, we investigate properties of countable stationary towers. We derive the regularity properties of sets of reals in $L(\mathbf{R})$ from some properties of countable stationary towers without explicit use of strong large cardinals such as Woodin cardinals. We also introduce the notion of semiprecipitousness and investigate its relation to precipitousness and presaturation of countable stationary towers. We show that precipitousness of countable stationary towers of weakly compact height implies the regularity properties of sets of reals in $L(\mathbf{R})$.

Keywords Stationary tower · Regularity properties of sets of reals

1 Introduction

In his expository article (written in Japanese) [8], Gaisi Takeuti called the development of a series of results concerning $L(\mathbf{R})$ and large cardinals, by Martin, Steel, and Woodin during the second half of the 1980s, a revolutionary leap in modern set theory. Long before the 1980s, Takeuti and Solovay pointed out that $L(\mathbf{R})$, the smallest inner model containing every real, is a natural candidate for an inner model of AD (axiom of determinacy) [3]. What Takeuti called a revolutionary leap has confirmed that their intuition was correct.

The first result in this revolutionary leap was proving various regularity properties, such as Lebesgue measurability, the Baire property, and the perfect set property, hold for every set of reals in $L(\mathbf{R})$ from the existence of a supercompact cardinal. Soon after, the large cardinal hypothesis was weakened and the conclusion was strength-

Y. Matsubara (✉)
Graduate School of Informatics, Nagoya University, Furo-cho, Chikusa-ku, Nagoya 464-8601, Japan
e-mail: yom@math.nagoya-u.ac.jp

T. Usuba
Faculty of Fundamental Science and Engineering, Waseda University, Okubo 3-4-1, Shinjyuku, Tokyo 169-8555, Japan
e-mail: usuba@waseda.jp

T. Arai et al. (eds.), *Advances in Mathematical Logic*, Springer Proceedings in Mathematics & Statistics 369, https://doi.org/10.1007/978-981-16-4173-2_6

ened drastically. For convenience, throughout this paper, by regularity properties, we mean Lebesgue measurability, the Baire property, and the perfect set property.

Woodin realized the existence of a certain generic elementary embedding would imply regularity properties of sets of reals in $L(\mathbf{R})$ [9].

Theorem 1 (Woodin) *Let λ be an inaccessible cardinal. Suppose there is a partial order $\mathbb{P}$ such that $\mathbb{P}$ forces the following:*

There is a generic elementary embedding $j : V \to M$ with

1. $\mathrm{crit}(j) = \aleph_1^V$ *and* $j(\aleph_1^V) = \lambda$
2. *M is closed under taking ω sequences in $V^{\mathbb{P}}$*
3. *every real is obtained by a small (size less than λ) forcing, i.e. in $V[G]$, where G is $\mathbb{P}$-generic, for each real x there are a poset $\mathbb{P}_x$ in the ground model V with cardinality less than λ and a $\mathbb{P}_x$-generic $H \in V[G]$ such that $x \in V[H]$.*

Then every set of reals in $L(\mathbf{R})$ satisfies the regularity properties. □

Woodin developed the notion of stationary tower forcing to produce a generic elementary embedding satisfying the hypothesis of the above theorem. Although Woodin's theory of stationary towers is more general, we will concentrate our attention on countable stationary towers. We present a brief sketch of the notion of countable stationary tower forcing. We refer the reader to Larson's book [5] for a more thorough treatment of stationary tower forcing.

Definition 1 Let γ be an uncountable limit ordinal. We denote the set of all $p \in V_\gamma$ such that p is a stationary subset of $[\bigcup p]^\omega$ by $\mathbb{Q}_\gamma$. For $p, q \in \mathbb{Q}_\gamma$, we say $p \leq_{\mathbb{Q}_\gamma} q$ if $\bigcup p \supseteq \bigcup q$ and $\forall x \in p(x \cap \bigcup q \in q)$.

The partial order $\langle \mathbb{Q}_\gamma, \leq_{\mathbb{Q}_\gamma} \rangle$ is known as a countable stationary tower of height γ. Throughout this paper, whenever we mention a stationary tower, we are assuming its height is an uncountable limit ordinal.

If G is a $\mathbb{Q}_\gamma$-generic filter over a ground model V, then we can take the generic ultrapower $\mathrm{Ult}(V; G)$ of V. We refer a reader to Larson's book [5] for the definitions and basic facts concerning generic ultrapowers induced by stationary tower forcing.

In general, there is no guarantee that the generic ultrapower $\mathrm{Ult}(V; G)$ is well-founded.

Definition 2 $\mathbb{Q}_\gamma$ is said to be precipitous if $\mathbb{Q}_\gamma$ forces that $\mathrm{Ult}(V; G)$ is well-founded.

Remark 1 It turns out that the precipitousness of $\mathbb{Q}_\gamma$ is equivalent to "$\mathbb{Q}_\gamma$ forces that $\mathrm{Ult}(V_\delta; G)$ is well-founded where $\delta = (|V_\gamma|^+)^V$". This motivated the next definition:

Definition 3 $\mathbb{Q}_\gamma$ is said to be semiprecipitous if $\mathbb{Q}_\gamma$ forces that $\mathrm{Ult}(V_\gamma; G)$ is well-founded.

If $\mathrm{Ult}(V; G)$ is well-founded, then, by taking the transitive collapse of $\mathrm{Ult}(V; G)$, we can define a generic elementary embedding $j : V \to M$ by $j(x) = [c_x]$, where c_x is the constant function whose domain is $[\aleph_1]^\omega$ and that takes the constant value x.

Woodin proved that if λ is a large cardinal, such as a supercompact cardinal, then $\mathbb{Q}_\lambda$ is precipitous. Actually, he derived a somewhat stronger property than precipitousness from the large cardinal in his proof.

Definition 4 $\mathbb{Q}_\gamma$ is said to be presaturated if $\mathbb{Q}_\gamma$ forces that $\mathrm{Ult}(V; G)$ is closed under taking ω sequences in generic extensions.

Remark 2 It is easy to see that presaturation of $\mathbb{Q}_\gamma$ implies precipitousness of $\mathbb{Q}_\gamma$. It turns out that the presaturation of $\mathbb{Q}_\gamma$ is equivalent to "$\mathbb{Q}_\gamma$ forces $\mathrm{cf}(\gamma) > \omega$". Furthermore, the presaturation of $\mathbb{Q}_\gamma$ implies that $\mathbb{Q}_\gamma$ forces $j(\aleph_1^V) = \gamma$, where j is the corresponding generic elementary embedding.

Woodin proved that if λ is a supercompact cardinal, then $\mathbb{Q}_\lambda$ is presaturated and satisfies the hypotheses of Theorem 1. Therefore, the existence of a supercompact cardinal implies that every set of reals in $L(\mathbf{R})$ satisfies the regularity properties.

Woodin later proved that if λ is a Woodin cardinal which is substantially weaker than supercompact, then $\mathbb{Q}_\lambda$ is presaturated. A generic elementary embedding generated by $\mathbb{Q}_\lambda$ with Woodin λ always satisfies properties 1 and 2 of the hypotheses of Theorem 1. But property 3 can fail. It is known that the existence of one Woodin cardinal, in fact even ω many Woodin cardinals, could not imply that every set of reals in $L(\mathbf{R})$ satisfies the regularity properties. In particular Steel [7] proved that if the existence of infinitely many Woodin cardinals is consistent, then it is consistent with the axiom of choice holding in $L(\mathbf{R})$. It turns out that if there are infinitely many Woodin cardinals and a measurable cardinal above them, then every set of reals in $L(\mathbf{R})$ satisfies the regularity properties.

In this paper, we investigate the following questions:

1. Can we derive the regularity properties of sets of reals in $L(\mathbf{R})$ from properties of countable stationary towers, such as precipitousness or presaturation, without explicit use of strong large cardinals such as supercompact cardinals or Woodin cardinals?
2. Are there any relations between presaturation, precipitousness, and semiprecipitousness other than obvious implications?

2 Results

Burke used the following game to characterize the precipitousness of a stationary tower [1].

For an uncountable limit ordinal γ, let Γ_γ denote the following two player game of length ω: At each inning, players I and II choose conditions $p_n, q_n \in \mathbb{Q}_\gamma$ alternately with $p_0 \geq q_0 \geq p_1 \geq q_1 \geq \cdots$. For the play $\langle p_n, q_n \mid n < \omega\rangle$, player II wins if there is some $x \in [V_\gamma]^\omega$ such that $x \cap \bigcup p_n \in p_n$ for every $n < \omega$. Otherwise I wins. The following theorem is due to Burke [1]:

Theorem 2 *Let γ be an uncountable limit ordinal. Then the following are equivalent:*

1. *$\mathbb{Q}_\gamma$ is precipitous.*
2. *Player I does not have a winning strategy in the game Γ_γ.* □

For a certain γ, semiprecipitousness of $\mathbb{Q}_\gamma$ is equivalent to precipitousness of $\mathbb{Q}_\gamma$.

Proposition 1 *Let λ be a singular cardinal such that $|V_\lambda| = \lambda$ and $\omega < \mathrm{cf}(\lambda) < \lambda$. If $\mathbb{Q}_\lambda$ is semiprecipitous, i.e. $\Vdash_{\mathbb{Q}_\lambda}$ "$\mathrm{Ult}(V_\lambda; \dot{G})$ is well-founded", then $\mathbb{Q}_\lambda$ is precipitous.*

Proof Suppose $\mathbb{Q}_\lambda$ is not precipitous. Then player I has a winning strategy σ in Γ_λ.

Let T be the set of all finite sequences $\langle\langle q_i, x_i\rangle \mid i \leq n\rangle$ such that:

1. $q_i \in \mathbb{Q}_\lambda$ and $x_i \in q_i$.
2. $x_{i+1} \cap \bigcup q_i = x_i$ for every $i < n$.
3. $\sigma(\emptyset) \geq q_0 \geq \sigma(q_0) \geq q_1 \geq \cdots \geq \sigma(q_0, \ldots, q_{n-1}) \geq q_n$.

For $s, t \in T$, we define $\leq_T$ by $s \leq_T t$ if s is an extension of t. Since σ is a winning strategy for player I, T is well-founded, i.e., there is no infinite descending sequence in T.

We note that $T \subseteq V_\lambda$. Consider the structure $\langle V_\lambda, \in, T\rangle$, where we identify T as a unary predicate. Since $|V_\lambda| = \lambda$, we can check the following:

1. $\langle V_\lambda, \in, T\rangle \models$"$T$ is well-founded".
2. Since $T \cap V_\alpha \in V_\lambda$ for every $\alpha < \lambda$, we have $\langle V_\lambda, \in, T\rangle \models$"$\forall\alpha\exists x(x = T \cap V_\alpha)$".
3. $\langle V_\lambda, \in, T\rangle \models$"$\forall\alpha$, $T \cap V_\alpha$ is well-founded, and the rank function $f : T \cap V_\alpha \to \beta$ exists for some ordinal β".

Take a $\mathbb{Q}_\lambda$-generic G with $\sigma(\emptyset) \in G$. In $V[G]$, we can take the ultrapower of the structure $\langle V_\lambda, \in, T\rangle$ by G. Since $\mathrm{Ult}(V_\lambda; G)$ is well-founded, the ultrapower of $\langle V_\lambda, \in, T\rangle$ is also well-founded. Let $M = \langle M, \in, j(T)\rangle$ be the transitive collapse of the ultrapower, and $j : V_\lambda \to M$ be the elementary embedding. Then, because $\omega < \mathrm{cf}(\lambda) < \lambda$, we have that $j``\lambda$ is bounded in $M \cap \mathrm{ON}$. Let $\gamma = \sup(j``\lambda) \in M \cap \mathrm{ON}$. Consider $T^* = j(T) \cap M_\gamma$. By (2) above, we have $T^* \in M$, and by (3), there exists the rank function $f : T^* \to \beta$ in M. Hence, T^* is also well-founded in $V[G]$, and there is no infinite descending sequence of T^* in $V[G]$.

By induction on $n < \omega$, we define a descending sequence $\langle p_n, q_n \mid n < \omega\rangle$ as follows: First, let $p_0 = \sigma(\emptyset) \in G$. Then $\{\sigma(q) \mid q \leq p_0\}$ is dense below p_0, so we can take $q_0 \leq p_0$ and $p_1 = \sigma(q_0) \leq q_0$ with $p_1 \in G$. We have $q_0 \in G$. Again, since $\{\sigma(q_0, q) \mid q \leq p_1\}$ is dense below p_1, we can choose $q_1 \leq p_1$ and $p_2 = \sigma(q_0, q_1) \in G$, and so on. We know $j(p_n) = j(\sigma)(j(q_0), \ldots, j(q_{n-1})) \geq j(q_n)$. For $n < \omega$, let $x_n = \bigcup q_n$. We know $j``x_n \in j(q_n)$, and $j``x_{n+1} \cap \bigcup(j(q_n)) = j``x_n$. Because $j(q_n), j``x_n \in M_\gamma$, we have that $\langle\langle j(q_i), j``x_i\rangle \mid i \leq n\rangle \in T^*$ for every $n < \omega$. Therefore, the sequence $\langle\langle j(q_n), j``x_n\rangle \mid n < \omega\rangle$ induces an infinite descending sequence in T^*. This is a contradiction. □

Remark 3 The reader might wonder if the hypothesis of the last proposition is vacuous. In our next paper [6], we prove the consistency of precipitousness of $\mathbb{Q}_\lambda$ where λ satisfies the conditions of Proposition 1 assuming a sufficiently strong large cardinal hypothesis.

Later in this paper, we will show that, in general, semiprecipitousness of $\mathbb{Q}_\lambda$ does not imply precipitousness.

We can also use Burke's game to prove the following proposition:

Proposition 2 *Suppose α and β are uncountable limit ordinals with $\alpha < \beta$. If $\mathbb{Q}_\alpha$ is precipitous, then, for each $p \in \mathbb{Q}_\alpha$, there is some $p^* \in \mathbb{Q}_\beta$ such that $p^* \leq p$ and $p^* \Vdash_{\mathbb{Q}_\beta}$ "$\dot{G} \cap \mathbb{Q}_\alpha$ is $\mathbb{Q}_\alpha$-generic" where $\dot{G}$ is the canonical $\mathbb{Q}_\beta$-name for a generic filter.*

This proposition is an immediate consequence of the next lemma. The referee informed us that this lemma is a "tower" analogue of Proposition 2.4 of [4].

Lemma 1 *Let α be an uncountable limit ordinal, and suppose $\mathbb{Q}_\alpha$ is precipitous. Then, for every $p \in \mathbb{Q}_\alpha$, the set $\{x \in [V_{\alpha+1}]^\omega \mid x \cap \bigcup p \in p, \forall D \in x$ (if $D \subseteq \mathbb{Q}_\alpha$ is dense, then $\exists r \in D \cap x\ (x \cap \bigcup r \in r))\}$ is stationary in $[V_{\alpha+1}]^\omega$.*

Proof Take $p \in \mathbb{Q}_\alpha$. Fix a well-ordering Δ on $V_{\alpha+1}$ and a function $f : [V_{\alpha+1}]^{<\omega} \to V_{\alpha+1}$. For a set $x \subseteq V_{\alpha+1}$, let $Sk(x)$ denote the Skolem hull of x under the structure $\langle V_{\alpha+1}, \in, \Delta, f\rangle$. We will find a set x such that $Sk(x)$ belongs to the required set.

Now we describe a strategy for player I in Γ_α. Fix a surjection $\pi : \omega \to \omega \times \omega$ such that if $\pi(n) = \langle i, j\rangle$ then $i \leq n$. For a countable $x \subseteq V_{\alpha+1}$, we also fix an enumeration $\langle D^x_j \mid j < \omega\rangle$ of all dense subsets of $\mathbb{Q}_\alpha$ that belong to $Sk(x)$.

First, let $p'_0 = p$. Take $j < \omega$ with $\pi(0) = \langle 0, j\rangle$. Since p'_0 is stationary in $[\bigcup p'_0]^\omega$, there is some dense set $D_0 \subseteq \mathbb{Q}_\alpha$ with $\{x \in p'_0 \mid D_0 = D^x_j\}$ stationary in $[\bigcup p'_0]^\omega$, so $\{x \in p'_0 \mid D_0 = D^x_j\}$ is a condition of $\mathbb{Q}_\alpha$. Take $p''_0 \in D_0$ with $p''_0 \leq \{x \in p'_0 \mid D_0 = D^x_j\}$. Then take $p_0 \leq p''_0$ such that $p''_0 \in x$ and $Sk(x) \cap \bigcup p_0 = x$ for every $x \in p_0$. This p_0 is the first move of player I.

Let $n > \omega$, and suppose $p \geq p'_0 \geq p''_0 \geq p_0 \geq q_0 \geq \cdots \geq q_{n-1}$ were chosen so that, for every $k < n$:

1. If $\pi(k) = \langle i, j\rangle$, then, for every $x \in p''_k$, we have $D^{x \cap \bigcup p'_i}_j = D_k$.
2. $p''_k \in D_k$.
3. $Sk(x) \cap \bigcup p_k = x$ and $p''_k \in x$ for every $x \in p_k$.

As in the $n = 0$ case, we can choose $p_n \leq p''_n \leq p'_n \leq q_{n-1}$ as required, and player I chooses p_n as his move. This completes the description of our strategy for player I.

Since $\mathbb{Q}_\alpha$ is precipitous, the strategy above is not a winning strategy for player I. Thus, we can find a descending sequence $p \geq p'_0 \geq p''_0 \geq p_0 \geq q_0 \geq \cdots \geq q_{n-1} \geq p'_n \geq \cdots$ such that each p_n, p'_n, p''_n is chosen according to the above strategy, but there is some $x \subseteq V_\alpha$ such that $x \cap \bigcup p_n \in p_n$ for every $n < \omega$. We may assume that $x = \bigcup_{n<\omega}(x \cap \bigcup p_n)$. Since $p'_n \geq p''_n \geq p_n \geq p'_{n+1}$, we also have $x \cap \bigcup p'_n \in p'_n$, $x \cap \bigcup p''_n \in p''_n$, and $x = \bigcup_{n<\omega}(x \cap \bigcup p'_n) = \bigcup_{n<\omega}(x \cap \bigcup p''_n)$. Hence $Sk(x) = \bigcup_{n<\omega} Sk(x \cap \bigcup p'_n)$. For $n < m$, we have $Sk(x \cap \bigcup p_m) \cap \bigcup p''_n = (x \cap \bigcup p_m) \cap \bigcup p''_n = x \cap \bigcup p''_n$, thus $Sk(x) \cap \bigcup p''_n = x \cap \bigcup p''_n$ for every $n < \omega$. We show that, if $D \in Sk(x)$ is dense in $\mathbb{Q}_\alpha$, then there is $r \in D \cap Sk(x)$ with $Sk(x) \cap \bigcup r \in r$, which completes our proof. Take a dense set $D \in Sk(x)$. Since $Sk(x) = \bigcup_{n<\omega} Sk(x \cap \bigcup p'_n)$, there is some $i < \omega$ with $D \in Sk(x \cap \bigcup p'_i)$. Take $j < \omega$ with

$D = D_j^{x\cap\bigcup p_i'}$, and pick $n < \omega$ with $\pi(n) = \langle i, j\rangle$. Then, since $x \cap \bigcup p_n'' \in p_n''$, we have that $p_n'' \in D_n = D_j^{x\cap\bigcup p_n'} = D$. $p_n'' \in x \cap \bigcup p_n \subseteq x$, so $p_n'' \in D \cap Sk(x)$ and $Sk(x) \cap \bigcup p_n'' = x \cap \bigcup p_n'' \in p_n''$. □

Using the above proposition we can obtain the next result concerning Question 1.

Proposition 3 *Suppose λ is an inaccessible cardinal. If $\mathbb{Q}_\lambda$ is presaturated and $\{\alpha < \lambda \mid \mathbb{Q}_\alpha$ is precipitous$\}$ is cofinal in λ, then every set of reals in $L(\mathbf{R})$ satisfies the regularity properties.*

Proof For each $\gamma < \lambda$, we let E_γ denote the set of all of $p \in \mathbb{Q}_\lambda$ such that p forces "$\exists\alpha(\gamma < \alpha < \lambda \wedge G \cap \mathbb{Q}_\alpha$ is $\mathbb{Q}_\alpha$-generic) where G is $\mathbb{Q}_\lambda$-generic". We note that Proposition 2 implies that, for each $\gamma < \lambda$, the set E_γ is dense in $\mathbb{Q}_\lambda$. Let G be a $\mathbb{Q}_\lambda$-generic filter over V. We work in $V[G]$. From the denseness of E_γ for all $\gamma < \lambda$, it is easy to see that $\{\alpha < \lambda \mid G \cap \mathbb{Q}_\alpha$ is $\mathbb{Q}_\alpha$-generic$\}$ is cofinal.

If $j : V \to M$ is the corresponding generic elementary embedding, then conditions (1) and (2) of Woodin's theorem are satisfied since $\mathbb{Q}_\lambda$ is presaturated. Now we check that condition (3) is also satisfied.

Suppose r is a real in $V[G]$. By presaturation of $\mathbb{Q}_\lambda$ we know that r belongs to M, the generic ultrapower. Hence there is some function f in the ground model with $\mathrm{dom}(f) \in G$ such that $[f] = r$. Choose large enough $\alpha < \lambda$ such that $\mathrm{dom}(f) \in \mathbb{Q}_\alpha$ and $G \cap \mathbb{Q}_\alpha$ is $\mathbb{Q}_\alpha$-generic. This would imply that $r \in V[G \cap \mathbb{Q}_\alpha]$. Hence the conditions of Woodin's theorem are satisfied by our generic elementary embedding. □

The stationarity of $\{\alpha < \lambda \mid \mathbb{Q}_\alpha$ is precipitous$\}$ has some interesting consequences.

Lemma 2 *Suppose λ is inaccessible. If the set $\{\alpha < \lambda \mid \mathbb{Q}_\alpha$ is precipitous$\}$ is stationary in λ, then $\mathbb{Q}_\lambda$ is presaturated.*

Proof Let $p \in \mathbb{Q}_\lambda$, and take $\mathbb{Q}_\lambda$-names $\dot{f}_n$ $(n < \omega)$ such that $p \Vdash_{\mathbb{Q}_\lambda}$"$\dot{f}_n \in V$ is a function whose domain is some element of $\mathbb{Q}_\lambda$". For each $n < \omega$, take a maximal antichain $A_n \subseteq \mathbb{Q}_\lambda$ such that, for each $q \in A_n$, either $q \perp p$ or $q \leq p$, and, if $q \leq p$, then there is some function f_n^q with $q \Vdash_{\mathbb{Q}_\lambda}$"$\dot{f}_n = f_n^q$". Since the set $\{\alpha < \lambda \mid \mathbb{Q}_\alpha$ is precipitous$\}$ is stationary in λ, there is $\alpha < \lambda$ such that $\mathbb{Q}_\alpha$ is precipitous, and, for every $n < \omega$, $A_n \cap \mathbb{Q}_\alpha$ is predense in $\mathbb{Q}_\alpha$, and $\mathrm{dom}(f_n^q) \in V_\alpha$ for every $q \in A_n \cap V_\alpha$. Let $D_n = \{r \in \mathbb{Q}_\alpha \mid \exists q \in A_n \cap \mathbb{Q}_\alpha\ (r \leq q)\}$. D_n is dense in $\mathbb{Q}_\alpha$.

By the previous lemma, the set $p' = \{x \in [V_{\alpha+1}]^\omega \mid x \cap \bigcup p \in p, \forall n < \omega \exists r \in D_n \cap x\ (x \cap \bigcup r \in r)\}$ is stationary in $[V_{\alpha+1}]^\omega$. We know $p' \leq p$. For $x \in p'$ and $n < \omega$, take $r_n^x \in D_n \cap x$ with $x \cap \bigcup r_n^x \in r_n^x$, and we may assume that there is a unique $q_n^x \in A_n \cap x$ with $r_n^x \leq q_n^x$. Then define $g : [V_{\alpha+1}]^\omega \to V$ by $g(x) = \{f_n^{q_n^x}(x \cap \bigcup q_n^x) \mid n < \omega\}$. It is routine to check that $p' \Vdash_{\mathbb{Q}_\lambda}$"$[g]$ represents the set $\{[\dot{f}_n] \mid n < \omega\}$". □

We use the last lemma to obtain some consequences of precipitousness of $\mathbb{Q}_\lambda$ where λ is a weakly compact cardinal.

Proposition 4 *Suppose λ is a weakly compact cardinal. If $\mathbb{Q}_\lambda$ is precipitous, then $\{\alpha < \lambda \mid \mathbb{Q}_\alpha$ is precipitous$\}$ is stationary in λ. Hence , under our hypothesis, $\mathbb{Q}_\lambda$ is presaturated.*

Proof The statement that "player I does not have a winning strategy for Burke's game Γ_λ" can be described by some Π^1_1-statement φ over $\langle V_\lambda, \in\rangle$. Hence, if $\mathbb{Q}_\lambda$ is precipitous, then $V_\lambda \models \varphi$, and, since λ is weakly compact, we have that $\{\alpha < \lambda \mid V_\alpha \models \varphi\}$ is stationary in λ. Therefore the set $\{\alpha < \lambda \mid \mathbb{Q}_\alpha$ is precipitous$\}$ is stationary in λ. □

By combining Propositions 3 and 4 we obtain the next result:

Corollary 1 *If $\mathbb{Q}_\lambda$ is precipitous for a weakly compact cardinal λ, then every set of reals in $L(\mathbf{R})$ satisfies the regularity properties.*

We note that the least Woodin cardinal is not a weakly compact cardinal. A weaker large cardinal could also reflect semistationarity.

Proposition 5 *If $\mathbb{Q}_\lambda$ is semiprecipitous for a Mahlo cardinal λ, then the set $\{\alpha < \lambda \mid \mathbb{Q}_\alpha$ is semiprecipitous where α is regular $\}$ is stationary in λ.*

Before we present the proof of this proposition, we state its corollary:

Corollary 2 *The semiprecipitousness of $\mathbb{Q}_\gamma$, in general, cannot imply the precipitousness of $\mathbb{Q}_\gamma$.*

Proof Suppose otherwise. Assume that λ is a Woodin cardinal. By Woodin's result, we know that $\mathbb{Q}_\lambda$ is presaturated.

Since a Woodin cardinal is Mahlo, by Proposition 5, $\{\alpha < \lambda \mid \mathbb{Q}_\alpha$ is semiprecipitous $\}$ is stationary. So, by our assumption, $\{\alpha < \lambda \mid \mathbb{Q}_\alpha$ is precipitous $\}$ is stationary. Then, by Proposition 3, every set of reals in $L(\mathbf{R})$ satisfies the regularity properties. But the existence of one Woodin cardinal cannot imply this conclusion. □

To prove Proposition 5, we use the following variant of Burke's game Γ_γ to characterize the precipitousness of $\mathbb{Q}_\gamma$.

Let Γ^*_γ denote the following two player game of length ω: At the nth inning of the game player I plays $\langle p_n, f_n\rangle$ with $p_n \in \mathbb{Q}_\gamma$, $f_n : p_n \to \mathrm{ON}$, and player II plays $q_n \in \mathbb{Q}_\gamma$ alternately with the requirements: $p_0 \geq q_0 \geq p_1 \geq q_1 \geq \cdots$ and $\forall x \in p_{n+1}(f_{n+1}(x) \in f_n(x \cap \bigcup p_n))$. Player I wins if and only if the game does not stop. By the proof of Burke's theorem, we see that $\mathbb{Q}_\gamma$ is precipitous if and only if player I does not have a winning strategy. This variant of Burke's game was inspired by a similar game due to Goldring [2] characterizing precipitousness of ideals over $\mathcal{P}_\kappa\lambda$. Note that one advantage of the game Γ^*_γ over Γ_γ is that, if player I does not have a winning strategy in Γ^*_γ, then, by the theorem of Gale and Stewart, player II does have a winning strategy.

Proof (*of Proposition* 5) Suppose that $\mathbb{Q}_\lambda$ is semiprecipitous for a Mahlo cardinal λ. For an uncountable limit ordinal γ, we consider the following game $\Gamma_\gamma^{\text{local}}$ which is a localized version of Γ_γ^*.

The game $\Gamma_\gamma^{\text{local}}$ is almost identical to Γ_γ^* with the added requirement that the function f_n, played by player I, must take its values in γ. It is easy to see that player I does not have a winning strategy in $\Gamma_\gamma^{\text{local}}$ is equivalent to $\mathbb{Q}_\gamma$ is semiprecipitous. Since player I does not have a winning strategy for the game $\Gamma_\lambda^{\text{local}}$, player II does have a winning strategy τ for that game. From the Mahloness of λ, by letting player II play the strategy τ, it is not too difficult to see that the set $\{\gamma < \lambda \mid \tau \text{ is a winning strategy for player II in } \Gamma_\gamma^{\text{local}} \text{ with } \gamma \text{ regular}\}$ is stationary in λ. □

It turns out that a large cardinal which is weaker than Mahlo suffices to reflect semiprecipitousness to points of cofinality ω.

Proposition 6 *If $\mathbb{Q}_\lambda$ is semiprecipitous for an inaccessible cardinal λ, then there is a club C in λ such that, for every $\gamma \in C$, if $\mathrm{cf}(\gamma) = \omega$ then $\mathbb{Q}_\gamma$ is semiprecipitous.*

Proof Since $\mathbb{Q}_\lambda$ is semiprecipitous, player II in $\Gamma_\lambda^{\text{local}}$ has a winning strategy τ.

Take a sufficiently large regular cardinal, and take an elementary submodel $M \prec H_\theta$ containing all relevant objects satisfying $M \cap \lambda \in \lambda$ and $\mathrm{cf}(M \cap \lambda) = \omega$. Let $\gamma = M \cap \lambda$. We shall show that player II in $\Gamma_\gamma^{\text{local}}$ has a winning strategy. Let us describe player II's strategy.

Suppose player I plays $\langle p_0, f_0 \rangle$. Note that $p_0 \in \mathbb{Q}_\gamma \subseteq M$ but f_0 may not be an element of M, since $\mathrm{range}(f_0)$ could be cofinal in γ. However, since $\mathrm{cf}(\gamma) = \omega$, there is $\gamma' < \gamma$ such that $\{x \in p_0 \mid f(x) \leq \gamma'\}$ is stationary. Let $\gamma*$ be the least ordinal such that $p_0' = \{x \in p_0 \mid f(x) \leq \gamma*\} \in \mathbb{Q}_\gamma$. Then we have $\langle p_0', f_0 \restriction p_0' \rangle \in M$, and we can identify $\langle p_0', f_0 \restriction p_0' \rangle$ as player I's play in $\Gamma_\lambda^{\text{local}}$. Let $q_0 = \tau(\langle p_0', f_0 \restriction p_0' \rangle) \in M \cap \mathbb{Q}_\lambda = \mathbb{Q}_\gamma$ be player II's response. Note that $q_0 \leq p_0' \leq p_0$ in $\mathbb{Q}_\gamma$. If $\langle p_1, f_1 \rangle$ is a response of player I in $\Gamma_\gamma^{\text{local}}$, by repeating the above procedure, we can find $p_1' \subseteq p_1$ with $\mathrm{range}(f_1 \restriction p_1')$ bounded in γ. Since $\bigcup p_0 = \bigcup p_0'$, for every $x \in p_1'$, we have $f_1(x) \in f_0(x \cap \bigcup p_0) = f_0(x \cap \bigcup p_0')$. Hence $\langle p_1', f_1 \restriction p_1' \rangle, q_0, \langle p_0', f_0 \restriction p_0' \rangle$ satisfy the requirements of a play in $\Gamma_\lambda^{\text{local}}$. Since $\langle p_0', f_0 \restriction p_0' \rangle, \langle p_1', f_1 \restriction p_1' \rangle \in M$, we let player II play $q_1 = \tau(\langle p_0', f_0 \restriction p_0' \rangle, \langle p_1', f_1 \restriction p_1' \rangle) \in M$, and so on. Since τ is a winning strategy of player II in $\Gamma_\lambda^{\text{local}}$, it is easy to see that the strategy above is a winning strategy of player II in $\Gamma_\gamma^{\text{local}}$. □

Remark 4 We note that, by the proof of Corollary 2, the last proposition indicates that the hypothesis "$\omega < \mathrm{cf}(\lambda)$" of Proposition 1 cannot be dropped.

In the late 1980s, the first author was in Los Angeles and received a letter from Gaisi Takeuti. He was inquiring about results concerning $L(\mathbf{R})$ and large cardinals. The first author still remembers the letter that professed Takeuti's passion and enthusiasm for set theory and logic. We regret that we did not have an opportunity to ask Professor Takeuti's view on the revolutionary leap in person.

Acknowledgements The first author's research was supported by JSPS KAKENHI Grant Nos. 18K03394. The second author's research was supported by JSPS KAKENHI Grant Nos. 18K03403 and 18K03404.

References

1. Burke, D.R.: Precipitous towers of normal filters. J. Symb. Logic **62**(3), 741–754 (1997)
2. Goldring, N.: Projecting precipitousness. Israel J. Math. **74**(1), 13–31 (1991)
3. Kanamori, A.: The higher infinite. In: Large Cardinals in Set Theory from their Beginnings. Perspectives in Mathematical Logic. Springer, Berlin (1994)
4. Ketchersid, R., Larson, P., Zapletal, J.: Increasing δ_2^1 and Namba-style forcing. J. Symb. Logic **72**(4), 1372–1378 (2007)
5. Larson, P.: The Stationary Tower: Notes on a Course by W. Hugh Woodin, University Lecture Series, vol. 32 (American Mathematical Society). Oxford University Press, Oxford (2004)
6. Matsubara, Y., Usuba, T.: Reflection principle and stationary towers. In: Preparation
7. Steel, J.R.: Inner models with infinitely many Woodin cardinals. Ann. Pure Appl. Logic **65**(2), 185–209 (1993)
8. Takeuti, G.: Gendai Shugoron Nyumon (Introduction to Modern Set Theory), revised Nippon-hyoronsha, Tokyo (1989)
9. Woodin, W.H.: Σ_1^2 absoluteness. Circulated Notes (1985)

Reforming Takeuti's Quantum Set Theory to Satisfy de Morgan's Laws

Masanao Ozawa

Abstract In 1981, Takeuti introduced set theory based on quantum logic by constructing a model analogous to Boolean-valued models for Boolean logic. He defined the quantum logical truth value for every sentence of set theory. He showed that equality axioms do not hold, while axioms of ZFC set theory hold if appropriately modified with the notion of commutators. Here, we consider the problem in Takeuti's quantum set theory that De Morgan's laws do not hold for bounded quantifiers. We construct a counter-example to De Morgan's laws for bounded quantifiers in Takeuti's quantum set theory. We redefine the truth value for the membership relation and bounded existential quantification to ensure that De Morgan's laws hold. Then, we show that the truth value of every theorem of ZFC set theory is lower bounded by the commutator of constants therein as quantum transfer principle.

Keywords Quantum logic · Commutators · Quantum set theory · De Morgan's laws · Transfer principle

1 Introduction

Since quantum logic is an intrinsic logic governing observational propositions of quantum mechanics, it is an intriguing problem to develop mathematics based on quantum logic. In 1981, Takeuti [15] introduced quantum set theory for this purpose. As a start, he constructed a model of set theory based on quantum logic represented by the complete orthomodular lattice of projections on a Hilbert space, which is isomorphic to the lattice of closed subspaces in the Hilbert space. He defined the truth values for all sentences of set theory on the model assuming the Sasaki arrow for implication. In order to make quantum counter part of ZFC axioms, he introduced

M. Ozawa (✉)
Graduate School of Informatics, Nagoya University, Chikusa-ku, Nagoya 464-8601, Japan

Centre for Mathematical Science and Artificial Intelligence and Chubu University Academy of Emerging Sciences, 1200 Matsumoto-cho, Kasugai 487-8501, Japan
e-mail: ozawa@is.nagoya-u.ac.jp; ozawa@isc.chubu.ac.jp

T. Arai et al. (eds.), *Advances in Mathematical Logic*, Springer Proceedings in Mathematics & Statistics 369, https://doi.org/10.1007/978-981-16-4173-2_7

the notion of commutator in quantum logic, and he showed that the axioms of ZFC hold in quantum set theory if appropriately modified by commutators of elements of the model, while equality axioms do not generally hold in quantum set theory. He showed that the real numbers in the model correspond to the observables of the system to be described.

Following Takeuti's work, we explored the question how theorems of ZFC hold in quantum set theory [9]. We showed that every theorem of ZFC holds in quantum set theory with truth value greater than or equal to the commutator of elements of the model appearing therein. This result was extended to general complete orthomodular lattices and to a general class of operations for implication in Ref. [12]. Quantum set theory was effectively applied to quantum mechanics to extend the probabilistic interpretation from observational propositions to relations between observables [10, 11].

In this paper, we consider the problem in Takeuti's quantum set theory that De Morgan's laws do not hold for bounded quantifiers. Let $\mathcal{H}$ be a Hilbert space. The quantum logic $\mathcal{Q}$ on $\mathcal{H}$ is represented by the lattice of projections on $\mathcal{H}$, which is a complete orthomodular lattice, called the quantum logic on $\mathcal{H}$. The classical definition of implication, $P \to Q = P^{\perp} \vee Q$, does not work since the relation $P \to Q = 1$ and the order relation $P \le Q$ are not equivalent, so that the implication in quantum logic is, according to the majority view [16], defined as the Sasaki arrow $\to$, a binary operation of $\mathcal{Q}$ defined by $P \to Q = P^{\perp} \vee (P \wedge Q)$.

Takeuti [15], applying the method of Boolean-valued models to quantum logic $\mathcal{Q}$, constructed the model $V^{(\mathcal{Q})}$ of quantum set theory. He defined the $\mathcal{Q}$-valued truth value $[\![\phi]\!]$ of a sentence ϕ in the language of set theory.

In particular, the truth values of bounded quantifications are directly defined as follows.

(1) $[\![(\forall x \in u)\, \phi(x)]\!] = \bigwedge_{u' \in \mathrm{dom}(u)} (u(u') \to [\![\phi(u')]\!])$.
(2) $[\![(\exists x \in u)\, \phi(x)]\!] = \bigvee_{u' \in \mathrm{dom}(u)} (u(u') \wedge [\![\phi(u')]\!])$.

Takeuti noted "In Boolean valued universes, $[\![(\forall x \in u)\phi(x)]\!] = [\![\forall x(x \in u \to \phi(x)]\!]$ and $[\![(\exists x \in u)\phi(x)]\!] = [\![\exists x(x \in u \wedge \phi(x)]\!]$. But this is not the case for $V^{(\mathcal{Q})}$" [15, p. 315]. However, it is problematic that he avoids the classical definition of implication $P \to Q = P^{\perp} \vee Q$ in the bounded universal quantification, whereas he still uses the classical definition of conjunction in the bounded existential quantification. Since the relation $P \wedge Q = (P \to Q^{\perp})^{\perp}$ does not hold for the conjunction $\wedge$ and the Sasaki arrow $\to$, so that De Morgan's laws,

(3) $[\![\neg(\forall x \in u)\, \phi(x)]\!] = [\![(\exists x \in u)\, \neg\phi(x)]\!]$,
(4) $[\![\neg(\exists x \in u)\, \phi(x)]\!] = [\![(\forall x \in u)\, \neg\phi(x)]\!]$,

do not hold. In fact, if $\mathcal{Q}$ is not a Boolean algebra, we can construct a predicate $\phi(x)$ such that $[\![(\exists x \in u)\, \neg\phi(x)]\!] = 0$ but $[\![\neg(\forall x \in u)\, \phi(x)]\!] > 0$.

In this paper, we introduce a new binary operation $*$ by $P * Q = (P \to Q^{\perp})^{\perp}$ and redefine the truth values of membership relation and bounded existential quantification as follows.

(5) $[\![u \in v]\!] = \bigvee_{v' \in \mathrm{dom}(v)} (v(v') * [\![v' = u]\!])$.
(6) $[\![(\exists x \in u)\, \phi(x)]\!] = \bigvee_{u' \in \mathrm{dom}(u)} (u(u') * [\![\phi(u')]\!])$.

Then, De Morgan's laws hold for bounded universal quantification and bounded existential quantification. Thus, for the language of quantum set theory we can assume only negation, conjunction, and bounded and unbounded universal quantification as primitive, while disjunction, bounded and unbounded existential quantification are considered to be introduced by definition.

The operator $*$ was found by Sasaki [14], and has been studied as the Sasaki projection in connection with residuation theory, whereas the operation $*$ has not been used for defining bounded quantifiers in quantum logic. Its intuitive meaning and significance will be discussed elsewhere.

We consider the commutator $\underline{\vee}(u_1, \ldots, u_n) \in \mathscr{Q}$ of elements $u_1, \ldots, u_n$ of $V^{(\mathscr{Q})}$ in order to explore how theorems of ZFC hold in the new interpretation for the model $V^{(\mathscr{Q})}$. Then the following quantum transfer principle holds: If a Δ_0-formula $\phi(x_1, \ldots, x_n)$ of the language $\mathscr{L}(\in)$ of set theory is provable in ZFC, for every $u_1, \ldots, u_n \in V^{(\mathscr{Q})}$ we have

$$[\![\phi(u_1, \ldots, u_n)]\!] \geq \underline{\vee}(u_1, \ldots, u_n).$$

This paper is organized as follows. Section 2 provides preliminaries on quantum logic, commutators, and conditionals. Section 3 introduces the model $V^{(\mathscr{Q})}$ and its interpretation that satisfies De Morgan's laws. We also discuss Takeuti's interpretation and construct the above mentioned counterexample. Section 4 derives the quantum transfer principle for the new interpretation.

2 Preliminaries

2.1 Quantum Logic

Let $\mathscr{H}$ be a Hilbert space. For any subset $S \subseteq \mathscr{H}$, we denote by $S^{\perp}$ the orthogonal complement of S. Then, $S^{\perp\perp}$ is the closed linear span of S. Let $\mathscr{C}(\mathscr{H})$ be the set of all closed linear subspaces in $\mathscr{H}$. With the set inclusion ordering, the set $\mathscr{C}(\mathscr{H})$ is a complete lattice. The operation $M \mapsto M^{\perp}$ is an orthocomplementation on the lattice $\mathscr{C}(\mathscr{H})$, with which $\mathscr{C}(\mathscr{H})$ is a complete orthomodular lattices [7, p. 65], *i.e.*, the orthocomplementation satisfies

(C1) if $P \leq Q$ then $Q^{\perp} \leq P^{\perp}$,
(C2) $P^{\perp\perp} = P$,
(C3) $P \vee P^{\perp} = 1$ and $P \wedge P^{\perp} = 0$,

where $0 = \bigwedge \mathscr{Q}$ and $1 = \bigvee \mathscr{Q}$, and the *orthomodular law*

(OM) if $P \leq Q$ then $P \vee (P^{\perp} \wedge Q) = Q$.

We refer the reader to Kalmbach [7] for a standard textbook on orthomodular lattices.

Denote by $\mathscr{B}(\mathscr{H})$ the algebra of bounded linear operators on $\mathscr{H}$ and $\mathscr{Q}(\mathscr{H})$ the set of projections on $\mathscr{H}$. We define the *operator ordering* on $\mathscr{B}(\mathscr{H})$ by $A \le B$ iff $(\psi, A\psi) \le (\psi, B\psi)$ for all $\psi \in \mathscr{H}$. For any $A \in \mathscr{B}(\mathscr{H})$, denote by $\mathscr{R}(A) \in \mathscr{C}(\mathscr{H})$ the closure of the range of A, i.e., $\mathscr{R}(A) = (A\mathscr{H})^{\perp\perp}$. For any $M \in \mathscr{C}(\mathscr{H})$, denote by $\mathscr{P}(M) \in \mathscr{Q}(\mathscr{H})$ the projection of $\mathscr{H}$ onto M. Then, $\mathscr{R}\mathscr{P}(M) = M$ for all $M \in \mathscr{C}(\mathscr{H})$ and $\mathscr{P}\mathscr{R}(P) = P$ for all $P \in \mathscr{Q}(\mathscr{H})$, and we have $P \le Q$ if and only if $\mathscr{R}(P) \subseteq \mathscr{R}(Q)$ for all $P, Q \in \mathscr{Q}(\mathscr{H})$, so that $\mathscr{Q}(\mathscr{H})$ with the operator ordering is also a complete orthomodular lattice isomorphic to $\mathscr{C}(\mathscr{H})$. We consider $\mathscr{Q}(\mathscr{H})$ as the *standard quantum logic* of $\mathscr{H}$, or the logic of observational propositions in quantum mechanics for the system described by $\mathscr{H}$ [2, 6]. The lattice operations are characterized by $P \wedge Q = \text{weak-lim}_{n\to\infty}(PQ)^n$, $P^{\perp} = 1 - P$ for all $P, Q \in \mathscr{Q}(\mathscr{H})$.

A non-empty subset of $\mathscr{Q}(\mathscr{H})$ is called a *subalgebra* iff it is closed under $\wedge$, $\vee$, and $\perp$. A subalgebra $\mathscr{A}$ of $\mathscr{Q}(\mathscr{H})$ is said to be *complete* iff it has the supremum and the infimum in $\mathscr{Q}(\mathscr{H})$ of an arbitrary subset of $\mathscr{A}$.

Let $\mathscr{A} \subseteq \mathscr{B}(\mathscr{H})$. We denote by $\mathscr{A}'$ the *commutant of* $\mathscr{A}$ in $\mathscr{B}(\mathscr{H})$, *i.e.*,

$$\mathscr{A}' = \{A \in \mathscr{B}(\mathscr{H}) \mid AB = BA \text{ for any } B \in \mathscr{A}\}.$$

A self-adjoint subalgebra $\mathscr{M}$ of $\mathscr{B}(\mathscr{H})$ is called a *von Neumann algebra* on $\mathscr{H}$ iff $\mathscr{M}'' = \mathscr{M}$. For any self-adjoint subset $\mathscr{A} \subseteq \mathscr{B}(\mathscr{H})$, $\mathscr{A}''$ is the von Neumann algebra generated by $\mathscr{A}$. We denote by $\mathscr{P}(\mathscr{M})$ the set of projections in a von Neumann algebra $\mathscr{M}$.

We say that P and Q in $\mathscr{Q}(\mathscr{H})$ *commute*, in symbols $P \,{\scriptstyle\downarrow\!\circ}\, Q$, iff $P = (P \wedge Q) \vee (P \wedge Q^{\perp})$. For any $P, Q \in \mathscr{Q}(\mathscr{H})$, we have $P \,{\scriptstyle\downarrow\!\circ}\, Q$ iff $[P, Q] = 0$, where $[P, Q] = PQ - QP$.

For any subset $\mathscr{A} \subseteq \mathscr{Q}(\mathscr{H})$, the *commutant* $\mathscr{A}^{!}$ of $\mathscr{A}$ in $\mathscr{Q}(\mathscr{H})$ [7, p. 23] is defined by

$$\mathscr{A}^{!} = \{P \in \mathscr{Q}(\mathscr{H}) \mid P \,{\scriptstyle\downarrow\!\circ}\, Q \text{ for all } Q \in \mathscr{A}\}. \tag{1}$$

Then, $\mathscr{A}^{!}$ is a complete subalgebra of $\mathscr{Q}(\mathscr{H})$. A *sublogic* of $\mathscr{Q}(\mathscr{H})$ is a subset $\mathscr{A}$ of $\mathscr{Q}(\mathscr{H})$ satisfying $\mathscr{A} = \mathscr{A}^{!!}$. Any sublogic of $\mathscr{Q}(\mathscr{H})$ will be called a *logic* on $\mathscr{H}$. For any subset $\mathscr{A}$ of a logic $\mathscr{Q}$, the smallest logic including $\mathscr{A}$ is $\mathscr{A}^{!!}$ called the *logic generated by* $\mathscr{A}$. Then, a subset $\mathscr{Q} \subseteq \mathscr{Q}(\mathscr{H})$ is a logic on $\mathscr{H}$ if and only if $\mathscr{Q} = \mathscr{P}(\mathscr{M})$ for some von Neumann algebra $\mathscr{M}$ on $\mathscr{H}$ [9, Proposition 2.1]. A logic $\mathscr{Q}$ on $\mathscr{H}$ is a Boolean algebra if and only if $P \,{\scriptstyle\downarrow\!\circ}\, Q$ for all $P, Q \in \mathscr{Q}$ [7, pp. 24–25]

The center of a logic $\mathscr{Q}$, denoted by $Z(\mathscr{Q})$, is the set of elements of $\mathscr{Q}$ commute with every element of $\mathscr{Q}$, i.e., $Z(\mathscr{Q}) = \mathscr{Q}^{!} \cap \mathscr{Q}$. Then, it is easy to see that a subset $\mathscr{A}$ is a Boolean sublogic, or equivalently a distributive sublogic, if and only if $\mathscr{A} = \mathscr{A}^{!!} \subseteq \mathscr{A}^{!}$. The center of $\mathscr{A}^{!!}$ is given by $Z(\mathscr{A}^{!!}) = \mathscr{A}^{!} \cap \mathscr{A}^{!!}$.

2.2 Commutators

Marsden [8] introduced the commutator $\perp\!\!\!\perp(P,Q)$ of two elements P and Q of an orthomodular lattice $\mathcal{Q}$ by

$$\perp\!\!\!\perp(P,Q)=(P\wedge Q)\vee(P\wedge Q^{\perp})\vee(P^{\perp}\wedge Q)\vee(P^{\perp}\wedge Q^{\perp}). \tag{2}$$

Bruns and Kalmbach [3] generalized this notion to finite subsets of $\mathcal{Q}$ by

$$\perp\!\!\!\perp(\mathscr{F})=\bigvee_{\theta:\mathscr{F}\to\{\mathrm{id},\perp\}}\ \bigwedge_{P\in\mathscr{F}}P^{\theta(P)} \tag{3}$$

for any finite subsets $\mathscr{F}$ of $\mathcal{Q}$, where $\{\mathrm{id},\perp\}$ stands for the set consisting of the identity operation id and the orthocomplementation $\perp$. Generalizing this notion to arbitrary subsets $\mathscr{A}$ of $\mathcal{Q}(\mathscr{H})$, Takeuti [15] defined $\perp\!\!\!\perp(\mathscr{A})$ by

$$\perp\!\!\!\perp(\mathscr{A})=\bigvee\{E\in\mathscr{A}^{!}\mid P_1\wedge E\ \downfree\ P_2\wedge E \text{ for all } P_1,P_2\in\mathscr{A}\}, \tag{4}$$

for any subset $\mathscr{A}$ of $\mathcal{Q}$. By Takeuti's definition it is not clear whether the commutator $\perp\!\!\!\perp(\mathscr{A})$ is determined inside the logic $\mathscr{A}^{!!}$ generated by $\mathscr{A}$ or not, unlike the definition of $\perp\!\!\!\perp(\mathscr{F})$ for finite subsets $\mathscr{F}$. To resolve this problem, it was shown in Ref. [10, Theorem 2.5] that $\perp\!\!\!\perp(\mathscr{A})\in\mathscr{A}^{!}\cap\mathscr{A}^{!!}$, and we obtain the relation

$$\perp\!\!\!\perp(\mathscr{A})=\bigvee\{E\in\mathscr{A}^{!}\cap\mathscr{A}^{!!}\mid P_1\wedge E\ \downfree\ P_2\wedge E \text{ for all } P_1,P_2\in\mathscr{A}\}, \tag{5}$$

as an alternative definition for $\perp\!\!\!\perp(\mathscr{A})$.

We have the following characterizations of commutators [9, Theorems 2.5, 2.6, Proposition 2.2]: For any $\mathscr{A}\subseteq\mathcal{Q}(\mathscr{H})$, we have the following relations.

$$\perp\!\!\!\perp(\mathscr{A})=\mathscr{P}\{\psi\in\mathscr{H}\mid[P_1,P_2]P_3\psi=0 \text{ for all } P_1,P_2,P_3\in\mathscr{A}^{!!}\} \tag{6}$$

$$\perp\!\!\!\perp(\mathscr{A})=\mathscr{P}\{\psi\in\mathscr{H}\mid[A,B]\psi=0 \text{ for all } A,B\in\mathscr{A}''\}. \tag{7}$$

We refer the reader to Pulmannová [13] and Chevalier [4] for further results about commutators in orthomodular lattices.

2.3 Conditionals

In classical logic, the conditional operation $\to$ is defined by negation $\perp$ and disjunction $\vee$ as $P\to Q=P^{\perp}\vee Q$. In quantum logic there is well-known arbitrariness in choosing a binary operation for conditional. Following Hardegree [5], we define a *quantum material conditional* on a logic $\mathcal{Q}$ as a binary operation $\rightarrow$ on $\mathcal{Q}$ definable by an ortholattice polynomial $p(x, y)$ as $P \rightarrow Q = p(P, Q)$ for all $P, Q \in \mathcal{Q}$ satisfying the following "minimum implicative conditions":

(E) $P \rightarrow Q = 1$ if and only if $P \leq Q$.
(MP) (*modus ponens*) $P \wedge (P \rightarrow Q) \leq Q$.
(MT) (*modus tollens*) $Q^{\perp} \wedge (P \rightarrow Q) \leq P^{\perp}$.

Hardegree [5] showed that there are exactly three polynomially definable material conditionals:

(S) (Sasaki conditional) $P \rightarrow_S Q := P^{\perp} \vee (P \wedge Q)$,
(C) (Contrapositive Sasaki conditional) $P \rightarrow_C Q := (P \vee Q)^{\perp} \vee Q$,
(R) (Relevance conditional) $P \rightarrow_R Q := (P \wedge Q) \vee (P^{\perp} \wedge Q) \vee (P^{\perp} \wedge Q^{\perp})$.

Following Takeuti [15] we adopt the Sasaki arrow, the most favorable according to the majority view [16], as the conditional $\rightarrow$ for a logic $\mathcal{Q}$, *i.e.*, $P \rightarrow Q = P^{\perp} \vee (P \wedge Q)$. The logical equivalence $\leftrightarrow$ is defined by

$$P \leftrightarrow Q = (P \rightarrow Q) \wedge (Q \rightarrow P). \tag{8}$$

In Boolean logic, implication and conjunction are associated by the relation $P \wedge Q = (P \rightarrow Q^{\perp})^{\perp}$, and this relation plays an essential role in the duality between bounded universal quantification $(\forall x \in A)\phi(x)$ and bounded existential quantification $(\exists x \in A)\phi(x)$. In order to keep the above duality in quantum set theory, we introduce the binary operation $*$ dual to $\rightarrow$ by

$$P * Q = (P \rightarrow Q^{\perp})^{\perp}. \tag{9}$$

We have the following relations

$$P \rightarrow Q = (P \wedge Q) \vee (P^{\perp} \wedge Q) \vee (P^{\perp} \wedge Q^{\perp}) \vee (P^{\perp} \wedge \perp\!\!\!\perp(P, Q)^{\perp}). \tag{10}$$

$$P * Q = (P \wedge Q) \vee (P \wedge \perp\!\!\!\perp(P, Q)^{\perp}). \tag{11}$$

The following proposition is useful in later discussions [9, Proposition 2.4], [12, Proposition 3.1].

Proposition 1 *Let $\mathcal{Q}$ be a logic on $\mathcal{H}$. The following hold.*

(i) If $P_\alpha \in \mathcal{Q}$ and $P_\alpha \,\natural\, Q$ for all α, then $(\bigvee_\alpha P_\alpha) \,\natural\, Q$, $(\bigwedge_\alpha P_\alpha) \,\natural\, Q$, and $Q \wedge (\bigvee_\alpha P_\alpha) = \bigvee_\alpha (Q \wedge P_\alpha)$.
(ii) If $P_1, P_2 \,\natural\, Q$, then $(P_1 \rightarrow P_2) \wedge Q = [(P_1 \wedge Q) \rightarrow (P_2 \wedge Q)] \wedge Q$.
(iii) If $P_1, P_2 \,\natural\, Q$, then $(P_1 * P_2) \wedge Q = [(P_1 \wedge Q) * (P_2 \wedge Q)] \wedge Q$.

3 Quantum Set Theory

We denote by V the universe of the Zermelo-Fraenkel set theory with the axiom of choice (ZFC). Let $\mathscr{L}(\in)$ be the language of first-order theory with equality augmented by a connective $\to$, a binary relation symbol $\in$, bounded quantifier symbols $\forall x \in y$, $\exists x \in y$, and no constant symbols. For any class U, the language $\mathscr{L}(\in, U)$ is the one obtained by adding a name for each element of U. We take the symbols $\neg$, $\wedge$, $\to$, $\forall x \in y$, and $\forall x$ as primitive, and the symbols $\vee$, $\exists x \in y$, and $\exists x$ as derived symbols by defining:

(i) $\phi \vee \psi = \neg(\neg\phi \wedge \neg\psi)$,
(ii) $\exists x \in y\, \phi(x) = \neg(\forall x \in y\, \neg\phi(x))$,
(iii) $\exists x \phi(x) = \neg(\forall x\, \neg\phi(x))$.

To each statement ϕ of $\mathscr{L}(\in, U)$, the satisfaction relation $\langle U, \in\rangle \models \phi$ is defined by the following recursive rules:

1. $\langle U, \in\rangle \models u \in v$ iff $u \in v$.
2. $\langle U, \in\rangle \models u = v$ iff $u = v$.
3. $\langle U, \in\rangle \models \neg\phi$ iff $\langle U, \in\rangle \models \phi$ does not hold.
4. $\langle U, \in\rangle \models \phi_1 \wedge \phi_2$ iff $\langle U, \in\rangle \models \phi_1$ and $\langle U, \in\rangle \models \phi_2$.
5. $\langle U, \in\rangle \models \phi_1 \to \phi_2$ iff if $\langle U, \in\rangle \models \phi_1$ then $\langle U, \in\rangle \models \phi_2$.
6. $\langle U, \in\rangle \models (\forall x \in u)\, \phi(x)$ iff $\langle U, \in\rangle \models \phi(u')$ for all $u' \in u$.
7. $\langle U, \in\rangle \models (\forall x)\, \phi(x)$ iff $\langle U, \in\rangle \models \phi(u)$ for all $u \in U$.

Our assumption that V satisfies ZFC means that if $\phi(x_1, \ldots, x_n)$ is provable in ZFC, i.e., ZFC $\vdash \phi(x_1, \ldots, x_n)$, then $\langle V, \in\rangle \models \phi(u_1, \ldots, u_n)$ for any formula $\phi(x_1, \ldots, x_n)$ of $\mathscr{L}(\in)$ and all $u_1, \ldots, u_n \in V$.

In what follows let $\mathscr{Q}$ be a logic on $\mathscr{H}$. For each ordinal α, let

$$V_\alpha^{(\mathscr{Q})} = \{u |\ u : \mathrm{dom}(u) \to \mathscr{Q} \text{ and } (\exists\beta < \alpha)\, \mathrm{dom}(u) \subseteq V_\beta^{(\mathscr{Q})}\}. \tag{12}$$

The *$\mathscr{Q}$-valued universe* $V^{(\mathscr{Q})}$ is defined by

$$V^{(\mathscr{Q})} = \bigcup_{\alpha \in \mathrm{On}} V_\alpha^{(\mathscr{Q})}, \tag{13}$$

where On is the class of all ordinals. For every $u \in V^{(\mathscr{Q})}$, the rank of u, denoted by rank(u), is defined as the least α such that $u \in V_{\alpha+1}^{(\mathscr{Q})}$. It is easy to see that if $u \in \mathrm{dom}(v)$ then $\mathrm{rank}(u) < \mathrm{rank}(v)$.

An induction on rank argument leads to the following [1].

Theorem 2 (Induction Principle for $V^{(\mathscr{Q})}$) *For any predicate* $\phi(x)$,

$$\forall u \in V^{(\mathscr{Q})}[\forall u' \in \mathrm{dom}(u)\phi(u') \to \phi(u)] \to \forall u \in V^{(\mathscr{Q})}\phi(u)$$

For any $u, v \in V^{(\mathcal{Q})}$, the $\mathcal{Q}$-valued truth values $[\![u = v]\!]_{\mathcal{Q}}$ and $[\![u \in v]\!]_{\mathcal{Q}}$ of atomic formulas $u = v$ and $u \in v$ are assigned by the following rules recursive in rank.

(iv) $[\![u = v]\!]_{\mathcal{Q}} = \bigwedge_{u' \in \mathrm{dom}(u)}(u(u') \to [\![u' \in v]\!]_{\mathcal{Q}}) \wedge \bigwedge_{v' \in \mathrm{dom}(v)}(v(v') \to [\![v' \in u]\!]_{\mathcal{Q}})$.
(v) $[\![u \in v]\!]_{\mathcal{Q}} = \bigvee_{v' \in \mathrm{dom}(v)}(v(v') * [\![u = v']\!]_{\mathcal{Q}})$.

To each statement ϕ of $\mathscr{L}(\in, V^{(\mathcal{Q})})$ we assign the $\mathcal{Q}$-valued truth value $[\![\phi]\!]_{\mathcal{Q}}$ by the following rules.

(vi) $[\![\neg\phi]\!]_{\mathcal{Q}} = [\![\phi]\!]_{\mathcal{Q}}^{\perp}$.
(vii) $[\![\phi_1 \wedge \phi_2]\!]_{\mathcal{Q}} = [\![\phi_1]\!]_{\mathcal{Q}} \wedge [\![\phi_2]\!]_{\mathcal{Q}}$.
(viii) $[\![\phi_1 \to \phi_2]\!]_{\mathcal{Q}} = [\![\phi_1]\!]_{\mathcal{Q}} \to [\![\phi_2]\!]_{\mathcal{Q}}$.
(ix) $[\![(\forall x \in u)\,\phi(x)]\!]_{\mathcal{Q}} = \bigwedge_{u' \in \mathrm{dom}(u)}(u(u') \to [\![\phi(u')]\!]_{\mathcal{Q}})$.
(x) $[\![(\forall x)\,\phi(x)]\!]_{\mathcal{Q}} = \bigwedge_{u \in V^{(\mathcal{Q})}}[\![\phi(u)]\!]_{\mathcal{Q}}$.

By the definitions of derived logical symbols, (i)–(iii), we have the following relations.

(xi) $[\![\phi_1 \vee \phi_2]\!]_{\mathcal{Q}} = [\![\phi_1]\!]_{\mathcal{Q}} \vee [\![\phi_2]\!]_{\mathcal{Q}}$.
(xii) $[\![(\exists x \in u)\,\phi(x)]\!]_{\mathcal{Q}} = \bigvee_{u' \in \mathrm{dom}(u)}(u(u') * [\![\phi(u')]\!]_{\mathcal{Q}})$.
(xiii) $[\![(\exists x)\,\phi(x)]\!]_{\mathcal{Q}} = \bigvee_{u \in V^{(\mathcal{Q})}}[\![\phi(u)]\!]_{\mathcal{Q}}$.

To see (xii), we have

$$\begin{aligned}
[\![(\exists x \in u)\,\phi(x)]\!]_{\mathcal{Q}} &= [\![\neg(\forall x \in u\, \neg\phi(x))]\!]_{\mathcal{Q}} \\
&= (\bigwedge_{u' \in \mathrm{dom}(u)} u(u') \to [\![\neg\phi(u')]\!]_{\mathcal{Q}})^{\perp} \\
&= (\bigwedge_{u' \in \mathrm{dom}(u)} u(u') \to [\![\phi(u')]\!]_{\mathcal{Q}}{}^{\perp})^{\perp} \\
&= \bigvee_{u' \in \mathrm{dom}(u)} (u(u') \to [\![\phi(u')]\!]_{\mathcal{Q}}{}^{\perp})^{\perp} \\
&= \bigvee_{u' \in \mathrm{dom}(u)} (u(u') * [\![\phi(u')]\!]_{\mathcal{Q}}).
\end{aligned}$$

Note that according to the above, we have the following relations

(xiv) $[\![u = v]\!]_{\mathcal{Q}} = [\![\forall x \in u(x \in v) \wedge \forall x \in v(x \in u)]\!]_{\mathcal{Q}}$,
(xv) $[\![u \in v]\!]_{\mathcal{Q}} = [\![\exists x \in v(x = u)]\!]_{\mathcal{Q}}$.

We also have the following relations satisfying De Morgan's laws:

(xvi) $[\![\neg(\phi_1 \wedge \phi_2)]\!]_{\mathcal{Q}} = [\![\neg\phi_1 \vee \neg\phi_2]\!]_{\mathcal{Q}}$,
(xvii) $[\![\neg(\phi_1 \vee \phi_2)]\!]_{\mathcal{Q}} = [\![\neg\phi_1 \wedge \neg\phi_2]\!]_{\mathcal{Q}}$,
(xviii) $[\![\neg(\forall x \in u\, \phi(x))]\!]_{\mathcal{Q}} = [\![\exists x \in u\, (\neg\phi(x))]\!]_{\mathcal{Q}}$,
(xix) $[\![\neg(\exists x \in u\, \phi(x))]\!]_{\mathcal{Q}} = [\![\forall x \in u\, (\neg\phi(x))]\!]_{\mathcal{Q}}$,
(xx) $[\![\neg(\forall x\, \phi(x))]\!]_{\mathcal{Q}} = [\![\exists x\, (\neg\phi(x))]\!]_{\mathcal{Q}}$,
(xxi) $[\![\neg(\exists x\, \phi(x))]\!]_{\mathcal{Q}} = [\![\forall x\, (\neg\phi(x))]\!]_{\mathcal{Q}}$.

A formula in $\mathscr{L}(\in)$ is called a Δ_0*-formula* iff it has no unbounded quantifiers $\forall x$ nor $\exists x$. The following theorem holds.

Theorem 3 (Δ_0-Absoluteness Principle) *For any Δ_0-formula $\phi(x_1, \ldots, x_n)$ of $\mathscr{L}(\in)$ and $u_1, \ldots, u_n \in V^{(\mathscr{Q})}$, we have*

$$[\![\phi(u_1, \ldots, u_n)]\!]_{\mathscr{Q}} = [\![\phi(u_1, \ldots, u_n)]\!]_{\mathscr{Q}(\mathscr{H})}.$$

Proof The assertion is proved by the induction on the complexity of formulas and the rank of elements of $V^{(\mathscr{Q})}$. Let $u, v \in V^{(\mathscr{Q})}$. By induction hypothesis, for any $u' \in \mathrm{dom}(u)$ and $v' \in \mathrm{dom}(v)$ we have $[\![u' \in w]\!]_{\mathscr{Q}} = [\![u' \in w]\!]_{\mathscr{Q}(\mathscr{H})}$, $[\![v' \in w]\!]_{\mathscr{Q}} = [\![v' \in w]\!]_{\mathscr{Q}(\mathscr{H})}$, and $[\![w = v']\!]_{\mathscr{Q}} = [\![w = v']\!]_{\mathscr{Q}(\mathscr{H})}$ for all $w \in V^{(\mathscr{Q})}$. Thus,

$$\begin{aligned}
[\![u = v]\!]_{\mathscr{Q}} &= \bigwedge_{u' \in \mathrm{dom}(u)} (u(u') \to [\![u' \in v]\!]_{\mathscr{Q}}) \wedge \bigwedge_{v' \in \mathrm{dom}(v)} (v(v') \to [\![v' \in u]\!]_{\mathscr{Q}}) \\
&= \bigwedge_{u' \in \mathrm{dom}(u)} (u(u') \to [\![u' \in v]\!]_{\mathscr{Q}(\mathscr{H})}) \wedge \bigwedge_{v' \in \mathrm{dom}(v)} (v(v') \to [\![v' \in u]\!]_{\mathscr{Q}(\mathscr{H})}) \\
&= [\![u = v]\!]_{\mathscr{Q}(\mathscr{H})},
\end{aligned}$$

and we also have

$$\begin{aligned}
[\![u \in v]\!]_{\mathscr{Q}} &= \bigvee_{v' \in \mathrm{dom}(v)} (v(v') * [\![u = v']\!]_{\mathscr{Q}}) \\
&= \bigvee_{v' \in \mathrm{dom}(v)} (v(v') * [\![u = v']\!]_{\mathscr{Q}(\mathscr{H})}) \\
&= [\![u \in v]\!]_{\mathscr{Q}(\mathscr{H})}.
\end{aligned}$$

Thus, the assertion holds for atomic formulas. Any induction step adding a logical symbol works easily, even when bounded quantifiers are concerned, since the ranges of the supremum and the infimum are common for evaluating $[\![\cdots]\!]_{\mathscr{Q}}$ and $[\![\cdots]\!]_{\mathscr{Q}(\mathscr{H})}$. □

Henceforth, for any Δ_0-formula $\phi(x_1, \ldots, x_n)$ and $u_1, \ldots, u_n \in V^{(\mathscr{Q})}$, we abbreviate $[\![\phi(u_1, \ldots, u_n)]\!] = [\![\phi(u_1, \ldots, u_n)]\!]_{\mathscr{Q}}$, which is the common $\mathscr{Q}(\mathscr{H})$-valued truth value for $u_1, \ldots, u_n \in V^{(\mathscr{Q})}$.

The universe V can be embedded in $V^{(\mathscr{Q})}$ by the following operation $\vee : v \mapsto \check{v}$ defined by the $\in$-recursion: for each $v \in V$, $\check{v} = \{\check{u} | \, u \in v\} \times \{1\}$. Then we have the following.

Theorem 4 (Δ_0-Elementary Equivalence Principle) *Let $\phi(x_1, \ldots, x_n)$ be a Δ_0-formula of $\mathscr{L}(\in)$. For any $u_1, \ldots, u_n \in V$, we have*

$$\langle V, \in \rangle \models \phi(u_1, \ldots, u_n) \text{ if and only if } [\![\phi(\check{u}_1, \ldots, \check{u}_n)]\!] = 1.$$

Proof Let **2** be the sublogic such that $\mathbf{2} = \{0, 1\}$. Then, by induction it is easy to see that $\langle V, \in\rangle \models \phi(u_1, \ldots, u_n)$ if and only if $[\![\phi(\check{u}_1, \ldots, \check{u}_n)]\!]_{\mathbf{2}} = 1$ for any $\phi(x_1, \ldots, x_n)$ in $\mathscr{L}(\in)$, and this is equivalent to $[\![\phi(\check{u}_1, \ldots, \check{u}_n)]\!] = 1$ for any Δ_0-formula $\phi(x_1, \ldots, x_n)$ by the Δ_0-absoluteness principle. □

Instead of (v) and (xii), Takeuti [15] defined the truth values of membership relation and existential quantification as follows.

(v') $[\![u \in v]\!] = \bigvee_{v' \in \mathrm{dom}(v)} (v(v') \wedge [\![u = v']\!])$.
(xii') $[\![(\exists x \in u)\, \phi(x)]\!] = \bigvee_{u' \in \mathrm{dom}(u)} (u(u') \wedge [\![\phi(u')]\!])$.

In this case, De Morgan's laws do not hold in general as follows.

Suppose that $\mathcal{Q}$ is not a Boolean algebra. Then, there exists a pair $P_0, Q_0 \in \mathcal{Q}$ such that P_0 does not commute with Q_0, so that $\perp\!\!\!\perp(P_0, Q_0)^\perp > 0$. Let $E = \perp\!\!\!\perp(P_0, Q_0)^\perp$, $P = P_0 \wedge E$, and $Q = Q_0 \wedge E$. If $P = 0$ then $P_0 = P_0 \wedge \perp\!\!\!\perp(P_0, Q_0)$ so that $P_0 \,{}_{\circ}^{|}\, Q_0$, a contradiction. Thus, $P \neq 0$. We also have that $P \wedge Q = P_0 \wedge Q_0 \wedge \perp\!\!\!\perp(P_0, Q_0)^\perp = 0$, so that $P \wedge Q = 0$. Let $u = \{\langle \check{0}, P\rangle\}$ and $v = \{\langle \check{0}, Q\rangle\}$. Consider the formula $\phi(x) = \neg(x \in v)$. Then, we can show

$$[\![\neg(\forall x \in u)\phi(x)]\!] > [\![(\exists x \in u)\neg\phi(x)]\!] = 0. \tag{14}$$

In fact, we have

$$\begin{aligned}
[\![(\exists x \in u)\neg\phi(x)]\!] &= \bigvee_{u' \in \mathrm{dom}(u)} (u(u') \wedge [\![\neg\phi(u')]\!]) \\
&= u(\check{0}) \wedge [\![\neg\neg(\check{0} \in v)]\!] \\
&= u(\check{0}) \wedge [\![\check{0} \in v]\!] \\
&= u(\check{0}) \wedge \bigvee_{v' \in \mathrm{dom}(v)} (v(v') \wedge [\![\check{0} = v']\!]) \\
&= u(\check{0}) \wedge (v(\check{0}) \wedge [\![\check{0} = \check{0}]\!]) \\
&= u(\check{0}) \wedge v(\check{0}) \\
&= P \wedge Q \\
&= 0.
\end{aligned}$$

Similarly we have

$$\begin{aligned}
[\![\neg(\forall x \in u)\phi(x)]\!] &= [\![(\forall x \in u)\phi(x)]\!]^\perp \\
&= \Big(\bigwedge_{u' \in \mathrm{dom}(u)} (u(u') \to [\![\phi(u')]\!])\Big)^\perp \\
&= (u(\check{0}) \to [\![\phi(\check{0})]\!])^\perp \\
&= u(\check{0}) * [\![\phi(\check{0})]\!]^\perp \\
&= u(\check{0}) * [\![\neg(\check{0} \in v)]\!]^\perp
\end{aligned}$$

$$
\begin{aligned}
&= u(\check{0}) * [\![\check{0} \in v]\!] \\
&= u(\check{0}) * \bigvee_{v' \in \mathrm{dom}(v)} (v(v') \wedge [\![\check{0} = v']\!]) \\
&= u(\check{0}) * (v(\check{0}) \wedge [\![\check{0} = \check{0}]\!]) \\
&= u(\check{0}) * v(\check{0}) \\
&= P * Q \\
&= (P \wedge Q) \vee (P \wedge \perp\!\!\!\perp(P, Q)^{\perp}) \\
&= P.
\end{aligned}
$$

Since $P \neq 0$, Eq. (14) follows.

Thus, if $\mathcal{Q}$ is not a Boolean algebra, there exists a predicate $\phi(x)$ such that $[\![(\exists x \in u)\, \neg\phi(x)]\!] = 0$ but $[\![\neg(\forall x \in u)\, \phi(x)]\!] > 0$.

4 Transfer Principle

In this section, we investigate the transfer principle that gives any Δ_0-formula provable in ZFC a lower bound for its truth value, which is determined by the degree of the commutativity of the elements of $V^{(\mathcal{Q})}$ appearing in the formula as constants. The results in this section was obtained in Ref. [9] for Takeuti's original formulation. Here, we extends the argument in a self-contained manner to the present formulation, in which De Morgan's laws hold for bounded quantifiers.

For $u \in V^{(\mathcal{Q})}$, we define the *support* of u, denoted by $L(u)$, by transfinite recursion on the rank of u by the relation

$$
L(u) = \bigcup_{x \in \mathrm{dom}(u)} L(x) \cup \{u(x) \mid x \in \mathrm{dom}(u)\} \cup \{0\}. \tag{15}
$$

For $\mathscr{A} \subseteq V^{(\mathcal{Q})}$ we write $L(\mathscr{A}) = \bigcup_{u \in \mathscr{A}} L(u)$ and for $u_1, \dots, u_n \in V^{(\mathcal{Q})}$ we write $L(u_1, \dots, u_n) = L(\{u_1, \dots, u_n\})$. Then, we obtain the following characterization of subuniverses of $V^{(\mathcal{Q}(\mathscr{H}))}$.

Proposition 5 *Let $\mathcal{Q}$ be a logic on $\mathscr{H}$ and α an ordinal. For any $u \in V^{(\mathcal{Q}(\mathscr{H}))}$, we have $u \in V_{\alpha}^{(\mathcal{Q})}$ if and only if $u \in V_{\alpha}^{(\mathcal{Q}(\mathscr{H}))}$ and $L(u) \subseteq \mathcal{Q}$. In particular, $u \in V^{(\mathcal{Q})}$ if and only if $u \in V^{(\mathcal{Q}(\mathscr{H}))}$ and $L(u) \subseteq \mathcal{Q}$. Moreover,* $\mathrm{rank}(u)$ *is the least α such that $u \in V_{\alpha+1}^{(\mathcal{Q}(\mathscr{H}))}$ for any $u \in V^{(\mathcal{Q})}$.*

Proof Immediate from transfinite induction on α. □

Let $\mathscr{A} \subseteq V^{(\mathcal{Q})}$. The *commutator of* $\mathscr{A}$, denoted by $\underline{\vee}(\mathscr{A})$, is defined by

$$
\underline{\vee}(\mathscr{A}) = \perp\!\!\!\perp(L(\mathscr{A})). \tag{16}
$$

For any $u_1, \ldots, u_n \in V^{(\mathcal{Q})}$, we write $\underline{\vee}(u_1, \ldots, u_n) = \underline{\vee}(\{u_1, \ldots, u_n\})$.

Let $u \in V^{(\mathcal{Q})}$ and $p \in \mathcal{Q}$. The *restriction* $u|_p$ of u to p is defined by the following transfinite recursion:

$$u|_p = \{\langle x|_p, u(x) \wedge p\rangle \mid x \in \mathrm{dom}(u)\} \cup \{\langle u, 0\rangle\}.$$

The last term $\{\langle u, 0\rangle\}$ has no essential role but ensures the well-definedness of the function $u|_p : \mathrm{dom}(u|_p) \to \mathcal{Q}$.

Proposition 6 *For any $\mathscr{A} \subseteq V^{(\mathcal{Q})}$ and $p \in \mathcal{Q}$, we have*

$$L(\{u|_p \mid u \in \mathscr{A}\}) = L(\mathscr{A}) \wedge p. \tag{17}$$

Proof By induction, it is easy to see the relation $L(u|_p) = L(u) \wedge p$, so that the assertion follows easily. □

Let $\mathscr{A} \subseteq V^{(\mathcal{Q})}$. The *logic generated by* $\mathscr{A}$, denoted by $\mathcal{Q}(\mathscr{A})$, is defined by

$$\mathcal{Q}(\mathscr{A}) = L(\mathscr{A})^{!!}. \tag{18}$$

For $u_1, \ldots, u_n \in V^{(\mathcal{Q})}$, we write $\mathcal{Q}(u_1, \ldots, u_n) = \mathcal{Q}(\{u_1, \ldots, u_n\})$.

Proposition 7 *For any Δ_0-formula $\phi(x_1, \ldots, x_n)$ in $\mathscr{L}(\in)$ and $u_1, \cdots, u_n \in V^{(\mathcal{Q})}$, we have $[\![\phi(u_1, \ldots, u_n)]\!] \in \mathcal{Q}(u_1, \ldots, u_n)$.*

Proof Let $\mathscr{A} = \{u_1, \ldots, u_n\}$. Since $L(\mathscr{A}) \subseteq \mathcal{Q}(\mathscr{A})$, it follows from Proposition 5 that $u_1, \ldots, u_n \in V^{(\mathcal{Q}(\mathscr{A}))}$. By the Δ_0-absoluteness principle, we have $[\![\phi(u_1, \ldots, u_n)]\!] = [\![\phi(u_1, \ldots, u_n)]\!]_{\mathcal{Q}(\mathscr{A})} \in \mathcal{Q}(\mathscr{A})$. □

Proposition 8 *For any Δ_0-formula $\phi(x_1, \ldots, x_n)$ in $\mathscr{L}(\in)$ and $u_1, \ldots, u_n \in V^{(\mathcal{Q})}$, if $p \in L(u_1, \ldots, u_n)^!$, then $p \,\text{\rotatebox{180}{δ}}\, [\![\phi(u_1, \ldots, u_n)]\!]$ and $p \,\text{\rotatebox{180}{δ}}\, [\![\phi(u_1|_p, \ldots, u_n|_p)]\!]$.*

Proof Let $u_1, \ldots, u_n \in V^{(\mathcal{Q})}$. If $p \in L(u_1, \ldots, u_n)^!$, then $p \in \mathcal{Q}(u_1, \ldots, u_n)^!$. From Proposition 7, $[\![\phi(u_1, \ldots, u_n)]\!] \in \mathcal{Q}(u_1, \ldots, u_n)$, so that $p \,\text{\rotatebox{180}{δ}}\, [\![\phi(u_1, \ldots, u_n)]\!]$. From Proposition 6, $L(u_1|_p, \ldots, u_n|_p) = L(u_1, \ldots, u_n) \wedge p$, and hence $p \in L(u_1|_p, \ldots, u_n|_p)^!$, so that $p \,\text{\rotatebox{180}{δ}}\, [\![\phi(u_1|_p, \ldots, u_n|_p)]\!]$. □

We define the binary relation $x_1 \subseteq x_2$ by $\forall x \in x_1 (x \in x_2)$. Then, by definition for any $u, v \in V^{(\mathcal{Q})}$ we have

$$[\![u \subseteq v]\!] = \bigwedge_{u' \in \mathrm{dom}(u)} (u(u') \to [\![u' \in v]\!]), \tag{19}$$

and we have $[\![u = v]\!] = [\![u \subseteq v]\!] \wedge [\![v \subseteq u]\!]$.

Proposition 9 *For any $u, v \in V^{(\mathcal{Q})}$ and $p \in L(u, v)^!$, we have the following relations.*

(i) $[\![u|_p \in v|_p]\!] = [\![u \in v]\!] \wedge p$.
(ii) $[\![u|_p \subseteq v|_p]\!] \wedge p = [\![u \subseteq v]\!] \wedge p$.
(iii) $[\![u|_p = v|_p]\!] \wedge p = [\![u = v]\!] \wedge p$

Proof We prove the relations by induction on the ranks of u, v. If $\mathrm{rank}(u) = \mathrm{rank}(v) = 0$, then $\mathrm{dom}(u) = \mathrm{dom}(v) = \emptyset$, so that the relations trivially hold. Let $u, v \in V^{(\mathcal{Q})}$ and $p \in L(u, v)^!$. To prove (i), let $v' \in \mathrm{dom}(v)$. Then, we have $p \,\natural\, v(v')$ by the assumption on p. By induction hypothesis, we have also $[\![u|_p = v'|_p]\!] \wedge p = [\![u = v']\!] \wedge p$. By Proposition 8, we have $p \,\natural\, [\![u = v']\!]$, so that $v(v'), [\![u = v']\!] \in \{p\}^!$. From Eq. (11) we have $(v(v') \wedge p) * [\![u|_p = v'|_p]\!] \le p$, and hence we have

$$\begin{aligned}(v(v') \wedge p) * [\![u|_p = v'|_p]\!] &= (v(v') \wedge p) * ([\![u|_p = v'|_p]\!] \wedge p)\\ &= (v(v') \wedge p) * ([\![u = v']\!] \wedge p)\\ &= (v(v') * [\![u = v']\!]) \wedge p.\end{aligned}$$

Thus, we have

$$\begin{aligned}[\![u|_p \in v|_p]\!] &= \bigvee_{v' \in \mathrm{dom}(v|_p)} (v|_p(v') * [\![u|_p = v']\!])\\ &= \bigvee_{v' \in \mathrm{dom}(v)} (v|_p(v'|_p) * [\![u|_p = v'|_p]\!])\\ &= \bigvee_{v' \in \mathrm{dom}(v)} [(v(v') \wedge p) * [\![u|_p = v'|_p]\!]]\\ &= \bigvee_{v' \in \mathrm{dom}(v)} [(v(v') * [\![u = v']\!]) \wedge p]\\ &= \left(\bigvee_{v' \in \mathrm{dom}(v)} (v(v') * [\![u = v']\!])\right) \wedge p,\end{aligned}$$

where the last equality follows from Proposition 1(i). Thus, by definition of $[\![u \in v]\!]$ we obtain the relation $[\![u|_p \in v|_p]\!] = [\![u \in v]\!] \wedge p$, and relation (i) has been proved. To prove (ii), let $u' \in \mathrm{dom}(u)$. Then, we have $[\![u'|_p \in v|_p]\!] = [\![u' \in v]\!] \wedge p$ by induction hypothesis. Thus, we have

$$\begin{aligned}[\![u|_p \subseteq v|_p]\!] &= \bigwedge_{u' \in \mathrm{dom}(u|_p)} (u|_p(u') \to [\![u' \in v|_p]\!])\\ &= \bigwedge_{u' \in \mathrm{dom}(u)} (u|_p(u'|_p) \to [\![u'|_p \in v|_p]\!])\\ &= \bigwedge_{u' \in \mathrm{dom}(u)} [(u(u') \wedge p) \to ([\![u' \in v]\!] \wedge p)].\end{aligned}$$

We have $p \mathrel{\text{\raisebox{0pt}{$\circ$}}\!\!\!|} u(u')$ by assumption on p, and $p \,{}_{\circ}\!\!\downarrow [\![u' \in v]\!]$ by Proposition 8, so that $p \,{}_{\circ}\!\!\downarrow u(u') \to [\![u' \in v]\!]$ and $p \,{}_{\circ}\!\!\downarrow (u(u') \wedge p) \to ([\![u' \in v]\!] \wedge p)$. Thus, by Proposition 1 we have

$$
\begin{aligned}
p \wedge [\![u|_p \subseteq v|_p]\!] &= p \wedge \bigwedge_{u' \in \mathrm{dom}(u)} [(u(u') \wedge p) \to ([\![u' \in v]\!] \wedge p)] \\
&= p \wedge \bigwedge_{u' \in \mathrm{dom}(u)} (u(u') \to [\![u' \in v]\!]) \\
&= p \wedge [\![u \subseteq v]\!].
\end{aligned}
$$

Thus, we have proved relation (ii). Relation (iii) follows easily from relation (ii). □

We have the following theorem.

Theorem 10 (Δ_0-Restriction Principle) *For any Δ_0-formula $\phi(x_1, \ldots, x_n)$ in $\mathscr{L}(\in)$ and $u_1, \ldots, u_n \in V^{(\mathscr{Q})}$, if $p \in L(u_1, \ldots, u_n)^!$, then $[\![\phi(u_1, \ldots, u_n)]\!] \wedge p = [\![\phi(u_1|_p, \ldots, u_n|_p)]\!] \wedge p$.*

Proof We shall write $\mathbf{u} = (u_1, \ldots, u_n)$ and $\mathbf{u}|_p = (u_1|_p, \ldots, u_n|_p)$. We prove the assertion by induction on the complexity of $\phi(x_1, \ldots, x_n)$. From Proposition 9, the assertion holds for atomic formulas. Thus, it suffices to consider the following induction steps: (i) $\phi \Rightarrow \neg\phi$, (ii) $\phi_1, \phi_2 \Rightarrow \phi_1 \wedge \phi_2$ (iii) $\phi_1, \phi_2 \Rightarrow \phi_1 \to \phi_2$, (iv) $\{\phi(x) \mid x \in \mathrm{dom}(u)\} \to \bigwedge_{x \in \mathrm{dom}(u)} \phi(x)$.

(i) If $a \,{}_{\circ}\!\!\downarrow p$, the relation

$$a^{\perp} \wedge p = (a \wedge p)^{\perp} \wedge p \tag{20}$$

follows easily. Let $p \in L(\mathbf{u})^!$. Suppose $[\![\phi(\mathbf{u})]\!] \wedge p = [\![\phi(\mathbf{u}|_p)]\!] \wedge p$. From Eq. (20) we have

$$
\begin{aligned}
[\![\phi(\mathbf{u})]\!]^{\perp} \wedge p &= ([\![\phi(\mathbf{u})]\!] \wedge p)^{\perp} \wedge p \\
&= ([\![\phi(\mathbf{u}|_p)]\!] \wedge p)^{\perp} \wedge p \\
&= [\![\phi(\mathbf{u}|_p)]\!]^{\perp} \wedge p,
\end{aligned}
$$

so that we have

$$[\![\neg\phi(\mathbf{u})]\!] \wedge p = [\![\neg\phi(\mathbf{u}|_p)]\!] \wedge p.$$

(ii) Let $p \in L(\mathbf{u})^!$. Suppose $[\![\phi_j(\mathbf{u})]\!] \wedge p = [\![\phi_j(\mathbf{u}|_p)]\!] \wedge p$ for $j = 1, 2$. Then, it follows easily from associativity of $\wedge$, we have

$$[\![\phi_1(\mathbf{u}) \wedge \phi_2(\mathbf{u})]\!] \wedge p = [\![\phi_1(\mathbf{u}|_p) \wedge \phi_2(\mathbf{u}|_p)]\!] \wedge p.$$

(iii) Recall the relation

$$(a \to b) \wedge p = [(a \wedge p) \to (b \wedge p)] \wedge p$$

for all $a, b \in \{p\}^!$ as shown in Proposition 1 (ii). Let $p \in L(\mathbf{u})^!$. Suppose $[\![\phi_j(\mathbf{u})]\!] \wedge p = [\![\phi_j(\mathbf{u}|_p)]\!] \wedge p$ for $j = 1, 2$. It follows from the above relation and the induction hypothesis that

$$\begin{aligned}[\![\phi_1(\mathbf{u}) \to \phi_2(\mathbf{u})]\!] \wedge p &= [([\![\phi_1(\mathbf{u})]\!] \wedge p) \to ([\![\phi_2(\mathbf{u})]\!] \wedge p)] \wedge p \\ &= [([\![\phi_1(\mathbf{u}|_p)]\!] \wedge p) \to ([\![\phi_2(\mathbf{u}|_p)]\!] \wedge p)] \wedge p \\ &= ([\![\phi_1(\mathbf{u}|_p)]\!] \to [\![\phi_2(\mathbf{u}|_p)]\!]) \wedge p,\end{aligned}$$

so that we have

$$[\![\phi_1(\mathbf{u}) \to \phi_2(\mathbf{u})]\!] \wedge p = [\![\phi_1(\mathbf{u}|_p) \to \phi_2(\mathbf{u}|_p)]\!] \wedge p.$$

(iv) Note that the relation

$$(\bigwedge_\alpha P_{1,\alpha} \to P_{2,\alpha}) \wedge Q = (\bigwedge_\alpha (P_{1,\alpha} \wedge Q) \to (P_{2,\alpha} \wedge Q)) \wedge Q$$

holds if $P_{j,\alpha} \downfootline Q$ for $j = 1, 2$, which follows from Proposition 1 (i) and (ii). Suppose $[\![\phi_j(u)]\!] \wedge p = [\![\phi_j(u|_p)]\!] \wedge p$ for $j = 1, 2$ for any $u \in V^{(\mathcal{Q})}$ and $p \in L(u)^!$. Suppose $u \in V^{(\mathcal{Q})}$ and $p \in L(u)^!$. Let $u' \in \mathrm{dom}(u)$. Since $L(u') \subseteq L(u)$, we have $p \in L(u')^!$. It follows that

$$[\![\phi_j(u')]\!] \wedge p = [\![\phi_j(u'|_p)]\!] \wedge p \quad \text{and} \quad p \downfootline [\![\phi(u')]\!], [\![\phi(u'|_p)]\!]$$

for all $u' \in \mathrm{dom}(u)$. Thus, we have

$$\begin{aligned}[\![(\forall x \in u)\phi(x)]\!] \wedge p &= \left(\bigwedge_{u' \in \mathrm{dom}(u)} (u(u') \to [\![\phi(u')]\!])\right) \wedge p \\ &= \bigwedge_{u' \in \mathrm{dom}(u)} [(u(u') \to [\![\phi(u')]\!]) \wedge p] \\ &= \bigwedge_{u' \in \mathrm{dom}(u)} \{[(u(u') \wedge p) \to ([\![\phi(u')]\!] \wedge p)] \wedge p\} \\ &= \bigwedge_{u' \in \mathrm{dom}(u|_p)} \{[u|_p(u') \wedge p \to ([\![\phi(u')]\!] \wedge p)] \wedge p\} \\ &= \bigwedge_{u' \in \mathrm{dom}(u|_p)} \{[u|_p(u') \to ([\![\phi(u')]\!])] \wedge p\} \\ &= \left(\bigwedge_{u' \in \mathrm{dom}(u|_p)} (u|_p(u') \to [\![\phi(u')]\!])\right) \wedge p.\end{aligned}$$

It follows that

$$[\![(\forall x \in u)\phi(x)]\!] \wedge p = [\![(\forall x \in u|_p)\phi(x)]\!] \wedge p.$$

□

Now, we obtain the following transfer principle for bounded theorems of ZFC in the new truth-value assignments for membership and existential quantifications to fully satisfy De Morgan's laws.

Theorem 11 (Δ_0-ZFC Transfer Principle) *For any Δ_0-formula $\phi(x_1, \ldots, x_n)$ of $\mathcal{L}(\in)$ and $u_1, \ldots, u_n \in V^{(\mathcal{Q})}$, if $\phi(x_1, \ldots, x_n)$ is provable in ZFC, then we have*

$$[\![\phi(u_1, \ldots, u_n)]\!] \geq \underline{\vee}(u_1, \ldots, u_n). \tag{21}$$

Proof Let $p = \underline{\vee}(u_1, \ldots, u_n)$. Then, we have $a \wedge p \downarrow\!\!\!\circ\, b \wedge p$ for any $a, b \in L(u_1, \ldots, u_n)$, and hence there is a Boolean sublogic $\mathcal{B}$ such that $L(u_1, \ldots, u_n) \wedge p \subseteq \mathcal{B}$. From Proposition 6, we have $L(u_1|_p, \ldots, u_n|_p) \subseteq \mathcal{B}$. From Proposition 5, we have $u_1|_p, \ldots, u_n|_p \in V^{(\mathcal{B})}$. By the ZFC transfer principle of the Boolean-valued universe [1, Theorem 1.33], we have $[\![\phi(u_1|_p, \ldots, u_n|_p)]\!]_{\mathcal{B}} = 1$. By the Δ_0-absoluteness principle, we have $[\![\phi(u_1|_p, \ldots, u_n|_p)]\!] = 1$. From Theorem 10, we have $[\![\phi(u_1, \ldots, u_n)]\!] \wedge p = [\![\phi(u_1|_p, \ldots, u_n|_p)]\!] \wedge p = p$, and the assertion follows. □

Acknowledgements The author acknowledges the support of the JSPS KAKENHI, No. 26247016, No. 17K19970, and the support of the IRI-NU collaboration. The author thanks the referee for calling his attention to the well-definedness of restrictions of quantum sets.

References

1. Bell, J.L.: Set Theory: Boolean-Valued Models and Independence Proofs, 3rd edn. Oxford UP, Oxford (2005). https://doi.org/10.1093/acprof:oso/9780198568520.001.0001
2. Birkhoff, G., von Neumann, J.: The logic of quantum mechanics. Ann. Math. **37**, 823–843 (1936). https://doi.org/10.2307/1968621
3. Bruns, G., Kalmbach, G.: Some remarks on free orthomodular lattices. In: Schmidt, J., (ed.), Proceedings of the Lattice Theory Conference, pp. 397–408. Houston, USA (1973)
4. Chevalier, G.: Commutators and decompositions of orthomodular lattices. Order **6**, 181–194 (1989). https://doi.org/10.1007/BF02034335
5. Hardegree, G.M.: Material implication in orthomodular (and Boolean) lattices. Notre Dame J. Formal Logic **22**, 163–182 (1981). https://doi.org/10.1305/ndjfl/1093883401
6. Husimi, K.: Studies on the foundation of quantum mechanics I. Proc. Physico-Math. Soc. Japan **19**, 766–778 (1937)
7. Kalmbach, G.: Orthomodular Lattices. Academic, London (1983)
8. Marsden, E.L.: The commutator and solvability in a generalized orthomodular lattice. Pacific J. Math **33**, 357–361 (1970). https://doi.org/10.2140/pjm.1970.33.357
9. Ozawa, M.: Transfer principle in quantum set theory. J. Symb. Log. **72**, 625–648 (2007). https://doi.org/10.2178/jsl/1185803627

10. Ozawa, M.: Quantum set theory extending the standard probabilistic interpretation of quantum theory. New Generat. Comput. **34**, 125–152 (2016). https://doi.org/10.1007/s00354-016-0205-2
11. Ozawa, M.: Operational meanings of orders of observables defined through quantum set theories with different conditionals. In: Duncan, R., Heunen, C., (eds.) Quantum Physics and Logic (QPL) 2016, EPTCS, vol. 236, pp. 127–144. Open Publishing Association (2017). https://doi.org/10.4204/EPTCS.236.9
12. Ozawa, M.: Orthomodular-valued models for quantum set theory. Rev. Symb. Log. **10**, 782–807 (2017). https://doi.org/10.1017/S1755020317000120
13. Pulmannová, S.: Commutators in orthomodular lattices. Demonstratio Math. **18**, 187–208 (1985). https://doi.org/10.1515/dema-1985-0116
14. Sasaki, U.: Orthocomplemented lattices satisfying the exchange axiom. J. Sci. Hiroshima Univ. A **17**, 293–302 (1954)
15. Takeuti, G.: Quantum set theory. In: Beltrametti, E.G., van Fraassen, B.C., (eds.), Current Issues in Quantum Logic: Proceedings of the Workshop on Quantum Logic, December 2–9, 1979, Erice, Sicily, Italy, pp. 303–322. Plenum, New York (1981). https://doi.org/10.1007/978-1-4613-3228-2_19
16. Urquhart, A.: Review. J. Symb. Log. **48**, 206–208 (1983). https://doi.org/10.2307/2273336

Choiceless Löwenheim–Skolem Property and Uniform Definability of Grounds

Toshimichi Usuba

Abstract In this paper, without the axiom of choice, we show that if a certain downward Löwenheim–Skolem property holds then all grounds are uniformly definable. We also prove that the axiom of choice is forceable if and only if the universe is a small extension of some transitive model of ZFC.

Keywords Axiom of choice · Forcing method · Set-theoretic geology

1 Introduction

Set-theoretic geology, which was initiated by Fuchs–Hamkins–Reitz [2], is a study of the structure of all ground models of the universe. In standard set-theoretic geology, the universe is assumed to be a model of ZFC, and all ground models are also supposed to satisfy ZFC. On the other hand, it is possible that the universe is a generic extension of some choiceless model, moreover in modern set theory, the forcing method over choiceless models has become a common tool, e.g., Woodin's $\mathbb{P}$-max forcing over $L(\mathbb{R})$. So it is natural to consider set-theoretic geology without the Axiom of Choice (AC). **The base theory in this paper is ZF unless otherwise specified**. Let us say that a transitive model W of ZF is a *ground* of V if there is a poset $\mathbb{P} \in W$ and a $(W, \mathbb{P})$-generic G with $V = W[G]$. V is a trivial ground of V. Again, **we do not assume that a ground satisfies AC, unless otherwise specified**.

A first problem in developing set-theoretic geology without AC is the uniform definability of all grounds. Laver [12], and independently Woodin, proved that, in ZFC, the universe V is definable in its forcing extension $V[G]$ by a first-order formula with parameters. Fuchs–Hamkins–Reitz [2] refined their result and showed that, in ZFC, all grounds are uniformly definable by a first-order formula:

Theorem 1 (Fuchs–Hamkins–Reitz [2], Reitz [14], in ZFC) *There is a first-order formula $\varphi(x, y)$ of set-theory such that:*

T. Usuba (✉)
Faculty of Fundamental Science and Engineering, Waseda University, Okubo 3-4-1, Shinjyuku, Tokyo 169-8555, Japan
e-mail: usuba@waseda.jp

T. Arai et al. (eds.), *Advances in Mathematical Logic*, Springer Proceedings in Mathematics & Statistics 369, https://doi.org/10.1007/978-981-16-4173-2_8

1. *For every set r, the class $W_r = \{x \mid \varphi(x,r)\}$ is a ground of V with $r \in W_r$ and satisfies* AC.
2. *For every ground W of V, if W satisfies* AC *then $W = W_r$ for some r.*

First-order definability is an important property, which allows us to treat all grounds within the first-order theory ZFC. However, their proofs heavily reply on AC, and it is still open if all grounds are uniformly definable without AC. Gitman-Johnstone [5] obtained a partial result under a fragment of AC. For instance, they showed that if DC_δ holds and a poset $\mathbb{P}$ has cardinality $\leq \delta$ ($\mathbb{P}$ is assumed to be well-orderable), then the universe V is definable in its forcing extension via $\mathbb{P}$. For this problem, we give another partial answer. We prove that if a certain downward Löwenheim–Skolem property holds, then all grounds are uniformly definable as in Theorem 1. We also show that such a downward Löwenheim–Skolem property holds in many natural models of ZF, or if V has many large cardinals. So we can start studying set-theoretic geology in many choiceless models.

We introduce the following notion, which corresponds to the downward Löwenheim–Skolem theorem in the context of ZFC:

Definition 1 An uncountable cardinal κ is a *Löwenheim–Skolem cardinal* (LS cardinal, for short) if for every $\gamma < \kappa \leq \alpha$ and $x \in V_\alpha$, there is $\beta > \alpha$ and an elementary submodel $X \prec V_\beta$ such that:

1. $V_\gamma \subseteq X$.
2. $x \in X$.
3. The transitive collapse of X belongs to V_κ.
4. ${}^{V_\gamma}(X \cap V_\alpha) \subseteq X$.

Clearly a limit of LS cardinals is also an LS cardinal, hence a singular LS cardinal can exist. In ZFC, a cardinal κ is LS if and only if $\kappa = \beth_\kappa$, so there are proper class many LS cardinals.

In ZF, every supercompact cardinal (see Definition 3 below) is an LS cardinal. We show that if there are proper class many LS cardinals, e.g., there are proper class many supercompact cardinals, then all grounds are uniformly definable.

Theorem 2 *Suppose there are proper class many LS cardinals. Then all grounds are uniformly definable, that is, there is a first-order formula $\varphi(x, y)$ of set-theory such that:*

1. *For every r, $W_r = \{x \mid \varphi(x,r)\}$ is a ground of V with $r \in W_r$.*
2. *For every ground W of V, there is r with $W_r = W$.*

We also prove that the statement "there are proper class many LS cardinals" is absolute between V and its forcing extensions. Hence under the assumption, in any grounds and generic extensions of V, we can define all its grounds uniformly.

In ZFC, there are proper class many LS cardinals, which is a consequence of the Löwenheim–Skolem theorem. This means that if there is a poset which forces AC, then we can conclude that V has proper class many LS cardinals. Let us say that AC is *forceable* if there is a poset which forces AC. This result lead us to the problem of when AC is forceable. Blass [1] already considered a necessary and sufficient condition for it. The principle SVC, *Small Violation of Choice*, is the assertion that there is a set X such that for every set Y, there is an ordinal α and a surjection $f : X \times \alpha \to Y$.

Theorem 3 (Blass [1]) *The following are equivalent:*

1. AC *is forceable.*
2. SVC *holds.*

Blass also showed that SVC holds in many choiceless models, such as symmetric models. So such models have proper class many LS cardinals, and all grounds are uniformly definable. We give another characterization, which tells us that AC is forceable if and only if V is a *small* extension of a model of ZFC. For a transitive model W of ZF and a set X, let $W(X)$ be the minimal transitive model of ZF with $W \subseteq W(X)$ and $X \in W(X)$ (see Definition 2 below).

Theorem 4 *The following are equivalent:*

1. AC *is forceable.*
2. *There is a transitive model W of* ZFC *and a set X such that W is definable in V with parameters from W and $V = W(X)$.*
3. *There is a transitive model W of* ZFC *and a set X such that W is definable in V with parameters from W, $V = W(X)$, and W is a ground of some generic extension of V.*

This characterization clarifies the structure of all grounds of V under AC.

Theorem 5 *Suppose V satisfies* AC. *Then for every transitive model M of* ZF, *M is a ground of V if and only if there is a ground W of V and a set X such that W satisfies* AC *and $M = W(X)$. In particular the collection of all grounds satisfying* AC *is dense in all grounds, with respect to $\subseteq$.*

2 Preliminaries

In this paper, we say that a collection M of sets is a *class* of V if $(V; \in, M)$ satisfies the collection scheme, that is, for every formula φ in the language $\{\in, M\}$ (where we identify M as a unary predicate) and all sets $a, v_0, \dots, v_n$, if the sentence $\forall b \in a \exists c\, \varphi(b, c, v_0, \dots, v_n)$ holds in V, then there is a set d such that the sentence $\forall b \in a \exists c \in d\, \varphi(b, c, \dots, v_n)$ holds. A class needs not be definable in V, but every definable collection of sets is a class in our sense. Note also that, by the forcing theorem, if W is a ground of V, then W is a class of V.

The following fact is well-known. See e.g. Jech [7] for the definitions and the proof.

Theorem 6 *Let M be a transitive class containing all ordinals. Then M is a model of* ZF *if and only if M is closed under the Gödel operations and M is almost universal, that is, for every set $x \subseteq M$, there is $y \in M$ with $x \subseteq y$.*

For a transitive model M of ZF and an ordinal α, let M_α be the set of all $x \in M$ with rank $< \alpha$.

We can develop a standard theory of the forcing method without AC. See e.g. Grigorieff [4] for the following facts:

Theorem 7 *Let $V[G]$ be a forcing extension of V via a poset $\mathbb{P} \in V$, and $V[G][H]$ of $V[G]$ via a poset $\mathbb{Q} \in V[G]$. Then there is a poset $\mathbb{R} \in V$ and a $(V, \mathbb{R})$-generic G' such that $V[G][H] = V[G']$.*

This fact shows that if M is a ground of V and W is of M, then W is a ground of V as well.

A poset $\mathbb{P}$ is *weakly homogeneous* if for every $p, q \in \mathbb{P}$, there is an automorphism $f : \mathbb{P} \to \mathbb{P}$ such that $f(p)$ is compatible with q.

Theorem 8 *Suppose $\mathbb{P}$ is a weakly homogeneous poset. For every $x_0, \dots, x_n \in V$ and formula φ, either $\Vdash_{\mathbb{P}} \varphi(x_0, \dots, x_n)$ or $\Vdash_{\mathbb{P}} \neg\varphi(x_0, \dots, x_n)$ in V.*

For a set S, let $\mathrm{Col}(S)$ be the poset consisting of all finite partial functions from ω to S ordered by reverse inclusion. $\mathrm{Col}(S)$ is weakly homogeneous, and if S is an ordinal definable set then so is $\mathrm{Col}(S)$.

Theorem 9 *Let $\mathbb{P}$ be a poset, and G be $(V, \mathbb{P})$-generic. Let α be a limit ordinal with $\alpha >$ rank$(\mathbb{P}) \cdot \omega$. Let H be $(V[G], \mathrm{Col}(V[G]_\alpha))$-generic. Then there is a $(V, \mathrm{Col}(V_\alpha))$-generic H' with $V[G][H] = V[H']$.*

Definition 2 For a transitive model M of ZF containing all ordinals and a set X, let $M(X) = \bigcup_{\alpha \in \mathrm{ON}} L(M_\alpha \cup \{X\})$.[1] If M is a class of V, then $M(X)$ is the minimal transitive class model of ZF with $M \subseteq M(X)$ and $X \in M(X)$.

The following useful fact will be applied frequently:

Theorem 10 (Theorem B in Grigorieff [4]) *Let $W \subseteq V$ be a ground of V. Let M be a transitive model of ZF and suppose $W \subseteq M \subseteq V$. Then the following are equivalent:*

1. *V is a generic extension of M.*
2. *M is of the form $W(X)$ for some $X \in M$.*

We also use the following fact due to Solovay.

Theorem 11 (Solovay, see Fuchs–Hamkins–Rietz [2]) *Let $\mathbb{P}$, $\mathbb{Q}$ be posets, and $G \times H$ be $(V, \mathbb{P} \times \mathbb{Q})$-generic. Then $V[G] \cap V[H] = V$.*

Lemma 1 (Folklore) *Let $\mathbb{P}$ be a poset, and $\alpha > \omega$ a limit ordinal with $\mathbb{P} \in V_\alpha$. Let G be $(V, \mathbb{P})$-generic. For a set $Y \in V$, let $Y[G] = \{\dot{a}_G \mid \dot{a} \in Y \text{ is a } \mathbb{P}\text{-name}\}$, where $\dot{a}_G$ is the interpretation of $\dot{a}$ by G. Then $V[G]_\alpha = V_\alpha[G]$.*

Proof (Sketch of proof) One can check that for every $\mathbb{P}$-name $\dot{a}$, we have rank$(\dot{a}_G) \leq$ rank$(\dot{a})$, hence $V_\alpha[G] \subseteq V[G]_\alpha$. For the converse, by induction on $\beta < \alpha$ with $\mathbb{P} \in V_\beta$, we can take a $\mathbb{P}$-name $\dot{\sigma}$ such that rank$(\dot{\sigma}) < \beta + \omega \leq \alpha$ and $\Vdash_{\mathbb{P}}$"$\dot{\sigma} = V[\dot{G}]_\beta$" (we do not need AC). Hence if $\mathbb{P} \in V_\alpha$ then $V[G]_\alpha \subseteq V_\alpha[G]$. □

[1] In [4], our $M(X)$ is referred to $M[X]$.

3 Löwenheim–Skolem Cardinals

In this section we shall observe some basic properties of LS cardinals, but results in this section are not required to prove the main theorems.

First we prove that in ZF, every supercompact cardinal is an LS-cardinal.

Definition 3 (*Woodin, Definition 220 in* [16]) An uncountable cardinal κ is *supercompact* if for every $\alpha \geq \kappa$, there is $\beta \geq \alpha$, a transitive set N, and an elementary embedding $j : V_\beta \to N$ such that the critical point of j is κ, $\alpha < j(\kappa)$, and ${}^{V_\alpha}N \subseteq N$.

If κ is supercompact, then κ is regular and V_κ is a model of ZF.

Lemma 2 *Every supercompact cardinal is an LS cardinal, and a limit of LS cardinals.*

This lemma is an immediate consequence of the following result of Woodin. For an ordinal γ, $V_\gamma \prec_{\Sigma_1^*} V$ means that $V_\gamma \prec_{\Sigma_1} V$ and for all $\alpha < \gamma$, $a \in V_\gamma$, and for all Σ_0-formula $\varphi(x, y)$, if there is $b \in V$ such that $\varphi(a, b)$ holds and ${}^{V_\alpha}b \subseteq b$ then there is $b \in V_\gamma$ such that $\varphi(a, b)$ holds and ${}^{V_\alpha}b \subseteq b$.

Theorem 12 (Woodin, Lemma 222 in [16]) *For an uncountable cardinal κ, the following are equivalent:*

1. *κ is supercompact.*
2. *For all $\gamma > \kappa$ such that $V_\gamma \prec_{\Sigma_1^*} V$, for all $a \in V_\gamma$, there exists $\overline{\gamma} < \kappa$, $\overline{a} \in V_{\overline{\gamma}}$, and an elementary embedding $j : V_{\overline{\gamma}+1} \to V_{\gamma+1}$ with critical point $\overline{\kappa} < \kappa$ such that $j(\overline{\kappa}) = \kappa$, $j(\overline{a}) = a$, and such that $V_{\overline{\gamma}} \prec_{\Sigma_1^*} V$.*

In ZFC, the existence of proper class many LS cardinals is provable, and LS cardinal is not a large cardinal. However we will see that the existence of an LS cardinal is not provable from ZF.

Definition 4 An uncountable cardinal κ is *weakly Löwenheim–Skolem* (weakly LS, for short) if for every $\gamma < \kappa \leq \alpha$ and $x \in V_\alpha$, there is $X \prec V_\alpha$ such that:

1. $V_\gamma \subseteq X$.
2. $x \in X$.
3. The transitive collapse of X belongs to V_κ.

Clearly every LS cardinal is weakly LS.

Lemma 3 *Let κ be a weakly LS cardinal.*

1. *For every $x \in V_\kappa$, there is no surjection from x onto κ.*
2. *For every cardinal $\lambda \geq \kappa$ and $x \in V_\kappa$, there is no cofinal map from x into λ^+. In particular* $\mathrm{cf}(\lambda^+) \geq \kappa$.

Proof (1). Suppose not. Then there is $\gamma < \kappa$ and a surjection f from V_γ onto κ. Take a large $\alpha \geq \kappa$ and $X \prec V_\alpha$ such that $V_\gamma \subseteq X$, $\gamma, f \in X$, and the transitive collapse of X is in V_κ. Clearly $|X \cap \kappa| < \kappa$, otherwise the transitive collapse of X cannot be in V_κ. However, since $V_\gamma \subseteq X$ and $f \in X$, we have $\kappa = f``V_\gamma \subseteq X$, this is a contradiction.

(2). Suppose to the contrary that there is a set $x \in V_\kappa$ and a cofinal map f from x into λ^+. Fix $\gamma < \kappa$ with $x \in V_\gamma$. Take a large $\alpha > \lambda^+$ and $X \prec V_\alpha$ such that:

1. $V_\gamma \subseteq X$.
2. The transitive collapse of X is in V_κ.
3. X contains all relevant objects.

Note that $x \subseteq X$, hence $f``x \subseteq X$ and $\lambda^+ = \sup(f``x) = \sup(X \cap \lambda^+)$.

Let Y be the transitive collapse of X, and $\pi : X \to Y$ be the collapsing map. Now define $g : Y \times \lambda \to \lambda^+$ as follows: For $\langle a, \beta \rangle \in Y \times \lambda$, if $\pi^{-1}(a)$ is a sujerction from λ onto some ordinal $< \lambda^+$, then $g(a, \beta) = \pi^{-1}(a)(\beta)$, and $g(a, \beta) = 0$ otherwise. Since $\sup(X \cap \lambda^+) = \lambda^+$, g is a sujerction from $Y \times \lambda$ onto λ^+. For $\beta < \lambda$, let $R_\beta = \{g(a, \beta) \mid a \in Y\} \subseteq \lambda^+$. We know $\lambda^+ = \bigcup_{\beta<\lambda} R_\beta$. If $\mathrm{ot}(R_\beta) < \kappa$ for every $\beta < \lambda$, we can take a canonical surjection from $\lambda \times \kappa$ onto λ^+. However we can prove $|\lambda \times \kappa| = \lambda$ in ZF, hence $|\lambda^+| = |\lambda \times \kappa| = \lambda$, so this is impossible. Thus there is $\beta < \lambda$ with $\mathrm{ot}(R_\beta) \geq \kappa$. This means that there is a surjection from Y onto κ via R_β, contradicting (1). □

Consider the model of ZF constructed by Gitik [3], which has no regular uncountable cardinals. By Lemma 3, there are no (weakly) LS cardinals in this model.

Corollary 1 *An uncountable cardinal κ is an LS-cardinal if and only if for every set x and $\gamma < \kappa$, there is $\alpha \geq \kappa$ and $X \prec V_\alpha$ such that $x \in X$, $V_\gamma \subseteq X$, ${}^{V_\gamma}X \subseteq X$, and the transitive collapse of X belongs to V_κ.*

Proof The "if" part is clear. For the converse, take a set x and $\gamma < \kappa$. By Lemma 3, we can find $\alpha > \kappa$ such that $x \in V_\alpha$, but there is no $y \in V_\kappa$ for which there is a cofinal map $f : y \to \alpha$. Since κ is LS, we can find $\beta > \alpha$ and $X' \prec V_\beta$ such that $x, \alpha \in X'$, $V_\gamma \subseteq X'$, ${}^{V_\gamma}(X' \cap V_\alpha) \subseteq X'$, and the transitive collapse of X' is in V_κ. Let $X = X' \cap V_\alpha$. We have that $X \prec V_\alpha$, $x \in X$, and its transitive collapse belongs to V_κ. Next take $f : V_\gamma \to X$. We know $f \in X'$. By the choice of α, the set $\{\mathrm{rank}(f(z)) \mid z \in V_\gamma\}$ is bounded in α. Hence $f \in V_\alpha$, and $f \in X' \cap V_\alpha = X$. □

Next we prove that if κ is weakly LS, then the club filter over λ^+ for $\lambda \geq \kappa$ is κ-complete.

Lemma 4 *Let κ be a weakly LS cardinal. Let $\lambda \geq \kappa$ be a cardinal and $x \in V_\kappa$. Let f be a function from x into the club filter over λ^+. Then $\bigcap f``x$ contains a club in λ^+. In particular, the club filter over λ^+ is κ-complete.*

Proof Take $\gamma < \kappa$ with $x \in V_\gamma$ and sufficiently large $\alpha > \lambda^+$. Take $X \prec V_\alpha$ such that $V_\gamma \subseteq X$, $f, \lambda^+ \in X$, and the transitive collapse of X is in V_κ. Put $\mathcal{C} = \{C \in X \mid C$ is a club in $\lambda^+\}$. We know that for every $a \in x$ there is a club $C \in \mathcal{C}$ with $C \subseteq f(a)$.

Let $D = \bigcap \mathcal{C} \subseteq \bigcap f``x$. It is enough to see that D is a club in λ^+. Closure is clear, so we check that D is unbounded in λ^+. Take $\xi < \lambda^+$. Fix $\delta < \kappa$ such that the transitive collapse of X is in V_δ. Again, take another large $\beta > \alpha$ and $Y \prec V_\beta$ such that $V_\delta \subseteq Y$, $X, \xi, D \in Y$, the transitive collapse of Y is in V_κ. There is a surjection from V_δ onto X, hence we have $X \subseteq Y$, and $\mathcal{C} \subseteq Y$. Note that $\sup(Y \cap \lambda^+) < \lambda^+$, otherwise we can take a cofinal map from the transitive collapse of Y into λ^+, which contradicts to Lemma 3. $\xi < \sup(Y \cap \lambda^+) < \lambda^+$, so it is sufficient to check that $\sup(Y \cap \lambda^+) \in D = \bigcap \mathcal{C}$. For each $C \in \mathcal{C}$, we have $C \in Y$. Since C is a club in λ^+ and $\sup(Y \cap \lambda^+) < \lambda^+$, we have $\sup(Y \cap \lambda^+) \in C$. □

We also show a variant of Fodor's lemma for weakly LS cardinals.

Lemma 5 *Let κ be a weakly LS cardinal. Let $\lambda \geq \kappa$ be a cardinal, and $f : \lambda^+ \setminus \{0\} \to \lambda^+$ be a regressive function. Then there is $\gamma < \lambda^+$ such that the set $\{\xi \in \lambda^+ \mid f(\xi) \leq \gamma\}$ is stationary in λ^+.*

Proof Let $f : \lambda^+ \setminus \{0\} \to \lambda^+$. Take a large $\alpha > \lambda^+$ and $X \prec V_\alpha$ such that $\lambda^+, f \in X$ and the transitive collapse of X is in V_κ. We know $\eta = \sup(X \cap \lambda^+) < \lambda^+$. Pick $\gamma \in X \cap \lambda^+$ with $f(\eta) \leq \gamma$. Then $S = \{\xi \in \lambda^+ \mid f(\xi) \leq \gamma\} \in X$. If S is non-stationary in λ^+, then there is a club $C \in X$ with $C \cap S = \emptyset$. But $\eta = \sup(X \cap \lambda^+) \in C$, so $\eta \in C \cap S$. This is a contradiction. □

That there are no LS cardinals holds in any model in which there are no regular uncountable cardinals, but this later statement is known to have large cardinals strength. So the following natural question arises: What is the consistency strength of "no (weakly) LS cardinals"? For this question, Asaf Karagila pointed out the following:

Theorem 13 (Karagila [8]) *Suppose V satisfies* AC *and* GCH. *Then there is an extension of V with same cofinalities as V, such that Fodor's lemma fails and the club filter is not σ-complete on every regular uncountable cardinal.*

In his model, there are no (weakly) LS cardinals by Lemmas 4 and 5.

Woodin (Theorem 227 in [16]) proved that if λ is a singular cardinal and a limit of supercompact cardinals, then λ^+ is regular, and the club filter over λ^+ is λ^+-complete. Now we can replace supercompact cardinals in Woodin's result by weakly LS cardinals:

Corollary 2 *Let λ be a singular weakly LS cardinal (e.g., a singular limit of weakly LS cardinals).*

1. *There is no cofinal map from V_λ into λ^+. In particular λ^+ is regular.*
2. *Let f be a function from V_λ into the club filter over λ^+. Then $\bigcap f``V_\lambda$ contains a club in λ^+". In particular the club filter over λ^+ is λ^+-complete.*
3. *For every regressive function $f : \lambda^+ \setminus \{0\} \to \lambda^+$, there is $\gamma < \lambda^+$ such that the set $\{\xi < \lambda^+ \mid f(\xi) = \gamma\}$ is stationary in λ^+.*

Proof Fix an increasing sequence $\langle \lambda_i \mid i < \mathrm{cf}(\lambda) \rangle$ with limit λ.

(1). Let $g : V_\lambda \to \lambda^+$. For $i < \mathrm{cf}(\lambda)$, let $\alpha_i = \sup(g``V_{\lambda_i})$. We have $\alpha_i < \lambda^+$ by Lemma 3. Again, since $\mathrm{cf}(\lambda) < \lambda$, we have $\sup(g``V_\lambda) = \sup\{\alpha_i \mid i < \mathrm{cf}(\lambda)\} < \lambda^+$ by Lemma 3.

(2). For a given f, we have that for every $i < \mathrm{cf}(\lambda)$, $\bigcap f``V_{\lambda_i}$ contains a club in $\lambda^+$ by Lemma 4. Then $\bigcap f``V_\lambda = \bigcap_{i<\mathrm{cf}(\lambda)} \bigcap f``V_{\lambda_i}$ contains a club by Lemma 4 again.

For (3), by Lemma 5, there is the minimal $\gamma < \lambda^+$ such that $\{\xi < \lambda^+ \mid f(\xi) \leq \gamma\}$ is stationary in λ^+. We will show that the set $\{\xi < \lambda^+ \mid f(\xi) = \gamma\}$ is stationary. Since λ is singular, we have $\mathrm{cf}(\gamma) < \lambda$. Take a sequence $\langle \gamma_i \mid i < \mathrm{cf}(\gamma) \rangle$ with limit γ, and let $S_i = \{\xi < \lambda^+ \mid f(\xi) \leq \gamma_i\}$. By the minimality of γ, each S_i is non-stationary. Then $\bigcup_{i<\mathrm{cf}(\gamma)} S_i$ is non-stationary by (2). This means that the set $\{\xi < \lambda^+ \mid f(\xi) = \gamma\}$ must be stationary. □

Question 1 1. Is the statement ZF+"there is a weakly LS cardinal which is not LS" consistent?
2. Suppose there is a supercompact cardinal (or an extendible cardinal). Then are there proper class many (weakly) LS cardinals?
3. Suppose λ is a singular weakly LS cardinal. Is the club filter over λ^+ normal?

4 Uniform Definability of Grounds

In this section, we prove that if there are proper class many LS cardinals, then all grounds are uniformly definable. For this purpose, we introduce a very rough measure on sets, which will be used instead of cardinality.

Definition 5 For a set x, the *norm of* x, $\|x\|$, is the least ordinal α such that there is a surjection from V_α onto x.

The following is easy to check:

Lemma 6 *1.* $\|x\| \leq \mathrm{rank}(x)$.
2. If $x \subseteq y$ then $\|x\| \leq \|y\|$.
3. If $M \subseteq V$ is a transitive model of (a sufficiently large fragment of) ZF *and $x \in M$, then $\|x\|^M \geq \|x\|$.*
4. If X is an extensional set (that is, for every $x, y \in X$, $x = y \iff \forall z \in X (z \in x \leftrightarrow z \in y)$) and its transitive collapse belongs to V_α for some α, then $\|X\| < \alpha$.

Definition 6 Let Z* be the theory Z, ZF−Replacement Scheme, with the conjunction of the following statements:

1. Every set x has transitive closure $\mathrm{trcl}(x)$.
2. Every set x has rank, that is, there is a surjection from $\mathrm{trcl}(x)$ onto some ordinal α such that $f(y) = \sup\{f(z) + 1 \mid z \in y\}$ for every $y \in \mathrm{trcl}(x)$. Such an α is the rank of x.

3. For every ordinal α, the collection of sets with rank $< \alpha$ forms a set.
4. Every extensional set has a (unique) transitive collapse and a collapsing map, that is, for every set X, if $\forall x, y \in X(x = y \iff \forall z \in X(z \in x \leftrightarrow z \in y))$, then there is a transitive set Y and an $\in$-isomorphism from X onto Y.

For a transitive model M of Z^* and $\alpha \in M \cap \mathrm{ON}$, let $M_\alpha = M \cap V_\alpha$. We know $M_\alpha \in M$.

Note 1 Let M be a transitive model of Z^*.

1. For $\alpha \in M \cap \mathrm{ON}$, $M_{\alpha+1} = \mathcal{P}(M_\alpha)^M = \mathcal{P}(M_\alpha) \cap M$, and $M_\alpha = \bigcup_{\beta<\alpha} M_\beta$ if α is limit.
2. $M = \bigcup_{\alpha \in M \cap \mathrm{ON}} M_\alpha$.
3. For $\gamma \in M \cap \mathrm{ON}$, if V_γ is a model of Z^* then M_γ is also a model of Z^*.

For models of Z^*, we define variants of the covering and approximation properties in Hamkins [6].

Definition 7 Let $M \subseteq V$ be a transitive model of Z^*. Let $\alpha \in M$ be an ordinal.

1. We say that M satisfies the *α-norm covering property* (for V) if for every $\beta \in M \cap \mathrm{ON}$ and set $x \subseteq M_\beta$, if $\|x\| < \alpha$ then there is $y \in M$ such that $x \subseteq y$ and $\|y\|^M < \alpha$.
2. We say that M satisfies the *α-norm approximation property* (for V) if for every $\beta \in M \cap \mathrm{ON}$ and set $x \subseteq M_\beta$, if $x \cap a \in M$ for every $a \in M$ with $\|a\|^M < \alpha$, then $x \in M$.

Note 2 1. M satisfies the α-norm covering property if and only if for every $\beta \in M \cap \mathrm{ON}$ and set $x \subseteq M_\beta$, if $\|x\| < \alpha$ then there is $y \in M_{\beta+1}$ such that $x \subseteq y$ and $\|y\|^M < \alpha$.
2. M satisfies the α-norm approximation property if and only if for every $\beta \in M \cap \mathrm{ON}$ and set $x \subseteq M_\beta$, if $x \cap a \in M$ for every $a \in M_{\beta+1}$ with $\|a\|^M < \alpha$, then $x \in M$.

Lemma 7 *Let $M \subseteq V$ be a transitive model of Z^*. Let $\gamma \in M \cap \mathrm{ON}$ be an ordinal, and suppose M satisfies the γ-norm covering and the γ-norm approximation properties for V. Fix $\alpha > \gamma$ with $\alpha \in M$, and let $\beta > \alpha$ and $X \prec V_\beta$ be such that:*

1. $V_\gamma \subseteq X$.
2. $M_{\alpha+1}, \gamma \in X$.
3. ${}^{V_\gamma}(X \cap V_\alpha) \subseteq X$.

Then $X \cap M_\alpha \in M$.

Proof By the γ-norm approximation property of M, it is enough to see that for every $a \in M_{\alpha+1}$, if $\|a\|^M < \gamma$ then $a \cap X \cap M_\alpha \in M$. Fix $a \in M_{\alpha+1}$ with $\|a\|^M < \gamma$. Note that $\|a\| < \gamma$. Since $a \cap X \cap M_\alpha \subseteq X$ and $\|a \cap X \cap M_\alpha\| \leq \|a\| < \gamma$, there is a surjection from V_γ onto $a \cap X \cap M_\alpha$. ${}^{V_\gamma}(X \cap V_\alpha) \subseteq X$, hence we have $a \cap X \cap M_\alpha \in X$. Because M satisfies the γ-norm covering property for V, there is some

$x \in M_{\alpha+1}$ such that $\|x\|^M < \gamma$ and $a \cap X \cap M_\alpha \subseteq x$. By the elementarily of X, we may assume that $x \in X$. Note that $x \subseteq X$ since $\|x\| \leq \|x\|^M < \gamma$ and $V_\gamma \subseteq X$. Then we have $a \cap X \cap M_\alpha = a \cap x \in M$. □

Lemma 8 *Let $M, N \subseteq V$ be transitive models of Z^* with $M \cap \mathrm{ON} = N \cap \mathrm{ON}$. Let κ be an LS cardinal with $\kappa \in M \cap N$, and suppose there is $\gamma < \kappa$ such that M and N satisfy the γ-norm covering and the γ-norm approximation properties for V. If $M_\kappa = N_\kappa$, then $M = N$.*

Proof We show $M_\alpha = N_\alpha$ by induction on $\alpha \in M \cap \mathrm{ON}$. The cases that $\alpha \leq \kappa$ and α is limit are clear. So suppose $\alpha = \bar{\alpha} + 1$ for some $\overline{\alpha} \geq \kappa$ and $M_{\bar{\alpha}} = N_{\bar{\alpha}}$.

First, we show that for every $x \in M_\alpha$, if $\|x\| < \gamma$ then $x \in N$. Since κ is LS, we can find a large $\beta > \alpha$ and $X \prec V_\beta$ such that:

1. $V_\gamma \subseteq X$,
2. ${}^{V_\gamma}(X \cap V_\alpha) \subseteq X$,
3. the transitive collapse of X is in V_κ, and
4. X contains all relevant objects.

Then, by Lemma 7, we have that $X \cap M_\alpha \in M$ and $X \cap N_\alpha \in N$. In particular $X \cap M_{\bar{\alpha}} \in M$ and $X \cap N_{\bar{\alpha}} \in N$. On the other hand, $M_{\bar{\alpha}} = N_{\bar{\alpha}}$ by the induction hypothesis. Hence we have $X \cap M_{\bar{\alpha}} = X \cap N_{\bar{\alpha}} \in M \cap N$. Since $\|x\| < \gamma$, we have $x \subseteq X$. Thus we have $x \subseteq X \cap M_{\bar{\alpha}} = X \cap N_{\bar{\alpha}}$. $X \cap M_{\bar{\alpha}}$ is extensional, so we can take the transitive collapse Y of $X \cap M_{\bar{\alpha}}$ and the collapsing map $\pi : X \cap M_{\bar{\alpha}} \to Y$. Note that Y and π are in $M \cap N$ because M and N are models of Z^*. The transitive collapse of X is in V_κ, hence Y is also in V_κ and thus $Y \in M_\kappa = N_\kappa$. Put $y = \pi``x \subseteq Y \in M_\kappa$. $y$ is in $M_\kappa$, and hence is also in $N_\kappa$. Now we have $x = \pi^{-1}``y \in N$.

The same argument shows that for every $x \in N_\alpha$, if $\|x\| < \gamma$ then $x \in M$. Finally, by the γ-norm approximation property of M and N, we have $\mathcal{P}(M_{\bar{\alpha}}) \cap M = \mathcal{P}(M_{\bar{\alpha}}) \cap N$, hence $M_\alpha = N_\alpha$. □

Corollary 3 *Suppose there are proper class many LS cardinals. Then there is a first-order formula $\varphi'(x, y)$ of set-theory such that:*

1. *$W'_r = \{x \mid \varphi'(x, r)\}$ is a transitive class model of ZF containing all ordinals such that $r \in W'_r$, and W'_r satisfies the α-norm covering and approximation properties for V for some α.*
2. *For every transitive class model $W \subseteq V$ of ZF containing all ordinals, if W satisfies the α-norm covering and approximation properties for V for some α, then there is r with $W'_r = W$.*

Proof For a set r, we define W'_r if r satisfies the following conditions:

1. $r = \langle X, \kappa, \alpha \rangle$ where κ is an LS cardinal, $\alpha < \kappa$, and X is a transitive set with $X \cap \mathrm{ON} = \kappa$.
2. For each cardinal $\lambda > \kappa$, if V_λ is a model of Z^* then there is a unique transitive model $X^{r,\lambda}$ of Z^* such that $X^{r,\lambda} \cap \mathrm{ON} = \lambda$, $(X^{r,\lambda})_\kappa = X$, and $X^{r,\lambda}$ satisfies the α-norm covering and the α-norm approximation properties for V.

In this case, let $W'_r = \bigcup\{X^{r,\lambda} \mid \lambda > \kappa$ is a cardinal, V_λ is a model of $\mathsf{Z}^*\}$. Otherwise, let $W'_r = V$. It is clear that the collection $\{W'_r \mid r \in V\}$ is a uniformly definable collection of classes. We see that $\{W'_r \mid r \in V\}$ is as required.

First we check condition (1). If $W'_r = V$, then it is clear. Suppose not, then r is of the form $\langle X, \kappa, \alpha\rangle$. It is clear that W'_r is transitive and contains all ordinals. Now fix cardinals $\lambda_0 > \lambda_1 > \kappa$ such that V_{λ_0} and V_{λ_1} are models of Z^*, and let X^{r,λ_0}, X^{r,λ_1} be transitive models of Z^*. It is routine to check that $(X^{r,\lambda_0})_{\lambda_1}$ is a model of Z^*, and satisfies the α-norm covering and approximation properties. Because $X = (X^{r,\lambda_0})_\kappa = (X^{r,\lambda_1})_\kappa$, we have $(X^{r,\lambda_0})_{\lambda_1} = X^{r,\lambda_1}$ by Lemma 8. This means that $(W'_r)_\lambda = X^{r,\lambda}$ for every cardinal $\lambda > \kappa$ with V_λ a model of Z^*, and that W'_r is almost universal and closed under the Gödel operations. Thus we have that W'_r is a model of ZF by Theorem 6. Moreover, by the definition of the norm covering and approximation properties, it is also easy to check that W'_r satisfies the α-norm covering and approximation properties. Finally, we have $X = (W'_r)_\kappa \in W'_r$, and $r \in W'_r$.

For (2), suppose W is a transitive class model of ZF and W satisfies the α-norm covering and approximation properties for V for some α. Fix an LS cardinal $\kappa > \alpha$, and let $X = M_\kappa$ and $r = \langle X, \kappa, \alpha\rangle$. For each cardinal $\lambda > \kappa$, if V_λ is a model of Z^* then W_λ is a transitive model of Z^*, satisfies the α-norm covering and approximation properties for V, and, by Lemma 8, W_λ is a unique transitive model M of Z^* satisfying the α-norm covering and approximation properties for V and $M_\kappa = X$. Then we have $W = W'_r$ by the definition of W'_r. □

Note 3 In item (2) of the previous corollary, W need not to be a class of V; In fact, it is sufficient that W is a transitive model of ZF satisfying the norm covering and approximation properties.

Lemma 9 *Let κ be a weakly LS cardinal. Let $M \subseteq V$ be a ground of V, and suppose there is a poset $\mathbb{P} \in M_\kappa$ and an $(M, \mathbb{P})$-generic G with $V = M[G]$. Then M satisfies the κ-norm covering and the κ-norm approximation properties for V.*

Proof First we show that M satisfies the κ-norm covering property for V. Take α and $x \subseteq M_\alpha$ with $\|x\| < \kappa$. Fix a limit $\gamma < \kappa$ and a surjection $f : V_\gamma \to x$. We may assume that $\mathbb{P} \in V_\gamma$. We know $V_\gamma = \{\dot{y}_G \mid \dot{y} \in M_\gamma$ is a $\mathbb{P}$-name$\}$ (see Lemma 1), where $\dot{y}_G$ is the interpretation of $\dot{y}$ by G. Hence we have a canonical surjection $\dot{y} \mapsto \dot{y}_G$ from all $\mathbb{P}$-names in M_γ onto V_γ, and so we can take a surjection g from M_γ onto x. Let $\dot{g}$ and $\dot{x}$ be $\mathbb{P}$-names for g and x respectively. We work in M. Fix $p_0 \in G$ such that $p_0 \Vdash_{\mathbb{P}}$"$\dot{g}$ is a surjection from M_γ onto $\dot{x} \subseteq M_\alpha$" in M. For $a \in M_\gamma$ and $p \in \mathbb{P}$ with $p \leq p_0$, take a unique $x_{a,p} \in M_\alpha$ with $p \Vdash_{\mathbb{P}}$"$\dot{g}(a) = x_{a,p}$" if it exists. If there is no such $x_{a,p}$, let $x_{a,p} = \emptyset$. Let $x' = \{x_{a,p} \mid a \in M_\gamma, p \in \mathbb{P}\} \in M$. Clearly $x \subseteq x'$. Moreover we can easily take a surjection from $M_\gamma \times \mathbb{P}$ onto x', hence $\|x'\|^M \leq \gamma + \omega < \kappa$.

For the κ-norm approximation property of M, take $\alpha \in M$, $A \subseteq M_\alpha$, and suppose $A \cap a \in M$ for every $a \in M_{\alpha+1}$ with $\|a\|^M < \kappa$. Take a $\mathbb{P}$-name $\dot{A} \in M$ for A. Take $p_0 \in G$ with $p_0 \Vdash_{\mathbb{P}}$"$\dot{A} \subseteq M_\alpha$, and $\dot{A} \cap a \in M$ for every $a \in M_{\alpha+1}$ with $\|a\|^M < \kappa$" in M. For $p \in \mathbb{P}$ with $p \leq p_0$, let $A_p = \{a \in M_\alpha \mid p \Vdash_{\mathbb{P}}$"$a \in \dot{A}$"$\} \in M$. We claim that there is $p \in \mathbb{P}$ with $A_p = A$, which completes our proof.

Suppose to the contrary that there is no $p \in \mathbb{P}$ with $A_p = A$. Take $\gamma < \kappa$ with $\mathbb{P} \in V_\gamma$. Since κ is a weakly LS cardinal, we can find $\beta > \alpha$ and $X \prec V_\beta$ such that:

1. $V_\gamma \subseteq X$.
2. The transitive collapse of X is in V_κ.
3. X contains all relevant objects.

Consider $M_\alpha \cap X$. Since the transitive collapse of X is in V_κ, we have $\|X\| < \kappa$, and $\|M_\alpha \cap X\| < \kappa$ as well. We know that M satisfies the κ-norm covering property, thus we can find $x \in M$ such that $\|x\|^M < \kappa$ and $M_\alpha \cap X \subseteq x$. We may assume that $x \in M_{\alpha+1}$. By the assumption, we have $A' = A \cap x \in M$. Thus there is $p \in G$ such that $p \leq p_0$ and $p \Vdash_{\mathbb{P}}$ "$\dot{A} \cap x = A'$", which means that $A \cap x = A_p \cap x$. Since $\mathbb{P} \in X$ and $\mathbb{P} \in V_\gamma$, we have $\mathbb{P} \subseteq X$, hence $p \in X$, and $A_p \in X$ as well. Since $A_p \neq A$, there is $a \in A \Delta A_p$. Because $A_p, A \in X$, we may assume $a \in X$, so $a \in X \cap M_\alpha \subseteq x$. Then $a \in (A \Delta A_p) \cap x = (A \cap x) \Delta (A_p \cap x)$, this is a contradiction. □

Now the uniform definability of grounds is immediate from Corollary 3 and Lemma 9:

Corollary 4 *Suppose there are proper class many LS cardinals. Then there is a formula $\varphi(x, y)$ of set-theory such that:*

1. *$W_r = \{x \mid \varphi(x, r)\}$ is a ground of V with $r \in W_r$.*
2. *For every ground W of V, there is r with $W_r = W$.*

Proof Let $\{W'_r \mid r \in V\}$ be the collection defined in Corollary 3. Then define $\{W_r \mid r \in V\}$ as follows: For a set r, if there are some poset $\mathbb{P} \in W'_r$ and an $(W'_r, \mathbb{P})$-generic G with $W'_r[G] = V$, then let $W_r = W'_r$. If otherwise, put $W_r = V$. By Corollary 3 and Lemma 9, the collection $\{W_r \mid r \in V\}$ is all grounds of V. □

Question 2 Suppose there is one supercompact cardinal (or one extendible cardinal). Are all grounds uniformly definable as in Theorem 2?

Finally we shall prove that the statement that "there are proper class many LS cardinals" is absolute between V and its forcing extensions.

Lemma 10 *Let $\mathbb{P}$ be a poset, and $\kappa < \lambda$ cardinals with $\mathbb{P} \in V_\kappa$. If $\Vdash_{\mathbb{P}}$ "κ and λ are LS", then λ is LS in V.*

Proof Take a set-forcing extension $V[G]$ of V via $\mathbb{P}$. In $V[G]$, since κ is an LS cardinal and $\mathbb{P} \in V_\kappa$, V satisfies the κ-norm covering and approximation properties for $V[G]$ by Lemma 9. We shall see that λ is LS in V. Take $\gamma < \lambda \leq \alpha$ and $x \in V_\alpha$. Take a large β and $X \prec V[G]_\beta$ such that $V[G]_\gamma \subseteq X$, X contains all relevant objects, the transitive collapse of X is in $V[G]_\lambda$, and ${}^{V[G]_\gamma}(X \cap V[G]_\alpha) \subseteq X$. Let $Y = X \cap V_{\alpha+1}$. We know $Y \prec V_{\alpha+1}$. Moreover $Y \in V$ by Lemma 7. Since the transitive collapse of X is in $V[G]_\lambda$, we have that Y is in V_λ. Since $V_\gamma \subseteq V[G]_\gamma \subseteq X$, we have $V_\gamma \subseteq X \cap V_{\alpha+1} = Y$. Finally we check that $({}^{V_\gamma}(Y \cap V_\alpha))^V \subseteq Y$. Take $f : V_\gamma \to Y \cap V_\alpha$ with $f \in V$. We know $V_\gamma \subseteq V[G]_\gamma$, $V_\alpha \subseteq V[G]_\alpha$, and ${}^{V[G]_\gamma}(X \cap V[G]_\alpha) \subseteq X$, so $f \in X \cap V_{\alpha+1} = Y$. □

Lemma 11 *(Folklore) Let $\mathbb{P}$ be a poset, and $\alpha > \omega$ a limit ordinal with $\mathbb{P} \in V_\alpha$. Suppose also that V_α satisfies the Σ_1-collection scheme. For every $X \prec V_\alpha$ with $\mathbb{P} \in X$ and $\mathbb{P} \subseteq X$, we have $X[G] \prec V[G]_\alpha$.*

Proof (Sketch of proof) If M is a transitive model of $\mathsf{Z}+\Sigma_1$-collection scheme, in M we can define the forcing relation and prove the forcing theorem,[2] that is, for every poset $\mathbb{P} \in M$, formula φ, $\mathbb{P}$-names $\dot{a}_0, \dots \dot{a} \in M$, and $(M, \mathbb{P})$-generic G, we have that $M[G] \models \varphi((\dot{a}_0)_G, \dots, (\dot{a}_n)_G)$ if and only if $p \Vdash_{\mathbb{P}} \varphi(\dot{a}_0, \dots, \dot{a}_n)$ holds in M for some $p \in G$.

Now suppose V_α satisfies the Σ_1-collection scheme. Note that $V[G]_\alpha = V_\alpha[G]$ by Lemma 8. To see that $X[G] \prec V[G]_\alpha = V_\alpha[G]$, by Tarski-Vaught criterion, it is enough to see that for every formula φ and $a_0, \dots, a_n \in X[G]$, if $V[G]_\alpha \models \exists x \varphi(a_0, \dots, a_n, x)$ then there is $b \in X[G]$ with $V[G]_\alpha \models \varphi(a_0, \dots, a_n, b)$. Let $\dot{a}_0, \dots, \dot{a}_n \in X$ be $\mathbb{P}$-names for $a_0, \dots, a_n$ respectively. By the forcing theorem, there is some $p \in G$ and $\mathbb{P}$-name $\dot{\sigma} \in V_\alpha$ such that $p \Vdash_{\mathbb{P}} \varphi(\dot{a}_0, \dots, \dot{a}_n, \dot{\sigma})$ in V_α. Hence the statement $\exists \dot{\sigma}(\dot{\sigma}$ is a $\mathbb{P}$-name and $p \Vdash_{\mathbb{P}} \varphi(\dot{a}_0, \dots, \dot{a}_n, \dot{\sigma}))$ holds in V_α. Because $X \prec V_\alpha$, we can find a witness $\dot{\tau} \in X$. Then $(\dot{\tau})_G \in X[G]$ and $V[G]_\alpha \models \varphi((\dot{a}_0)_G, \dots, (\dot{a}_n)_G, (\dot{\tau})_G)$, as required. □

Lemma 12 *Let κ be a cardinal limit of LS cardinals (hence κ itself is an LS cardinal) and $\mathbb{P} \in V_\kappa$ be a poset. Let G be $(V, \mathbb{P})$-generic. Then κ is LS in $V[G]$.*

Proof In $V[G]$, fix an ordinal $\alpha > \kappa$, $x \in V[G]_\alpha$, and $\gamma < \kappa$. We will find some $\beta > \alpha$ and $X \prec V[G]_\beta$ such that $V[G]_\gamma \subseteq X$, $x \in X$, the transitive collapse of X is in $V[G]_\kappa$, and ${}^{V[G]_\gamma}(X \cap V[G]_\alpha) \subseteq X$. Let $\dot{x}$ be a name for x.

In V, take a limit $\beta > \alpha$ such that V_β satisfies the Σ_1-collection scheme. Take an LS cardinal $\delta < \kappa$ with $\gamma < \delta$, and a submodel $Y \prec V_\beta$ such that $V_\delta \subseteq Y$, $\dot{x} \in Y$, the transitive collapse of Y is in V_κ, and ${}^{V_\delta}(Y \cap V_\alpha) \subseteq Y$. We may assume that $\mathbb{P} \subseteq Y$. We will show that $Y[G]$ is as required.

We have $x \in Y[G] \prec V_\beta[G] = V[G]_\beta$ and $V[G]_\gamma \subseteq V[G]_\delta = V_\delta[G] \subseteq Y[G]$. To show that ${}^{V[G]_\gamma}(Y[G] \cap V[G]_\alpha) \subseteq Y[G]$, take $f : V[G]_\gamma \to Y[G] \cap V[G]_\alpha$. We will find $y \in Y[G]$ with $\mathrm{range}(f) \subseteq y$ and $\|y\|^{V[G]} < \delta$. Then we will have $f \in Y[G]$ since $\mathcal{P}(y) \subseteq Y[G]$.

Let $\dot{f}$ be a $\mathbb{P}$-name for f. In V, since δ is LS, there is $Z \prec V_\beta$ such that $V_\gamma \subseteq Z$, $Y, \dot{f}, \dots \in Z$ and the transitive collapse of Z is in V_δ. Let $R = \{\dot{a} \in Z \cap Y \mid \exists p \in \mathbb{P} \exists \dot{b} \in V_\gamma\, (p \Vdash_{\mathbb{P}} \text{“}\dot{f}(\dot{b}) = \dot{a})\}$.

Claim $\mathrm{range}(f) \subseteq \{\dot{a}_G \mid \dot{a} \in R\}$.

Proof (Proof of Claim) Take $a \in \mathrm{range}(f)$. Then there are $\mathbb{P}$-names $\dot{b} \in V_\gamma$ and $\dot{a} \in Y \cap V_\alpha$ such that $f(\dot{b}_G) = a = \dot{a}_G$. Take $p \in \mathbb{P}$ with $p \Vdash_{\mathbb{P}} \text{“}\dot{f}(\dot{b}) = \dot{a}\text{”}$. Then the statement $\exists \dot{a}' \in Y (p \Vdash_{\mathbb{P}} \text{“}\dot{f}(\dot{b}) = \dot{a}\text{”})$ holds in V_β. Because $\mathbb{P} \subseteq V_\gamma \subseteq Z$ and $V_\gamma, Y \in Z$, we can find $\dot{a}' \in Z \cap Y$ with $p \Vdash_{\mathbb{P}} \text{“}\dot{f}(\dot{b}) = \dot{a}'\text{”}$, then $\dot{a}' \in R$ and $a = \dot{a}'_G$. □

---

[2] Actually Kripke-Platek set-theory is sufficient.

Now $R \subseteq Y$ and $\|R\| \leq \|Z\| < \delta$. Since ${}^{V_\delta}(Y \cap V_\alpha) \subseteq Y$, we have $R \in Y$. Let $y = \{\dot{a}_G \mid \dot{a} \in R\} \in Y[G]$. We have range$(f) \subseteq y$, and, since $\|R\|^V < \delta$, we know $\|y\|^{V[G]} < \delta$ as well. □

Corollary 5 *Let $V[G]$ be a generic extension of V. Then the statement that "there are proper class many LS cardinals" is absolute between V and $V[G]$.*

We say that all grounds are *uniformly definable in the generic multiverse* if there is a first-order formula $\varphi(x, y)$ of set-theory such that, in all grounds and generic extensions of V, all its grounds are uniformly definable by φ as in Theorem 2.[3]

By Corollary 5 and the proofs of Corollaries 3 and 4, we have:

Corollary 6 *Suppose there are proper class many LS cardinals. Then all grounds are uniformly definable in the generic multiverse.*

If AC is forceable, then there are proper class many LS cardinals in some generic extension of V, and hence also in V by Corollary 5. Hence we also have:

Corollary 7 *Suppose* AC *is forceable. Then there are proper class many LS cardinals, and all grounds are uniformly definable in the generic multiverse.*

In the next section, we discuss when AC is forceable.

By Corollary 7, we can easily construct a model V such that V does not satisfy AC but V has proper class many LS cardinals; For instance, AC is forceable over $L(\mathbb{R})$, so $L(\mathbb{R})$ has proper class many LS cardinals. On the other hand it is possible that $L(\mathbb{R})$ does not satisfy AC (e.g., see Theorem 15 below). However the author does not know if the converse direction of Corollary 7 fails:

Question 3 Is it consistent that there are proper class many LS cardinals but AC is never forceable?

This question might be connected with Woodin's Axiom of Choice Conjecture:

Conjecture 14 (Axiom of Choice Conjecture, Woodin, Definition 231 in [16]) *If V has a large cardinal, e.g., extendible cardinal, then* AC *is forceable.*

We know some notable models of ZF in which AC is never forceable, for instance:

1. A model of ZF which has no regular uncountable cardinals (Gitik [3]).
2. A model of ZF which has proper class many infinite but Dedekind-finite sets (Monro [13]).
3. A model of ZF in which Fodor's lemma fails everywhere and every club filter is not σ-complete (Karagila [8]).
4. The Bristrol model M, a transitive model of ZF which lies between L and the Cohen forcing extension $L[c]$, definable in $L[c]$ (Karagila [9]).

Daisuke Ikegami pointed out that Chang's model $L(\mathrm{ON}^\omega)$ is also an example. Kunen [10] showed that, in ZFC, AC fails in $L(\mathrm{ON}^\omega)$ if there are uncountably many measurable cardinals.

[3] Of course this definition cannot be formalized within ZF, so we will use it informally.

5. Chang's model $L(\mathrm{ON}^\omega)$ assuming that there are proper class many measurable cardinals.

If there are proper class many measurable cardinals, we can check that AC is not forcesable over $L(\mathrm{ON}^\omega)$ by a similar argument used in [10].

Question 4 Do these models have proper class many LS cardinals?

As stated before, models (1) and (3) have no LS cardinals.

Question 5 What does the geology of these models looks like? For instance, are all grounds uniformly definable in these models?

We know few things about the geology of these models.

5 The Mantle and the Generic Mantle

In this section we briefly discuss the mantle and the generic mantle of the universe.

Definition 8 Suppose all grounds are uniformly definable as in Theorem 2. The *mantle* $\mathbb{M}$ is the intersection of all grounds, that is, $\mathbb{M} = \bigcap_r W_r$.

The mantle is a parameter-free definable transitive class containing all ordinals. In ZFC, the intersection of all grounds satisfying AC is a model of ZFC [2, 15], so a natural and important question is:

Question 6 Is the mantle a model of ZF or ZFC?

Note that if all grounds of V are downward directed, that is, every two grounds of V have a common ground, then we can prove that the mantle is a model of ZF as in the context of ZFC (see [2]). In the ZFC-context, it is known that all grounds are downward directed (see Theorem 16 below). However, in the ZF-context, this downward directedness can fail. Now let us sketch the proof.

For sets X and Y, let $Fn(X, Y)$ be the poset consisting of all finite partial functions from X into Y with the reverse inclusion order. The following is known, e.g., see Exercise E in Chap. VII in Kunen [11]:

Theorem 15 (Millar) *Suppose $V = L$. Let G be $(V, Fn(\omega_1, 2))$-generic. Then, in $V[G]$, $L(\mathbb{R}^{V[G]})$ does not satisfy* AC.

Now suppose $V = L$. Let $\mathbb{P} = \mathrm{Fn}(\omega_1, 2)$. Take a $(V, \mathbb{P} \times \mathbb{P})$-generic $G \times H$, and work in $V[G \times H]$. Let $M_G = L(\mathbb{R}^{V[G]})$, and $M_H = L(\mathbb{R}^{V[H]})$. By Theorem 15, M_G and M_H do not satisfy AC. Hence $V = L$ is not a common ground of M_G and M_H. M_G and M_H are grounds of $V[G \times H]$ by Theorem 10. On the other hand, by Theorem 11, we have $V[G] \cap V[H] = V = L$. Because $M_G \subseteq V[G]$ and $M_H \subseteq V[H]$, we have $M_G \cap M_H = V = L$. This shows that M_G and M_H cannot have a common ground.

Again, suppose all grounds are uniformly definable in the generic multiverse. Then every forcing extension of V can define its mantle in the same way. The following is immediate from Theorem 8 and the weak homogeneity of $\mathrm{Col}(V_\alpha)$:

Lemma 13 *Suppose all grounds are uniformly definable in the generic multiverse.*

1. *For every limit ordinal α and $(V, \mathrm{Col}(V_\alpha))$-generic G_0, G_1, we have $\mathbb{M}^{V[G_0]} = \mathbb{M}^{V[G_1]} \subseteq \mathbb{M}^V \subseteq V$. Hence, for some/any $(V, \mathrm{Col}(V_\alpha))$-generic G, $\mathbb{M}^{V[G]}$ can be denoted as $\mathbb{M}^{\mathrm{Col}(V_\alpha)}$.*
2. *The collection $\{\mathbb{M}^{\mathrm{Col}(V_\alpha)} \mid \alpha$ is a limit ordinal$\}$ is uniformly definable in V.*
3. *Let $V[G]$ be a forcing extension of V. Then there is a limit ordinal α such that $\mathbb{M}^{V^{\mathrm{Col}(V_\alpha)}} \subseteq \mathbb{M}^{V[G]}$.*

Thus we can define the *generic mantle* $g\mathbb{M} = \bigcap\{\mathbb{M}^{V^{\mathrm{Col}(V_\alpha)}} \mid \alpha \in \mathrm{ON}\}$, which is the intersection of all mantles of all generic extensions. As in the context of ZFC (see [2]), we can check that $g\mathbb{M}$ is a parameter-free definable transitive model of ZF containing all ordinals. Clearly $g\mathbb{M} \subseteq \mathbb{M}$. In the ZFC-context, the mantle coincides with the generic mantle [2, 15]. How is it in ZF?

Question 7 Does $\mathbb{M} = g\mathbb{M}$?

6 When AC is Forceable

In this section, we discuss when AC is forceable. For this purpose, we use the DDG, downward directedness of the grounds.

Theorem 16 (Usuba [15], in ZFC) *Let $\{W_r \mid r \in V\}$ be the uniformly definable collection of all grounds satisfying* AC *as in Theorem 1. Let X be a set. Then there is a ground W of V such that W satisfies* AC, *and W is a ground of each W_r $(r \in X)$.*

Proposition 1 *Suppose* AC *is forceable, and let $V[G]$ be a generic extension of V such that $V[G]$ satisfies* AC. *Then there is a ground W of $V[G]$ and a set X such that $V = W(X)$ and W satisfies* AC.

Proof Let $\mathbb{P}$ be a poset, G be $(V, \mathbb{P})$-generic, and suppose $V[G]$ satisfies AC. Take a $(V[G], \mathbb{P})$-generic H. We may assume that $V[H]$ satisfies AC. Then $G \times H$ is $(V, \mathbb{P} \times \mathbb{P})$-generic, and $V[G \times H]$ is a common forcing extension of $V[G]$ and $V[H]$. Because both $V[G]$ and $V[H]$ satisfy AC, $V[G \times H]$ also satisfies AC. Then, by Theorem 16, there is a model W of ZFC which is a common ground of $V[G]$ and $V[H]$. We know $V = V[G] \cap V[H]$ by Theorem 11, hence $W \subseteq V \subseteq V[G]$. $V[G]$ is a forcing extension of W, so V must be of the form $W(X)$ for some $X \in V$ by Theorem 10. □

Now we have the following characterization.

Corollary 8 *The following are equivalent:*

1. AC *is forceable.*
2. *There is a transitive model W of* ZFC *and a set X such that W is definable in V with parameters from W and $V = W(X)$.*
3. *There is a transitive model W of* ZFC *and a set X such that W is definable in V with parameters from W, $V = W(X)$, and W is a ground of some generic extension of V.*

Proof (3) $\Rightarrow$ (2) is trivial.

(2) $\Rightarrow$ (1). Suppose $V = W(X)$. Let Y be the transitive closure of X, and $\mathbb{P} = \mathrm{Col}(Y)$. Take a $(V, \mathbb{P})$-generic G. In $V[G]$, Y is well-orderable. Because W satisfies AC and Y is well-orderable in $V[G]$, every element of $V = W(X)$ is well-orderable in $V[G]$. Then every element of $V[G]$ is well-orderable, because there is a canonical class surjection from V onto $V[G]$.

(1) $\Rightarrow$ (3). By Proposition 1, we can find a set-forcing extension $V[G]$ of V and a ground W of $V[G]$ such that $V[G]$ satisfies AC, W is a model of ZFC, and $V = W(X)$ for some $X \in V$. We have to check that W is definable in V. Because W is a ground of $V[G]$, W satisfies the α-norm covering and approximation properties for $V[G]$ for some large α. Then it is easy to check that W also satisfies the α-norm covering and approximation properties for V. Since AC is forceable over V, V has proper class many LS cardinals. Then, by Corollary 3, W is of the form W'_r for some $r \in W$, hence W is definable in V. □

Corollary 9 *Suppose* AC *is forceable. Then for every ground W of V, there is a transitive model M of* ZFC *and a set $X \in W$ such that M is definable in V with parameters from M and $W = M(X)$.*

Proof If W is a ground of V, then AC is forceable over W. Then the assertion follows from the previous corollary. □

Corollary 10 *Suppose V satisfies* AC*. Then for every transitive model M of* ZF*, M is a ground of V if and only if there is a ground W of V and a set X such that W satisfies* AC *and $M = W(X)$.*

Proof If M is a ground of V, then AC is forceable over M. We can find required W and X by Proposition 1. For the converse, suppose $M = W(X)$ for some ground W of V satisfying AC and a set X. Then M is a ground by Theorem 10. □

Corollary 11 *Suppose* AC *is forceable. Then the generic mantle is a model of* ZFC.

Proof Since AC is forceable, for every sufficiently large α and $(V, \mathrm{Col}(V_\alpha))$-generic G, we have that $V[G]$ satisfies AC. By Corollary 10, if $M \subseteq V[G]$ is a ground of $V[G]$, then there is a ground W of $V[G]$ such that $W \subseteq M$ and W satisfies AC. Hence the mantle of $V[G]$ coincides with the intersection of all grounds of $V[G]$ satisfying AC. In ZFC, it is known that the intersection of all grounds satisfying AC is a model of ZFC [15]. Therefore we have that for every large α, $\mathbb{M}^{\mathrm{Col}(V_\alpha)}$ is a model of ZFC. Since the generic mantle is the intersection of the $\mathbb{M}^{\mathrm{Col}(V_\alpha)}$'s, we can check that the generic mantle is a model of ZFC. □

A *symmetric model* is a type of choiceless model constructed as a submodel of a generic extension. See Grigorieff [4] for the definition of symmetric models. We use the following characterization of symmetric models.

For a class M of V, let $\mathrm{OD}(M)$ be the collection of all sets x such that x is definable with parameters from M and ordinals. $\mathrm{HOD}(M)$ is the collection of all sets x such that the transitive closure of x is a subset of $\mathrm{OD}(M)$. If a class M is a transitive model of ZF containing all ordinals, then $\mathrm{HOD}(M)$ is a transitive model of ZF with $M \subseteq \mathrm{HOD}(M)$.

Theorem 17 (Theorem C in [4]) *For transitive models M and N of* ZF *and a generic extension $M[G]$ of M, suppose $M \subseteq N \subseteq M[G]$ and N is a class of $M[G]$. Then N is a symmetric submodel of $M[G]$ if and only if N is of the form $\mathrm{HOD}(M(X))^{M[G]}$ for some $X \in N$.*

Proposition 2 *Suppose* AC *if forceable. Then there is a transitive model M of* ZFC *and a generic extension $M[G]$ of M such that M is definable in V with parameters from M and V is a symmetric submodel of $M[G]$.*

Proof By Corollary 8, there is a definable transitive model M of ZFC such that $V = M(X)$ for some $X \in V$, and M is a ground of some generic extension $V[H]$ of V. By appealing to Theorem 7, we may assume that H is $(V, \mathrm{Col}(V_\alpha))$-generic for some α, and there is β and $(M, \mathrm{Col}(M_\beta))$-generic G with $V[H] = M[G]$. Let us consider $\mathrm{HOD}(M(X))^{M[G]}$, which is a symmetric submodel of $M[G]$. Since $\mathrm{Col}(V_\alpha)$ is weakly homogeneous, ordinal definable, and $M \subseteq V$, we have $M(X) \subseteq \mathrm{HOD}(M(X))^{M[G]} = \mathrm{HOD}(M(X))^{V[H]} \subseteq \mathrm{HOD}(M(X))^V \subseteq V$, hence $M(X) = \mathrm{HOD}(M(X))^{M[G]} = V$. □

Corollary 12 *The following are equivalent:*

1. AC *is forceable.*
2. *There is a transitive model M of* ZFC *and a generic extension $M[G]$ of M such that M is definable in V with parameters from M, and V is a symmetric submodel of $M[G]$.*

Proof (1) $\Rightarrow$ (2) follows from the previous proposition, and (2) $\Rightarrow$ (1) follows from the result of Blass [1]. □

Acknowledgements The author would like to thank Asaf Karagila for his many valuable comments. The author also thank the referee who gives the author many corrections, and Daisuke Ikegami who pointed out the failure of SVC in Chang's model. This research was supported by JSPS KAKENHI Grant Nos. 18K03403 and 18K03404.

References

1. Blass, A.: Injectivity, projectivity, and the axiom of choice. Trans. Am. Math. Soc. **255**, 31–59 (1979)
2. Fuchs, G., Hamkins, J.D., Reitz, J.: Set-theoretic geology. Ann. Pure Appl. Logic **166**(4), 464–501 (2015)
3. Gitik, M.: All uncountable cardinals can be singular. Israel J. Math. **35**(1–2), 61–88 (1980)
4. Grigorieff, S.: Intermediate submodels and generic extensions in set theory. Annal. Math. **101**(3), 447–490 (1975)
5. Gitman, V., Johnstone, T.: On ground model definability. In: Infinity, Computability, and Metamathematics, vol. 23, pp. 205–227, Tributes. Colloquium Publications, London (2014)
6. Hamkins, J.D.: Extensions with the approximation and cover properties have no new large cardinals. Fund. Math. **180**(3), 257–277 (2003)
7. Jech, T.: Set theory. In: The Third Millennium Edition, Revised and Expanded. Springer, Berlin (2003)
8. Karagila, A.: Fodor's lemma can fail everywhere. Acta Math. Hungarica **154**(1), 231–242 (2018)
9. Karagila, A.: The Bristol model: an abyss called a Cohen real. J. Math. Logic **18**(02), 1850008 (2018)
10. Kunen, K.: A model for the negation of the axiom of choice. In: Cambridge Summer School in Mathematical Logic (Cambridge, 1971), pp. 489–494. Lecture Notes in Math. vol. 337. Springer, Berlin (1973)
11. Kunen, K.: Set theory. In: An Introduction to Independence Proofs. Studies in Logic and the Foundations of Mathematics, vol. 102. North-Holland (1980)
12. Laver, R.: Certain very large cardinals are not created in small forcing extensions. Ann. Pure Appl. Logic **149**(1–3), 1–6 (2007)
13. Monro, G.P.: Independence results concerning Dedekind-finite sets. J. Austral. Math. Soc. **19**, 35–46 (1975)
14. Reitz, J.: The ground axiom. J. Symb. Logic **72**(4), 1299–1317 (2007)
15. Usuba, T.: The downward directed grounds hypothesis and very large cardinals. J. Math. Log. 17(2), 1750009, 24 pp (2017)
16. Woodin, W.H.: Suitable extender models I. J. Math. Log. **10**(1–2), 101–339 (2010)

Irrational-Based Computability of Functions

Mariko Yasugi, Yoshiki Tsujii, and Takakazu Mori

Abstract We investigate a sort of a unifying theory of computability of real functions, continuous or discontinuous, called here "irrational-based" (IB-) computability. All the examples which are computable in our various theories presented previously are IB-computable. The basic requirements of computability, the sequential computability and the effective continuity, are defined relative to computable irrational real sequences. The family of IB-computable functions is closed under IB-effective convergence. In order to certify the fruitfulness of IB-computability, quite a number of examples are presented.

Keywords Computable irrational sequence · Irrational-based (IB-) computable real function · IB-effective convergence

1 Introduction

We present here a theory of a computable real functions taking the computable irrational sequences as basic. Computable functions of this theory include all the functions which are computable in the various theories we have studied previously.

We will confine the domain of our discourse to real numbers, real number sequences and real functions. We assume the traditional theory of the computability structure on the real line as well as that of continuous computable functions (cf. [6]).

Of continuous real functions, the notion of computability has been long defined and generally agreed (cf. [6], for example). It has proved its value in many works related to computability theories. The basis of the computability theory of functions

M. Yasugi (✉)
Kyoto Sangyo University Emerita, Kyoto, Japan
e-mail: yasugi@cc.kyoto-su.ac.jp

Y. Tsujii · T. Mori
Kyoto Sangyo University Emeritus, Kyoto, Japan
e-mail: tsujiiy@cc.kyoto-su.ac.jp

T. Mori
e-mail: morita@cc.kyoto-su.ac.jp

T. Arai et al. (eds.), *Advances in Mathematical Logic*, Springer Proceedings in Mathematics & Statistics 369, https://doi.org/10.1007/978-981-16-4173-2_9

is the theory of recursive functions as modulus of continuity, modulus of convergence etc. Everything else can be treated just as in classical mathematics. However, there are many discontinuous functions of interest and importance in science and technology which appeal to human mind as computable. A very simple and familiar example is the floor function. (Examples will be seen in Sect. 4.) Extending the notion of computability to such functions was a necessity, welling up from mathematics itself as well as from the mind of mathematicians working in the area of computability in mathematics.

There have been successful works on this theme. See, for example, [1, 6, 9, 10, 16] for general expositions. In particular, the type two Turing machine approach, originated by Weihrauch, has manifested one successful unifying logical principle for the notion of computability of functions (cf. [9]).

We have thus studied computability structures of real functions which are not necessarily continuous, following first the functional space approach developed by Pour-El and Richards [6]. In their method, the computability is controlled with recursive moduli of continuity and of convergence, and otherwise one can work within mathematics of functional spaces, such as Banach spaces [6, 10, 16]. We have also worked with the Fine space [3–5], uniform topological spaces [7, 13, 14, 17] and Fréchet spaces [8, 11]. In its essence, some discontinuous functions are made computable by changing the topology of the domain, the real line. Use of the limiting recursive modulus of convergence instead of recursive one is also treated in some of our works [11, 12].

We selected those background mathematical theories in order to extend the notion of computability to the respective functions of our concern.

On its surface, our treatments above may appear somewhat ad hoc, though mathematically sound and significant. So, we wished to devise a more comprehensive mathematical framework for some of the theories of computability of discontinuous functions, and we have been led to a notion of computability of functions which will be called here IB-computability. It is defined by requiring effective properties of functions with respect to computable *irrational sequences*. It turns out that all the (discontinuous) functions which we have made computable in our preceding works are IB-computable. Some more examples of IB-computable functions will also be included.

The family of IB-computable functions is not convenient to do elementary mathematics, since irrational numbers (and their sequences) are not closed with respect to elementary arithmetic, and IB-computable functions are not closed under compositions. The role of IB-computability may be hence regarded as the provision of a conceptual framework of computability for a wide range of real functions. It also suggests the usefulness of irrational numbers in computable mathematics.

Further, IB-computability has the advantage over the functional space approach in the following sense. In the functional space approach, a function is a point or an element of a space, and the computability is defined in terms of an environment, that is, in terms of a computability structure in the space, while IB-computability is defined for individual functions according to their behaviors.

We came to the idea of IB-computability while studying continued fractions. It is an easy practice to show that an *irrational sequence* is computable if and only if its (unique) regular continued fraction representation is recursive. That is, the computability of the former is characterized by a recursive (double) sequence of natural numbers. This fact alluded us that we might regard the family of irrational sequences as the basis in computable mathematics.

We present precise treatments of mathematical results and supply with examples which comprise all the computable functions treated earlier and some more, in order to certify that our theory is mathematically fruitful.

This article is composed as follows. Some basic known facts about computable numbers and functions are provided in Sect. 2. In Sect. 3, IB-computability of functions is defined, and the fundamental properties thereof follow. Section 4 is devoted to the examples which give substance to our theory. A counterexample is also given to show that the theory maintains the natural sense of effectivity. Convergence of IB-computable function sequences under a certain condition will be studied in Sect. 5, and the closure property is proved. Examples of converging sequences are included. As a final remark, Sect. 6 explains a similarity of IB-computability to a classical extension of continuity: it is a kind of effectivization of almost everywhere continuity, though there is a gap between them.

2 Preliminaries

Throughout this article, "computability" (without any modifier) of a real number or a real sequence as well as of real functions will mean the traditional computability. (See Definition 1 below.) Other notions of computability will be distinguished with some prefixes.

We will here review and analyze some basic relationships between the computabilities of rational number sequences, real number sequences and irrational number sequences, from our present viewpoint.

Definition 1 (*Computability of real number sequences:* [6]) (i) A recursive sequence of rational numbers is *computable*.

(ii) A sequence of real numbers is *computable* if it is effectively approximated by a (double) recursive sequence of rational numbers. ("Effective approximation" means that there is a recursive modulus of convergence.)

Remark 1 (1) Integers are included in rational numbers, and an integer sequence is computable if and only if it is recursive.

(2) (ii) in the definition above creates non-recursive but computable rational sequences.

It is the "computable irrational sequence" that plays the main role in this article.

Definition 2 (*Irrational sequence*) A sequence of real numbers is called an *irrational sequence* if all its terms are irrational. An irrational sequence is called a *computable irrational sequence* if it is computable.

Proposition 1 *Let* $\{s_n\}$ *be a recursive sequence of rational numbers. Then there is a computable double irrational sequence, say* $\{y_{nk}\}$, *which effectively converges to* $\{s_n\}$.

Proof Put

$$y_{nk} = s_n + \frac{\sqrt{2}}{2^{k+1}}.$$

$\{y_{nk}\}$ is a computable double irrational sequence, which effectively converges to $\{s_n\}$ with a recursive modulus of convergence $\alpha(n, k) = k$.

Proposition 2 Let $\{x_n\}$ be any computable sequence of real numbers. Then there is a computable double irrational sequence $\{w_{np}\}$ which effectively converges to $\{x_n\}$.

Proof By the definition of computability, there is a recursive double sequence of rational numbers, say $\{r_{nk}\}$, which converges to $\{x_n\}$ with a recursive modulus of convergence $\alpha(n, p)$. Applying Proposition 1 to the double sequence $\{r_{nk}\}$, we obtain a computable triple irrational sequence, say $\{z_{nkl}\}$, which converges to $\{r_{nk}\}$ with a recursive modulus of convergence $\beta(n, k, p)$. Put $w_{np} = z_{n,\alpha(n,p+1)\beta(n,\alpha(n,p+1),p+1)}$. $\{w_{np}\}$ is a computable irrational sequence which approximates $\{x_n\}$ with a recursive modulus of convergence $\gamma(n, p) = p$.

Definition 3 (*Classical notion of computable function:* [6]) A real function f is computable if

(i) (Sequential computability) for any computable sequence of real numbers $\{x_n\}$, $\{f(x_n)\}$ is computable;
(ii) (Effective continuity) there is a recursive function α such that

$$\forall M, p \forall x, y \in [-M, M]. |x - y| < \frac{1}{2^{\alpha(M,p)}} \Rightarrow |f(x) - f(y)| < \frac{1}{2^p},$$

where M and p are positive integers.

The definition can be easily extended to the computability of a sequence of functions.

Remark 2 The condition (ii) in Definition 3 implies that a computable function is continuous.

3 Irrational-Based Computable Function

In this section, we propose an extended notion of computability of real functions which applies to many of discontinuous functions. It will be called irrational-based computability, whose intended meaning is "relative to computable irrational numbers (number sequences)".

Definition 4 (*IB-computability*) (1) A real function f will be called *irrational-based computable* (abbreviated to IB-computable) if the following (i) and (ii) hold for f.

(i) (IB-sequential computability) f is *irrational-based sequentially computable* (IB-sequentially computable). That is, for any computable *irrational* sequence $\{x_m\}$, the sequence $\{f(x_m)\}$ is a computable sequence of real numbers.

(ii) (IBE-continuity) f is *irrational-based effectively* (IBE-) *continuous*. That is, for any computable *irrational* sequence $\{x_m\}$, there is a recursive function $\gamma(m, p)$ such that

$$\forall m, p \forall y. |x_m - y| < \frac{1}{2^{\gamma(m,p)}} \Rightarrow |f(x_m) - f(y)| < \frac{1}{2^p}.$$

γ will be called an effective modulus of continuity for (f and) $\{x_m\}$.

(2) A function sequence $\{f_n\}$ will be called IB-computable if the following (i') and (ii') hold.

(i') $\{f_n(x_m)\}$ is a computable double sequence of real numbers for any computable irrational sequence $\{x_m\}$.

(ii') There is an effective modulus of continuity $\gamma(n, m, p)$ for $\{f_n\}$ and $\{x_m\}$ as in (ii) above.

Remark 3 (1) Notice that, in (ii), y is an arbitrary real number.

(2) IB-computability is closed under the operations of addition, subtraction, multiplication and division of functions (presuming that the division is well-defined), but is not generally preserved under the composition of functions. It is also closed under "effective limit" (cf. Theorem 2 in Sect. 5).

(3) IBE-continuity of a function may seem far from the effective continuity in the sense of [6]. In [6], effective continuity is defined for continuous functions, while IBE-continuity applies also to many discontinuous functions. We will see what sort of a condition may supplement the gap, and show that IB-computability is an extension of (continuous) computability (cf. Theorem 1 below).

Theorem 1 *A (continuous) computable function is IB-computable.*

To show this, we define RE-continuity of a function and prove some propositions, from which the theorem is derived.

Definition 5 (*RE-continuity*) A function f is called *relatively effectively continuous* (RE-continuous) if it is effectively continuous relative to computable irrational numbers; that is, if there is a recursive function $\delta(M, p)$ such that, for every *computable irrational number* $u \in (-M, M)$,

$$\forall y \in (-M, M).|u - y| < \frac{1}{2^{\delta(M,p)}} \Rightarrow |f(u) - f(y)| < \frac{1}{2^p}.$$

Proposition 3 *An RE-continuous function is IBE-continuous ((ii) of Definition 4).*

Proof Suppose f is RE-continuous with δ.

Suppose $\{x_m\}$ is a computable irrational sequence. Then one can effectively find a recursive sequence $\{M_m\}$ so that $x_m \in (-M_m, M_m)$. One can then effectively evaluate $q = \gamma(m, p)$ such that

$$\frac{1}{2^q} < \min(\frac{x_m - (-M_m)}{2}, \frac{1}{2^{\delta(M_m,p)}}, \frac{M_m - x_m}{2}).$$

γ is recursive, and is obtained relative to $\{x_m\}$.

To see that this γ works, suppose $|x_m - y| < \frac{1}{2^{\gamma(m,p)}}$. Then $y \in (-M_m, M_m)$ and $|x_m - y| < \frac{1}{2^{\delta(M_m,p)}}$. So, by RE-continuity, $|f(x_m) - f(y)| < \frac{1}{2^p}$.

The next proposition uses the lemma below.

Lemma 1 *For any real numbers x and y with $x < y$, there is a computable irrational number u such that $x < u < y$. (No effectivity of finding such a u is claimed.)*

Proof There are rational numbers, say s and t so that $x < s < t < y$. Put $u = s + \frac{t-s}{2\sqrt{2}}$.

Proposition 4 *A function f is effectively continuous in the sense of [6] (cf. (ii) of Definition 3) if and only if it is RE-continuous in the sense of Definition 5.*

Proof "If" Part: Suppose f is RE-continuous with δ. Suppose also that $x, y \in [-N, N]$ and $x < y$. Then, putting $M = N + 1$, $x, y \in (-M, M)$.

By Lemma 1, there is a computable irrational number u satisfying $x < u < y$, hence $u \in (-M, M)$. For any q, if $|x - y| < \frac{1}{2^q}$, then $|x - u|, |u - y| < \frac{1}{2^q}$. If we put $q = \alpha(N, p) = \delta(N + 1, p + 1) = \delta(M, p + 1)$, then $|x - u|, |u - y| < \frac{1}{2^{\delta(M,p+1)}}$ holds. Applying RE-continuity to u and M, we obtain

$$|f(x) - f(y)| \leq |f(x) - f(u)| + |f(u) - f(y)| < \frac{1}{2^{p+1}} + \frac{1}{2^{p+1}} = \frac{1}{2^p},$$

which means that f is effectively continuous with a recursive modulus of continuity α.

The "only if" part is trivial, since RE-continuity is a restricted case of the effective continuity in the sense of [6].

Remark 4 (1) By virtue of Proposition 4, we can view that effective continuity is determined with the due property on computable irrational numbers.

(2) The converse of Proposition 3 does not hold, since RE-continuity is equivalent to effective continuity in the sense of [6] by Proposition 4 (hence applies only to

continuous functions), while there are many IBE-continuous functions which are not continuous, as will be seen in the examples in Sect. 4.

(3) Proposition 3 shows that IBE-continuity is a natural extension of RE-continuity, and hence, equivalently, of effective continuity in the sense of [6] by virtue of Proposition 4.

(4) Notice that the notions of effective continuity in Propositions 4 and 3 are treated independent of the notions of sequential computability.

Proposition 5 *An IB-sequentially computable function ((i) of Definition 4) is sequentially computable in the sense of [6] (cf. (i) of Definition 3) if it is effectively continuous (or equivalently RE-continuous: Proposition 4).*

Proof Let f be (a) IB-sequentially computable and (b) effectively continuous. Let $\{y_l\}$ be any computable sequence of real numbers. We are to show that $\{f(y_l)\}$ is computable.

From the conditions of f and $\{y_l\}$, the following properties hold.

1. For any computable irrational (double) sequence, say $\{z_{lm}\}, \{f(z_{lm})\}$ is computable (from (a)).
2. There is a recursive function γ such that

$$\forall N, p \forall x, y \in [-N, N]. |x - y| < \frac{1}{2^{\gamma(N,p)}} \Rightarrow |f(x) - f(y)| < \frac{1}{2^p}$$

(from (b)).

3. We can find effectively a recursive sequence $\{N_l\}$ such that $y_l \in (-N_l + 1, N_l - 1)$, since $\{y_l\}$ is computable.
4. By Proposition 2, there is a computable irrational double sequence $\{z_{lm}\}$ and a recursive function $\beta(l, q)$ such that

$$\forall m \geq \beta(l, q). |y_l - z_{lm}| < \frac{1}{2^q}.$$

Using those properties, we proceed to the proof.

5. Put $q = \gamma(N_l, p)$ in 4 with γ and N_l of 2 and 3 above. Then

$$\forall m \geq \beta(l, \gamma(N_l, p)). |y_l - z_{lm}| < \frac{1}{2^{\gamma(N_l,p)}},$$

and $y_l, z_{lm} \in [-N_l, N_l]$ for those m, since then $|y_l - z_{lm}| < 1$. Applying 2 to N_l, y_l and z_{lm}, we obtain from 5,

$$\forall m \geq \beta(l, \gamma(N_l, p)). |f(y_l) - f(z_{lm})| < \frac{1}{2^p}.$$

This means that $\{f(z_{ml})\}$ effectively approximates $\{f(y_l)\}$ with modulus of convergence $\alpha(l, p) = \beta(l, \gamma(N_l, p))$, which is recursive by 3, and hence by 1 $\{f(y_l)\}$ is computable, since computability is closed under effective convergence.

Propositions 4 and 5 imply the following.

Proposition 6 *A real function f is computable (in the sense of [6]) if and only if it is IB-sequentially computable and RE-continuous.*

Proof (1) Sequential computability: If f is sequentially computable in the sense of [6], then, in particular, it is IB-sequentially computable (cf. (i) of Definition 4). If f is IB-sequentially computable and RE-continuous, then it is sequentially computable in the sense of [6] by Proposition 5.

(2) Continuity: f is effectively continuous in the sense of [6] if and only if it is RE-continuous by Proposition 4.

Proof of Theorem 1 Immediate from Propositions 3 and 6.

Remark 5 (1) Among IB-computable functions, there are also discontinuous functions: all the individual discontinuous functions we have treated in our studies of computability in analysis based on various different theories (cf. [4, 5, 7, 11, 13–15], for example) are IB-computable as will be seen in the next section.

(2) So, IB-computability can be regarded as a unifying principle according to which both continuous and discontinuous functions that have been treated with different theories can be endowed with a certain effective property.

4 Examples

In order to show that the theory of IB-computability is a significant one, we will present here several examples. Most of them have been shown computable in our preceding theories. The point is that here they are all treated unanimously in one theory, the theory of IB-computability.

In the proofs of the examples below, (i) and (ii) will mean the respective numberings (i) (IB-sequential computability) and (ii) (IBE-continuity) in Definition 4.

Example 1 (*An effective uniform space:* [7]) Let $\{t_l\}_{l \in \mathbf{Z}}$ be a recursive sequence of rational numbers which are enumerated with all integers. Suppose it satisfies that $t_l < t_{l+1}$ for all l and that $\bigcup_l [t_l, t_{l+1}) = \mathbf{R}$ (the set of all real numbers). The sequence of intervals $\{S_l\}$, where $S_l = [t_l, t_{l+1})$, induces an effective uniform topology $\mathcal{U}$ (cf. [7]): for each positive integer p, and for any real number $x \in S_l$, put $U_p(x) = S_l \cap (x - \frac{1}{2^p}, x + \frac{1}{2^p})$, and let $\mathcal{U}$ consist of all those $U_p(x)$'s.

A $\mathcal{U}$-computable function f is IB-computable. (See [7] for the notion of computability of a function with respect to an effective uniform topology.)

Proof Let $\{x_m\}$ be an arbitrary computable irrational sequence.

(i) It can be easily proved that, $\{x_m\}$ is $\mathcal{U}$-computable. Since f is assumed to be $\mathcal{U}$-sequentially computable, $\{f(x_m)\}$ is computable. So, f is IB-sequentially computable.

(ii) We can determine a recursive sequence of integers $\{N_m\}$ such that $x_m \in S_{N_m}$. Since f is effectively $\mathcal{U}$-continuous, there is a recursive function γ (depending on $\{x_m\}$) such that

$$y \in U_{\gamma(m,p)}(x_m) \Rightarrow |f(x_m) - f(y)| < \frac{1}{2^p}. \tag{1}$$

Put $J_{m,p} = (x_m - \frac{1}{2^{\gamma(m,p)}}, x_m + \frac{1}{2^{\gamma(m,p)}})$. $U_{\gamma(m,p)}(x_m) = S_{N_m} \cap J_{m,p}$. If we put

$$z_{mp} = \min\left(\frac{1}{2^{\gamma(m,p)}}, x_m - t_{N_m}, t_{N_m+1} - x_m\right),$$

then $\{z_{mp}\}$ is a computable sequence of positive numbers, and hence we can find effectively positive integers $L(m, p)$ such that $\frac{1}{2^{L(m,p)}} < z_{mp}$. Then $|y - x_m| < \frac{1}{2^{L(m,p)}}$ implies $y \in U_{\gamma(m,p)}(x_m)$, and hence by (1) $|f(y) - f(x_m)| < \frac{1}{2^p}$, that is, f is IBE-continuous.

Remark 6 (1) Needless to say, there are many functions which are discontinuous at $\{t_l\}$ and yet $\mathcal{U}$-computable.

(2) For a piecewise continuous function with discontinuous points $\{a_m\}_{m\in\mathbf{Z}}$, a computable sequence of real numbers (which are not necessarily rational numbers), where $a_m < a_{m+1}$ and $\cup_m[a_m, a_{m+1}) = \mathbf{R}$, we may apply the theory of IB-computability by regarding a_m as a kind of "rational point" and defining $\{a_m\}$-irrational numbers, with a similar technique as in [13].

Example 2 (*Floor function:* [1, 11, 17]) As a corollary of Example 1, the floor function (the integer part of a number x, customarily denoted by $[x]$) is IB-computable. (This function was originally treated in [11] with the Frechét space, with upper semi-computability and with limiting recursion.)

Example 3 (*Fine computable function:* [3, 4]) Fine computability of real functions was originally treated in the above references regarding the Fine space as an effective uniform space. For simplicity, we will consider the Fine space on the domain [0, 1), although it can easily be extended to the entire real line.

For the reader's convenience, we quote the definition of Fine computability of a function f on [0, 1) from [3]. f is Fine computable if (F1) and (F2) below hold. (F1) (Fine sequential computability) For any Fine computable sequence of real numbers $\{x_m\}$, $\{f(x_m)\}$ is a computable sequence of real numbers.

(F2) (Effective Fine continuity) There is a recursive function $\beta(p, i)$ such that (with a recursive enumeration of all dyadic rationals $\{e_i\}$ in [0, 1))

(a) $x \in J(e_i, \beta(p, i))$ implies $|f(x) - f(e_i)| < \frac{1}{2^p}$;

(b) $\bigcup_{i=0}^{\infty} J(e_i, \beta(p, i)) = [0, 1)$ for all p.

($J(e_i, q)$ denotes the interval of the form $I(q, l) = [\frac{l}{2^q}, \frac{l+1}{2^q})$ which contains e_i. For any q, $\bigcup_{l=0}^{2^q-1} I(q, l)$ is a decomposition of $[0, 1)$.)

It holds that

$$\forall q \forall i \exists ! l (\leq 2^q - 1). e_i \in J(e_i, q) = I(q, l) = [\frac{l}{2^q}, \frac{l+1}{2^q}). \tag{2}$$

Now, a Fine computable function is IB-computable.

Proof (i) An irrational sequence is Fine computable if and only if it is computable, and a Fine computable function maps a Fine computable sequence to a computable sequence.

(ii) Let f be effectively Fine continuous.

Suppose $e_i = \frac{p}{2^h}$, where p is an odd number. p and h are determined by i. Then, $l(p, q, h) = [\frac{p}{2^{h-q}}]$, the integer part of $\frac{p}{2^{h-q}}$, is computable as a function of p, h and q, and, with $l = l(p, q, h)$, (2) holds.

Denote the left edge of the interval $I(q, l)$ with $\lambda(e_i, q)$ and the right one with $\rho(e_i, q)$. Those are computable as functions of i and q. Using those facts and notations, we show that a Fine computable function is IB-computable.

Now, let $\{x_m\}$ be a computable irrational sequence from $[0, 1)$. There is a recursive function ι so that $x_m \in J(e_{\iota(m,p)}, \beta(p+1, \iota(m, p)))$. By (a), $|f(x_m) - f(e_{\iota(m,p)})| < \frac{1}{2^{p+1}}$. Put

$$z_{mp} = \min\{x_m - \lambda(e_{\iota(m,p)}, \beta(p+1, \iota(m, p)), \rho(e_{\iota(m,p)}, \beta(p+1, \iota(m, p)) - x_m\}.$$

$\{z_{mp}\}$ is a computable sequence of positive irrational numbers. There is a recursive function $\gamma(m, p)$ so that $\frac{1}{2^{\gamma(m,p)}} < z_{mp}$. If $|y - x_m| < \frac{1}{2^{\gamma(m,p)}}$, then $y \in J(e_{\iota(m,p)}, \beta(p+1, \iota(m, p)))$. Using (a) again, $|f(y) - f(e_{\iota(m,p)})| < \frac{1}{2^{p+1}}$. Combining two inequalities, we obtain $|f(x_m) - f(y)| < \frac{1}{2^p}$.

Example 4 (*Examples in* [11]) In [11], computability notions of various discontinuous functions have been developed in terms of the limiting recursion, upper semi-computability, the theory of locally convex space and the theory of Frechét space with computability structures. All the examples there are proven to be IB-computable. We will treat some of those examples here.

In the following, $\{x_m\}$ is assumed to be a computable irrational sequence.

(1) Let h be the function defined by $h(x) = x - [x]$. h is not computable in the sense of [6].

(i) The integral part of $\{x_m\}$, $\{[x_m]\}$, is effectively computed. So $\{h(x_m)\}$ is a computable sequence of real numbers.

(ii) If we put $u_{mp} = \min(\frac{1}{2^p}, h(x_m), 1 - h(x_m))$, then $\{u_{mp}\}$ is a computable double sequence of positive real numbers, from which one can obtain an effective modulus of continuity for $\{h(x_m)\}$ as follows. One can effectively find a double sequence of positive integers $\{r_{mp}\}$ so that $\frac{1}{2^{r_{mp}}} < u_{mp}$. Put $\gamma(m, p) = r_{mp}$. This

γ serves as an effective modulus of continuity for h. (If $|y - x_m| < \frac{1}{2^{\gamma(m,p)}}$, then $|h(y) - h(x_m)| = |y - x_m| < \frac{1}{2^{\gamma(m,p)}} < \frac{1}{2^p}$.)
(2) Let $]x[$ denote the following real function.

$$]x[= n \text{ if and only if } n < x \leq n + 1$$

$]x[$ is lower semi-computable but is not upper semi-computable.

(i) The condition in the definition of the function can be replaced by $n < x_m < n + 1$, and hence $n_m =]x_m[$ can be effectively computed.

(ii) Put $u_{mp} = \min(x_m - n_m, n_m + 1 - x_m)$. There is a recursive function $\gamma(m, p)$ so that $\frac{1}{2^{\gamma(m,p)}} < u_{mp}$. This will do.
(3) Define a function σ as follows:

$$\sigma(x) = 0(x \in (-\infty, 0)), = w(x = 0), = 1(x \in (0, \infty)),$$

where w is any real number (computable or non-computable) different from 0 and 1. σ is a modified version of the signum function, and is neither upper nor lower semi-computable.

(i) For a computable irrational number, it can be effectively determined whether it be positive or negative, hence $\{\sigma(x_m)\}$ is computable.

(ii) For any computable irrational number, the function value is constant in its small enough neighborhood, hence IBE-continuity is obvious. ($x_m < 0$ or $x_m > 0$. So, for example, assume the latter. Then, one can effectively find integers j_m so that $0 < \frac{1}{2^{j_m}} < x_m$, and $\gamma(m, p) = j_m$ for all p will do.)
(4) Define a function ρ as follows. $\rho(x) = \tan \frac{\pi}{2}x$ if $2n + 1 < x < 2n + 3$; $\rho(x) = w$ if $x = 2n + 1$ for all n, where w is any real number. Unlike the preceding examples, the function ρ is unbounded.

Since $\tan \frac{\pi}{2}x$ is sequentially computable and effectively continuous between $2n + 1$ and $2n + 3$, the IB-computability of ρ follows.

Example 5 (*Thomae's function*) Thomae's function, known also with other names such as the popcorn function, is IB-computable.

Proof Let us review Thomae's function (named τ here).

$$\tau(x) = 0 \quad \text{if } x \text{ is irrational}; \quad = \frac{1}{q} \quad \text{if} \quad x = \frac{p}{q}$$

for coprime integers p and q, where $q > 0$. (If x is an integer n, it is expressed as $\frac{n}{1}$, and hence $\tau(x) = 1$.)

(i) If $\{x_m\}$ is an irrational (and computable) sequence, then $\tau(x_m) = 0$ for all m, and hence $\{\tau(x_m)\}$ is computable.

(ii) Let $\{x_m\}$ be a computable irrational sequence. We must find a recursive γ such that

$$|y - x_m| < \frac{1}{2^{\gamma(m,p)}} \Rightarrow |\tau(y) - \tau(x_m)| < \frac{1}{2^p}.$$

We will first work on one computable irrational number x. For a positive integer q, we can effectively find an $l = l(q)$ such that $\frac{l}{q} < x < \frac{l+1}{q}$. Put

$$\mu = \mu(q) = \min(x - \frac{l}{q}, \frac{l+1}{q} - x).$$

μ is computable and $0 < \mu(q) < \frac{1}{q}$. Put then

$$\delta(h) = \min\{\mu(q); 1 \leq q \leq h\}.$$

δ is computable and $0 < \delta(h) < \frac{1}{h}$ follows, and we can effectively find an $N = N(h)$ which satisfies

$$\frac{1}{2^N} < \delta(h).$$

Suppose now $|y - x| < \frac{1}{2^{N(h)}}$, hence $|y - x| < \delta(h) < \frac{1}{h}$. Since $\tau(x) = 0$,

$$|\tau(y) - \tau(x)| = |\tau(y)|.$$

If y is irrational, then $|\tau(y)| = 0$. If y is rational and expressed with an irreducible fraction $\frac{s}{t}$, then $t > h$ (since $|y - x| < \delta(h)$) and hence $\tau(y) = \frac{1}{t} < \frac{1}{h}$.

In either case, $|\tau(y) - \tau(x)| < \frac{1}{h}$. Taking $h = 2^p$ and putting $\gamma(p) = N(2^p)$, the condition of (ii) is satisfied. (Note that the objects defined here are obtained effectively, dependent on x.)

The argument above can be extended to a sequence. So, for a computable irrational sequence $\{x_m\}$, we can find effectively $l = l(m, q)$, $\mu = \mu(m, q)$, $\delta(m, h)$, $N = N(m, h)$, and $\gamma(m, p)$ satisfying the condition in (ii).

Remark 7 For each real number x, $\tau(x)$ is a rational number, and hence is computable, but τ is not sequentially computable. This can be seen as follows. Let α be a recursive injection whose range A is not recursive, and define

$$x_n = \frac{1}{2^{m+2}} \quad \text{if} \quad n = \alpha(m); \ = 0 \quad \text{if} \quad n \notin A.$$

Then $\{x_n\}$ is a computable sequence of rational numbers (cf. Example 3 of Chapter 0 in [6]), and

$$\tau(x_n) = \frac{1}{2^{m+2}} \quad \text{if} \quad n = \alpha(m); \ = 1 \quad \text{if} \quad n \notin A.$$

This shows that

$$n \in A \quad \text{if} \quad \tau(x_n) < \frac{1}{2}; \ n \notin A \quad \text{if} \quad \tau(x_n) > \frac{1}{2}. \tag{3}$$

Suppose τ were sequentially computable. Then we would be able to effectively decide "$n \in A$?" from the relations in (3), contradicting non-recursiveness of A.

Example 6 (*Probability distribution function G: Example 3.4 of* [5]) Let α be as in Remark 7 (assuming that the value does not take 0). If we put

$$d = \Sigma_{i=1}^{\infty} \frac{1}{2^{\alpha(i)}} (< 1),$$

d is not computable (cf. [6]). Put

$$G = (1 - d)\chi_{[0,1)} + \Sigma_{i=1}^{\infty}(\Sigma_{j=i+1}^{\infty} \frac{1}{2^{\alpha(j)}})\chi_{[\frac{1}{2^i}, \frac{1}{2^{i-1}})} + \chi_{[1,\infty)}. \tag{4}$$

$G = G_\nu$ (defined on the whole real line) is the distribution function of a computable probability distribution, called ν, of Example 3.4 in [5]. As a reference, we present the definition of ν, quoted from [5].

$$\nu = (1 - d)\delta_0 + \Sigma_{i=1}^{\infty} \frac{1}{2^{\alpha(i)}} \delta_{\frac{1}{2^{\alpha(i)}}}.$$

ν is computable in the sense that for every computable function with compact support, say f, $\nu(f)$ is a computable real number.

As for G itself, G is not computable with respect to any of our preceding methods, for $G(0) = 1 - d$ is not computable. (In [5], G was presented as a counter example to Fine computability.) Also, G is not effectively right continuous at 0, though right continuous. However, G is IB-computable. That is, G is an example of a function which is not Fine computable but is IB-computable. So, the notion of computability is in a sense conserved in the transfer of the distribution ν to its distribution function.

IB-computability of G can be shown as follows.

(i) For a computable irrational number x, it can be decided to which of the intervals in the formula (4) it belongs. Notice that it is not equal with the edge of any interval. If $x < 0$, then $G(x) = 0$, and if $x > 1$, then $G(x) = 1$. If $x \in [\frac{1}{2^i}, \frac{1}{2^{i-1}})$ for some i, then

$$G(x) = (1 - d) + \Sigma_{j=i+1}^{\infty} \frac{1}{2^{\alpha(j)}} = (1 - d) + (d - \Sigma_{j=1}^{i} \frac{1}{2^{\alpha(j)}}) = 1 - \Sigma_{j=1}^{i} \frac{1}{2^{\alpha(j)}},$$

which is computable. The argument can be extended to a computable irrational sequence, and hence IB-sequential computability holds.

(ii) The IBE-continuity can be shown as follows. For a computable irrational sequence $\{x_m\}$, it can be effectively determined which of the intervals $(-\infty, 0)$, $[\frac{1}{2^i}, \frac{1}{2^{i-1}})(i = 1, 2, \cdots)$ and $[1, \infty)$ each x_m belongs to. One can effectively find a positive integer l_m so that the open interval $(x_m - \frac{1}{2^{l_m}}, x_m + \frac{1}{2^{l_m}})$ lies within the relevant interval for each case as below. If the first is the case, take an l_m satisfying $\frac{1}{2^{l_m}} < |x_m|$. If the last is the case, take one satisfying $\frac{1}{2^{l_m}} < x_m - 1$. If $x_m \in [\frac{1}{2^i}, \frac{1}{2^{i-1}})$, then take the one satisfying $\frac{1}{2^{l_m}} < \min(x_m - \frac{1}{2^i}, \frac{1}{2^{i-1}} - x_m)$. $\{l_m\}$ is recursive. Now put $\gamma(m, p) = l_m$ for all p. This will serve as an effective modulus of continuity, since, within each interval, the value of G is constant.

Example 7 (*Uniformly continuous but not effectively uniformly continuous function:* [6]) Theorem 6 in Chap. 1 of [6] states the following.

> There exists a continuous function f on $[0, 1]$ such that f is sequentially computable but is not computable. Furthermore, f is differentiable (but not continuously differentiable).

In fact f is uniformly continuous but is not effectively uniformly continuous. We will show that f is IB-computable.

First, since f is sequentially computable, (i) trivially holds. So, we will concentrate on IBE-continuity, i.e., (ii) of Definition 4.

For the reader's convenience, we will cite a lemma and the construction of f from [6].

Effective modulus lemma (EML: Theorem 5 in Chap. 1 of [6]) There exists a computable sequence of rational numbers $\{r_m\}$ such that:

(1) the sequence $\{r_m\}$ is strictly increasing, $0 < r_m < 1$, and $\{r_m\}$ converges to a noncomputable real number α.
(2) the differences $(r_m - r_{m-1})$ do not approach zero effectively.
(3) for any computable sequence of reals $\{\gamma_k\}$, there exist recursive functions $d(k)$ and $e(k)$ such that

$$\forall k \forall m \geq e(k). |\gamma_k - r_m| \geq \frac{1}{d(k)}. \tag{5}$$

Using this lemma, f is constructed as follows.

Let $\{r_m\}$ be as in the lemma above. Put

$$a_m = r_m - r_{m-1};\ b_m = \min(a_m, a_{m-1})\ (m \geq 2).$$

From the construction in the lemma, $0 < r_m - r_{m-1} < 1$, and hence $0 < a_m, b_m < 1$. Define a function φ by

$$\varphi(x) = e^{\frac{-x^2}{(1-x^2)}}\ (|x| < 1); = 0\ (|x| \geq 1).$$

$\varphi(x)$ is a computable C^∞ pulse function with compact support $[-1, 1]$. Now f is defined by

$$f(x) = \sum_{m=2}^{\infty} a_m^2 \varphi[2^m b_m^{-1}(x - r_{m-1})].$$

$\sum a_m^2$ converges (not effectively), hence the series for f is uniformly convergent (not effectively). So, f is continuous, hence is uniformly continuous on the compact interval $[-1, 1]$. It is proved in [6] that f is C^∞ except at the point $\alpha = \lim r_m$, but that $f'(\alpha)$ exists and is zero. So, f is differentiable. It is also shown that f is not effectively uniformly continuous.

Note that, if we put $z_m(x) = 2^m b_m^{-1}(x - r_m)$, then

$$\varphi(z_m(x)) > 0 \Leftrightarrow |z_m(x)| < 1; \varphi(z_m(x)) = 0 \Leftrightarrow |z_m(x)| \geq 1. \tag{6}$$

In case x is irrational, the latter necessary condition is > 1.

Now we show that f is IBE-continuous: for any computable irrational sequence $\{\gamma_k\}$, there is a recursive function $\delta(k, p)$ such that

$$\forall y. |\gamma_k - y| < \frac{1}{2^{\delta(k,p)}} \Rightarrow |f(\gamma_k) - f(y)| < \frac{1}{2^p}.$$

In (3) of EML, let $\{\gamma_k\}$ be a computable irrational sequence (in $(0, 1)$). Then there are recursive functions $d(k)$ and $e(k)$ for which (5) holds.

We will first apply EML to one computable irrational number $\gamma \in (0, 1)$ and see that all the objects involved can be effectively obtained from the information on γ.

(3) in EML asserts that positive integers d and e can be effectively found so that

$$m \geq e \Rightarrow |\gamma - r_m| \geq \frac{1}{d}.$$

Put $M = \max(d, e)$. Then the sum in the definition of $f(\gamma)$ is bounded by M. Namely

$$f(\gamma) = \sum_{m=2}^{M} a_m^2 \varphi[2^m b_m^{-1}(\gamma - r_{m-1})]. \tag{7}$$

For, suppose $m > M$. Then $\frac{1}{2^m} < \frac{1}{2^d} < \frac{1}{d}$ and $m - 1 \geq M \geq e$, hence $m - 1 \geq e$. By (3) of EML, it then holds

$$|\gamma - r_{m-1}| \geq \frac{1}{d} > \frac{1}{2^m} > \frac{b_m}{2^m}.$$

That is, $|z_m(\gamma)| > 1$. So, by (6), $\varphi(z_m(\gamma)) = 0$.

Next, define a function f_M on $[0, 1]$ by

$$f_M(x) = \sum_{m=2}^{M} a_m^2 \varphi[2^m b_m^{-1}(x - r_{m-1})].$$

With these preparations, we show that f is IBE-continuous.

Suppose $|\gamma - y| < \frac{1}{2d}$. Then for $m > M$,

$$|y - r_{m-1}| \geq |\gamma - r_{m-1}| - |y - \gamma| \geq \frac{1}{d} - \frac{1}{2d} = \frac{1}{2d}$$

$$\geq \frac{1}{2^d} > \frac{1}{2^m} > \frac{b_m}{2^m}.$$

So, $|z_m(y)| > 1$, hence by (6) $\varphi(z_m(y)) = 0$. This implies $f(y) = f_M(y)$.

Since f_M is effectively uniformly continuous, there is a recursive function $\xi(p)$ such that

$$|\gamma - x| < \frac{1}{2^{\xi(p)}} \Rightarrow |f_M(\gamma) - f_M(x)| < \frac{1}{2^p}.$$

One can select a $\delta(p)$ effectively so that

$$\frac{1}{2^{\delta(p)}} < \frac{1}{2^{\xi(p)}}; \frac{1}{2^{\delta(p)}} < \frac{1}{2d}.$$

Noting that $|\gamma - y| < \frac{1}{2^{\delta(p)}}$ implies $|\gamma - y| < \frac{1}{2d}$, we then have

$$|\gamma - y| < \frac{1}{2^{\delta(p)}} \Rightarrow |f(\gamma) - f(y)| = |f_M(\gamma) - f_M(y)| < \frac{1}{2^p},$$

proving the effective continuity at γ.

In extending the above argument to a computable irrational sequence $\{\gamma_k\}$, one should mind that $M(k)$ can be recursive, and we define

$$f_{M(k)}(x) = \sum_{m=2}^{M(k)} a_m^2 \varphi[2^m b_m^{-1}(x - r_{m-1})].$$

Following the case for a single γ, we obtain that

$$|\gamma_k - y| < \frac{1}{2d(k)} \Rightarrow f(y) = f_{M(k)}(y).$$

Also, by examining the definition of $f_{M(k)}$, we obtain a modulus of continuity $\xi(k, p)$ which is recursive in k and p, from which a recursive function $\delta(k, p)$ for f at $\{\gamma_k\}$ can be found.

Remark 8 The function f in Example 7 was chosen in [6] as an example of a very nice and smooth function which is nevertheless "not effectively" uniformly continuous. For an example which is sequentially computable, uniformly continuous and not effectively uniformly continuous on a compact interval, it suffices to use a function simpler than φ, such as a tent function.

We have shown that the typical functions which have been treated in our preceding articles with various notions of computability are all IB-computable, and that there are functions which are not computable in the preceding theories but are IB-computable. Among the IB-computable functions, the function G in Example 6 is not Fine-computable, and the function f in Example 7 is, though a very smooth function in the classical sense, not effectively uniformly continuous. We have also treated here Thomae's function and a few others.

At the end, we present a counterexample.

Example 8 (*Counterexample*) Let g be the function which distinguishes rational numbers and irrational numbers: $g(x) = 1$ if x is rational, and $= 0$ if it is irrational.

g is IB-sequentially computable but is not IBE-continuous.

Obviously g is IB-sequentially computable. However, for any irrational number x (computable or not), there can be no neighborhood of x in which the value of g remains the same as $g(x)$.

5 Irrational-Based Effective Convergence of Function Sequence

In this section, $\{x_m\}$ will denote a computable irrational sequence, and $\{f_n\}$ and f will denote a sequence of real functions and a real function respectively.

Definition 6 (*IBE-convergence*) $\{f_n\}$ will be said to irrational-based effectively converge (IBE-converge) to f if the following condition (Cond) holds.
(Cond): For any $\{x_m\}$, there are recursive functions α_1 and α_2 such that,

$$\forall n \geq \alpha_1(m, p)(\forall y.|y - x_m| < \frac{1}{2^{\alpha_2(m,p)}} \Rightarrow |f_n(y) - f(y)| < \frac{1}{2^p}). \tag{8}$$

Remark 9 Notice that $\alpha_2(m, p)$ depends on p, and hence this notion of convergence is not even locally uniform.

In mathematics, a sequence (of real numbers or functions) plays an important role, especially, for example, with regard to the limit operation. The sequence version of IB-computability as in (2) of Definition 4 is a simple extension of the single function version. For mathematical development of the theory of IB-computability, however, it is more appropriate to somewhat restrict the function sequence as is seen below in Definition 7.

Definition 7 (*IBAQ-computability of function sequence*) A function sequence $\{f_n\}$ is called *irrational-based asymptotically equi-computable* (IBAQ-computable) if it satisfies the following 1° and 2°.
1°. (i') of (2) in Definition 4 (sequential computability) holds.
2°. (ii') is modified to *asymptotical equi-continuity* (IBAQ-continuity) as follows: for any $\{x_m\}$, there are recursive functions $\beta_1(m, p)$ and $\beta_2(m.p)$ such that

$$\forall n \geq \beta_1(m, p)\forall y.|x_m - y| < \frac{1}{2^{\beta_2(m,p)}} \Rightarrow |f_n(x_m) - f_n(y)| < \frac{1}{2^p} \tag{9}$$

β_2 may be called a recursive asymptotic modulus of continuity for $\{f_n\}$ and $\{x_m\}$.

Theorem 2 (Closure property) *If $\{f_n\}$ is an IBAQ-computable sequence (cf. Definition 7) and IBE-converges to f, then f is IB-computable.*

Proof We have the following (a), (b) and (c) at our disposal.

(a) (Cond) in Definition 6 (the existence of α_1 and α_2).
(b) The condition 1° in Definition 7 (computability of the double sequence $\{f_n(x_m)\}$).
(c) The condition 2° in Definition 7 (the existence of β_1 and β_2).

(i) of Definition 4: From (a) with $y = x_m$ and (b), $\{f(x_m)\}$ is a computable sequence of real numbers by virtue of "the closure under effective limit" of a computable real number sequence (Proposition 1 in Chapter 0 of [6]).
(ii) of Definition 4: What to be proved is, for every $\{x_m\}$, there is a recursive function γ satisfying,

$$|x_m - y| < \frac{1}{2^{\gamma(m,p)}} \Rightarrow |f(x_m) - f(y)| < \frac{1}{2^p}.$$

By (c), there are recursive β_1 and β_2 for which (9) holds. From (a), there are recursive α_1 and α_2 so that (8) holds. Now put

$$\delta_1(m, p) = \max(\alpha_1(m, p), \beta_1(m, p)); \delta_2(m, p) = \max(\beta_2(m, p), \alpha_2(m, p)),$$

and suppose $n \geq \delta_1(m, p)$. If $|x_m - y| < \frac{1}{2^{\delta_2(m,p)}}$, then the premises of (8) and (9) both hold. So, for $n = \delta_1(m, p)$, for example,

$$|f(x_m) - f(y)| \leq |f(x_m) - f_n(x_m)| + |f_n(x_m) - f_n(y)| + |f_n(y) - f(y)|$$

$$< \frac{1}{2^p} + \frac{1}{2^p} + \frac{1}{2^p} = \frac{3}{2^p},$$

where the three evaluations are due to respectively (8) with $y = x_m$, (9) and (8). Now put $\gamma(m, p) = \delta_2(m, p + 2)$.

Remark 10 Theorem 2 assures us that the definitions of IBAQ-computability of a function sequence (cf. Definition 7) and its IBE-convergence (cf. Definition 6) are natural ones. To reinforce this claim, we will present some examples.

Example 9 (*Powers of x on the unit interval*) Consider, on the compact domain $I = [0, 1]$, the functions

$$f_n(x) = x^n,$$

$$f(x) = 1 - \chi_{[0,1)}(x).$$

$\{f_n\}$ is IBAQ-computable, and IBE-converges to f, which is indeed IB-computable.

Proof IBAQ-computability of $\{f_n\}$ is easy to see.

For IBE-convergence, put

$$a_m = \frac{\min(1 - x_m, x_m)}{2} (> 0).$$

Then $\{a_m\}$ is a computable sequence of irrational numbers and $(x_m - a_m), (x_m + a_m) \in [0, 1)$. Since each f_n is an increasing function,

$$f_n(x_m - a_m) < f_n(x_m) < f_n(x_m + a_m).$$

An n satisfying $(x_m + a_m)^n < \frac{1}{2^p}$, or $n > \frac{-p \log 2}{\log(x_m + a_m)}$ can be effectively computed, depending on m and p. Now put $\alpha_1(m, p) =$ "this n". Also, a k satisfying $\frac{1}{2^k} < a_m$, or $k > -\frac{\log a_m}{\log 2}$, can be effectively computed, depending on m. Put $\alpha_2(m, p) =$ "this k" (uniformly in p).

It is obvious that, if $n \geq \alpha_1(m, p)$ and $|y - x_m| < \frac{1}{2^{\alpha_2(m,p)}}$, then $0 \leq y < x_m + a_m < 1$, and hence $f(y) = 0$. So,

$$|f_n(y) - f(y)| = f_n(y) < (x_m + a_m)^n < \frac{1}{2^p}.$$

Example 10 (*Convergence to probability distribution function G: Example 6*) Let α and d be as in Example 6. We consider the following functions on $[0, 1)$.

$$G_n = 1 - (\sum_{i=1}^{n} (\sum_{j=1}^{i} \frac{1}{2^{a(j)}}) \chi_{[\frac{1}{2^i}, \frac{1}{2^{i-1}})} + \sum_{j=1}^{n} \frac{1}{2^{a(j)}} \chi_{[0, \frac{1}{2^n})});$$

$$G = (1 - d)\chi_{[0,1)} + \Sigma_{i=1}^{\infty} (\Sigma_{j=i+1}^{\infty} \frac{1}{2^{\alpha(j)}}) \chi_{[\frac{1}{2^i}, \frac{1}{2^{i-1}})}.$$

$\{G_n\}$ is an IBAQ-computable sequence of functions, and IBE-converges to G on $[0, 1)$. (G has been proved IB-computable in Example 6. Here the last term $\chi_{[1,\infty)}$ in (4) is irrelevant.)

Proof We first consider the two conditions on IB-computability for one computable irrational number $x \in [0, 1)$, paying attention to the effectivity of the procedure to obtain certain numbers.

For each n, it can be effectively decided to which of the intervals $[\frac{1}{2^i}, \frac{1}{2^{i-1}})$ $(1 \leq i \leq n)$ or $[0, \frac{1}{2^n})$ x belongs. In the former case, $G_n(x) = 1 - \sum_{j=1}^{i} \frac{1}{2^{\alpha(j)}}$, and in the latter case, $G_n(x) = 1 - \sum_{j=1}^{n} \frac{1}{2^{\alpha(j)}}$. $G_n(x)$ is a computable number and the value can be effectively obtained. This process can be easily extended to a computable irrational sequence, hence $\{G_n\}$ is IB-sequentially computable (1° of Definition 7).

As for IBAQ-continuity (2° of Definition 7), suppose first that $x \in [\frac{1}{2^i}, \frac{1}{2^{i-1}})$ holds for an i, $1 \leq i \leq n$. Put $w_x = \min(x - \frac{1}{2^i}, \frac{1}{2^{i-1}} - x)$. w_x is computable from x. Then a positive integer β_2 can be effectively found so that $\frac{1}{2^{\beta_2}} < w_x$.

If $x \in [0, \frac{1}{2^n})$, then put $w_x = \min(x, \frac{1}{2^n} - x)$, and find a β_2 as above. In either case, if $|x - y| < \frac{1}{2^{\beta_2}}$, then y belongs to the same interval as x. So, $G_n(x) - G_n(y) = 0$. This process applies to any n (and x), hence we can take $\beta_1 = 1$. β_2 depends effectively on x, and hence for a sequence $\{x_m\}$ we should write $\beta_2(m, p)$, although it is independent of p.

Next, we show that $\{G_n\}$ IBE-converges to G.

Suppose $x \in [\frac{1}{2^i}, \frac{1}{2^{i-1}})$ for some i $(i = 1, 2, 3, \cdots)$. (An irrational number $x \in [0, 1)$ belongs to one of such intervals and i can be effectively obtained.) Then $G(x) = 1 - \sum_{j=1}^{i} \frac{1}{2^{\alpha(j)}}$. If $n \geq i$, then $G(x) = G_n(x)$. Put $w_x = \min(x - \frac{1}{2^i}, \frac{1}{2^{i-1}} - x)$. If $|x - y| < w_x$, then y belongs to the same interval as x. Now, take $\alpha_1 = i$ and α_2 to satisfy $\frac{1}{2^{\alpha_2}} < w_x$. Then, $G(x) = G(y) = G_n(x) = G_n(y)$.

The argument can be extended to a computable irrational sequence. Here α_1 and α_2 depend on m, but are independent of p.

Classically, the convergence is locally uniform.

Example 11 (*Convergence to Thomae's function τ: Example 5*) Define $\tau_n(x) = 0$ if x is irrational. Suppose $x = \frac{s}{t}$, an irreducible fraction with $t > 0$. Then $\tau_n(x) = \frac{1}{t}$ if $t \leq n$; $= 0$ if $t > n$.

$\{\tau_n\}$ is IBAQ-computable and IBE-converges to τ.

Proof IBAQ-computability: 1° of Definition 7 trivially holds.

For 2°, suppose x is a computable irrational number. Following the proof of (ii) in Example 5, reading τ as τ_n, we obtain $\mu = \mu(q)$, $\delta(h)$ and $N(h)$ effectively, depending on n.

Suppose $|y - x| < \frac{1}{2^{N(h)}}$. Then $|y - x| < \delta(h) < \frac{1}{h}$. Since $\tau_n(x) = 0$, $|\tau_n(y) - \tau_n(x)| = \tau_n(y)$. If y is irrational, then the value is 0. Suppose y is rational and expressed with an irreducible fraction $\frac{s}{t}$. Then $t > h$. If $t > n$, then $\tau_n(y) = 0$. If $t \leq n$, then $\tau_n(y) = \frac{1}{t}$. Since $t > h$, $\tau_n(y) < \frac{1}{h}$. In any case, $|\tau_n(y) - \tau_n(x)| < \frac{1}{h}$. This holds for all n. As before, taking $h = 2^p$ and putting $\beta_1(p) = 1$ and $\beta_2(p) = N(2^p)$, we obtain that $|y - x| < \frac{1}{2^{\beta_2(p)}}$ implies $|\tau_n(y) - \tau_n(x)| < \frac{1}{2^p}$.

Extend the argument to a sequence $\{x_m\}$, putting $\beta_1(m, p) = 1$ and $\beta_2(m, p) = N(m, 2^p)$.

IBE-convergence: We should notice that, for any x,

$$\forall x. |\tau_n(x) - \tau(x)| < \frac{1}{n}$$

holds, that is, the convergence is uniform. (Suppose $x = \frac{s}{t}$, and $t \leq n$. Then $\tau_n(x) = \tau(x) = \frac{1}{t}$. If $t > n$, then $\tau_n(x) = 0$, while $\tau(x) = \frac{1}{t}$, hence $|\tau_n(x) - \tau(x)| \leq \frac{1}{t} < \frac{1}{n}$.) Then $\alpha_1(p) = 2^p$ and $\alpha_2(p) = 1$ will do.

Remark 11 The function sequences $\{G_n\}$ and $\{\tau_n\}$ have not been ad hoc devised for the sake of examples. In fact one often draws approximate graphs of G and τ respectively in terms of those sequences.

Example 12 (*Fine computable sequence: Example 3*)

In order to deal with this example, we first introduce a condition on a Fine computable sequence of functions. (We refer the reader to [3, 4] for notations and notions with regards to Fine computable sequences of functions.)

Definition 8 (*AQ-Fine computability*) A function sequence $\{f_n\}$ on $[0, 1)$ is called *asymptotically equi-Fine computable* (AQ-Fine computable) if the following (i) and (ii) hold.

(i) It is sequentially Fine computable.

(ii) It is *effectively asymptotically equi-Fine continuous* (EAQ-Fine continuous). That is, there are recursive functions γ_1 and γ_2 such that

(a) for $n \geq \gamma_1(p, i)$, $x \in J(e_i, \gamma_2(p, i))$ implies $|f_n(x) - f_n(e_i)| < \frac{1}{2^p}$;

(b) $\bigcup_{i=0}^{\infty} J(e_i, \gamma_2(p, i)) = [0, 1)$ for all p.

Definition 9 (*Effective Fine convergence:* [3]) A function sequence $\{f_n\}$ on $[0, 1)$ is said to Fine converge effectively to a function f if there are recursive functions M and N for which the following (c) and (d) hold.

(c) For any i, p and any $n \geq M(p, i)$, if $x \in J(e_i, N(p, i))$, then $|f_n(x) - f(x)| < \frac{1}{2^p}$.

(d) For any p, $\bigcup_{i=0}^{\infty} J(e_i, N(p, i)) = [0, 1)$.

Proposition 7 *An AQ-Fine computable sequence of functions on* $[0, 1)$ *is IBAQ-computable on* $[0, 1)$ *(cf. Definition 7).*

Proof Let $\{f_n\}$ be an AQ-Fine computable function sequence on $[0, 1)$. Sequential computability (1°) is obvious (cf. Example 3).

For 2°, we can first find effectively $i = \iota(m, p)$ satisfying $x_m \in J(e_i, \gamma_2(p, i))$ (by (b) of Definition 8), and next $l = l(m, p)$ such that, putting $q = \gamma_2(p, i)$, $J(e_i, q) = [\frac{l}{2^q}, \frac{l+1}{2^q})$ holds. Now if $n \geq \gamma_1(p, i)$, then, by (a) of Definition 8, , $|f_n(x_m) - f_n(e_i)| < \frac{1}{2^p}$. Put

$$z_{mp} = \min(x_m - \frac{l}{2^q}, \frac{l+1}{2^q} - x_m).$$

$\{z_{mp}\}$ is a positive computable sequence, and a positive integer $u = u(m, p)$ can be found effectively so that $\frac{1}{2^u} < z_{mp}$.

Suppose $|x_m - y| < \frac{1}{2^u}$. Then $y \in J(e_i, \gamma_2(p, i))$, and hence by (a) $|f_n(y) - f_n(e_i)| < \frac{1}{2^p}$. Replacing p by $p + 1$, we then obtain $|f_n(x_m) - f_n(y)| < \frac{1}{2^p}$. We can now take $\beta_1(m, p) = \gamma_1(p + 1, \iota(m, p + 1))$ and $\beta_2(m, p) = u(m, p + 1)$.

Proposition 8 *If an AQ-Fine computable function sequence* $\{f_n\}$ *effectively Fine-converges to a function* f *on* $[0, 1)$*, then it IBE-converges, and hence* f *is IB-computable by Theorem 2.*

Proof From the premise follows that there are recursive functions M and N satisfying (c) and (d) in Definition 9. We can follow the proof for 2° of Proposition 7 with M in the place of γ_1 and N in the place of γ_2, thus obtaining recursive functions ι and u so that, with $u = u(m, p)$ and $i = \iota(m, p)$, if $|x_m - y| < \frac{1}{2^u}$, then $y \in J(e_i, N(p, i))$. Then by (c), for $n \geq M(p, i)$, $|f_n(y) - f(y)| < \frac{1}{2^p}$. So, $\alpha_1(m, p) = M(p, \iota(m, p))$ and $\alpha_2(m, p) = u(m, p)$ will do (cf. Definition 6).

6 Final Note: Classical Account

It is generally agreed that the existing computability notion of continuous functions (cf. [6]) is a natural effectivization of continuity.

There are many different definitions of continuity which apply also to some discontinuous functions (in the Euclidean topology), but we feel "almost everywhere continuity" closer to the "standard" continuity compared with many others. Our counterexample g (Example 8) can be a sort of test function to see this: g is not almost everywhere continuous, while it is "continuous" in the sense of almost continuity, for instance. (See Note below.) We view that g is too far from continuity, especially from the computability point of view.

Our present theory, IB-computability, has a close similarity to the notion of almost everywhere continuity.

For the reader's convenience, we present the definition of almost everywhere continuity on the real line with regard to a subset of real numbers, say $\mathbf{J}$, where the Lebesgue measure of $\mathbf{J}$ restricted to each compact interval between two consecutive integers, say $[M, M+1]$ is 1, hence the complement of $\mathbf{J}$ is of measure 0. Now, a real function f is almost everywhere continuous with respect to $\mathbf{J}$ if f is continuous at every point of $\mathbf{J}$. That is,

$$\forall x \in \mathbf{J} \forall \varepsilon > 0 \exists \delta > 0 \forall y \in (x-\delta, x+\delta).|f(y) - f(x)| < \varepsilon. \tag{10}$$

Notice that $\delta = \delta_{x,\varepsilon}$ depends on x and ε, and that the continuity at a point in $\mathbf{J}$ is the usual one. The condition on $\mathbf{J}$ implies that it contains "most of" irrational numbers.

So, we assume the set $\mathbf{J}$ to be the set of all irrational numbers to make the discussion simpler and clearer.

Then, one idea of effectivizing the condition in (10) can be stated as follows.

(a) There is an effective method to compute the value $f(x)$ for any computable irrational number x.

(b) Given any computable irrational number x and a positive rational number ε, a positive number δ can be effectively computed so that

$$\forall y.|x-y| < \delta \Rightarrow |f(x) - f(y)| < \varepsilon.$$

Notice that the effective method of computing δ depends on x (and ε).

Instead of referring to "an effective method to compute" each time, we require the properties (a) and (b) for computable sequences of irrational numbers. (This is a standard practice in such a situation: see [6].)

We thus come to Definition 4, IB-computability of f, in Sect. 3.

Remark 12 (1) In (a) above, we could require the computability of the value $f(x)$ for any computable real number x. The theory of IB-computability of functions holds sound with this setting. However, this generalization does not fit our image. Almost everywhere continuity is endowed with the word "continuity" by excluding the rational points from the continuity requirement, except for the smoothness of the function in the whole of a neighborhood of each irrational number. It is therefore more natural not to question the nature of the values at rational numbers.

(2) The condition (b) does not necessarily imply almost everywhere continuity of f, since it does not guarantee the continuity at non-computable irrational numbers. The function θ in Example 13 below provides such an example.

Example 13 (*IB-computable but not almost everywhere continuous function*) We assume the $\{r_m\}$, d, e, α of EML in Sect. 4. Recall that $\alpha = \lim r_m$.

The function θ defined on $[0, 1]$ by

$$\theta(x) = 1(x = \alpha); = 0(x \neq \alpha)$$

is IB-computable, but is not almost everywhere continuous with respect to the set of all irrational numbers.

Proof It is obvious that θ is well defined, and is IB-sequentially computable since its values at computable irrational numbers are all 0.

Let $\{\gamma_k\}$ be a computable irrational sequence. By EML, there are recursive functions $d(k)$ and $e(k)$ such that

$$\forall k \forall m \geq e(k).|\gamma_k - r_m| \geq \frac{1}{d(k)}.$$

From this follows that $|\gamma_k - \alpha| \geq \frac{1}{d(k)}$. Now select $\delta(k, p) = \delta(k)$ so that

$$\frac{1}{2^{\delta(k)}} \leq \frac{1}{2d(k)}.$$

If $|\gamma_k - y| < \frac{1}{2^{\delta(k)}}$, then

$$|y - \alpha| \geq |\gamma_k - \alpha| - |y - \gamma_k| \geq \frac{1}{d(k)} - \frac{1}{2d(k)} = \frac{1}{2d(k)} > 0.$$

So, $\theta(y) = 0$. Since $\theta(\gamma_k) = 0$, $|\theta(y) - \theta(\gamma_k)| = 0$, and hence the above $\delta(k, p) = \delta(k)$ serves as an effective modulus of continuity for θ. θ is thus IB-computable.

On the other hand, θ is discontinuous at the non-computable irrational number α, which implies that θ is not almost everywhere continuous.

Note A real function f is said to be almost continuous if the following holds.

$$\forall x \forall \varepsilon > 0 \exists \delta > 0.\{y : |f(x) - f(y)| < \varepsilon\} \text{ is dense in } (x - \delta, x + \delta).$$

Almost continuity of our g is shown in Example 1 of [2].

References

1. Brattka, V., Hertling, P., Weihrauch, K.: A Tutorial on Computable Analysis. New Computational Paradigms, pp. 425–491. Springer, Berlin (2008)
2. Hussain, H., Dwivedi, T.D.: A characterization of real almost continuous functions. Can. Math. Bull. **10**(3), 361–369 (1967)
3. Mori, T., Tsujii, Y., Yasugi, M.: Fine computable functions and effective Fine convergence. In: CCA2005 Proceedings (Informatik Berichte 326-7/2005 FernUniversiat in Hagen, 2005), pp. 177–198
4. Mori, T., Yasugi, M., Tsujii Y.: Fine convergence of functions and its effectivization. In: Automata, Formal Languages and Algebraic Systems, pp. 139–162. World Scientific, Singapore (2010)
5. Mori, T., Tsujii, Y., Yasugi, M.: Fine Computability of Probability Distribution Functions and Computability of Probability Distributions on the Real Line, "Acta humanistica et scientifica", Universitatis Sangyo Kyotiensis, Natural Science Series, vol. 44, pp. 107–119. http://hdl.handle.net/10965/1180
6. Pour-El, M.B., Richards, J.I.: Computability in Analysis and Physics. Springer, Berlin (1989): Available also at Projecteuclid
7. Tsujii, Y., Yasugi, M., Mori, T., Some properties of the effectively uniform topological space. In: Computability and Complexity in Analysis, LNCS, vol. 2064, pp. 336–356. Springer, Berlin (2001)
8. Washihara, M., Yasugi, M.: Computability and metrics in a Fréchet space. Math. Japonica **43**, 1–13, Japanese Assoc. of Math. Sci. (1996)
9. Weihrauch, K.: Computable Analysis. Springer, Berlin (2000)
10. Yasugi, M., Washihara, M.: Computability structures in analysis. Sugaku Expositions (AMS) **13**(2), 215–235 (2000)
11. Yasugi, M., Brattka, V., Washihara, M.: Computability aspects of some discontinuous functions. Sci. Math. Jpn. (SCMJ) **5**, 405–419 (2001)
12. Yasugi, M.: Limiting computability of discontinuous functions-significance and problems. J. Jpn. Assoc. Phil. Sci. **30**(2), 13–18 (2003) (in Japanese)
13. Yasugi, M., Tsujii, Y.: Computability of a function with jumps-Effective uniformity and limiting recursion. Topol. Appl. (Elsevier Science) **146–147**, 563–582 (2005)
14. Yasugi, M., Tsujii, Y., Mori, T.: Sequential computability of a function - effective fine space and limiting recursion. JUCS **11-12**, 2179–2191 (2005). http://www.jucs.org/
15. Yasugi, M., Washihara, M.: A note on Rademacher functions and computability. In: Words, Languages and Combinatorics III, pp. 466–475. World Scientific, Singapore (2003)
16. Yasugi, M.: On notions of computation over the continuum - Beyond recursive functions. Kyoto Grad. J. Philos. (Tetsugakuronso) XXXV, 199–209 (2008). (in Japanese): http://hdl.handle.net/2433/96269
17. Yasugi, M., Washihara, M.: Sequential computability of a function -limiting recursion versus effective uniformity. Sci. Math. Jpn. **71**(3), 331–341 (2010); Online, e-2010-16, No. 23, 153–163 (2010): http://www.jams.or.jp/scm/contents/e-2010-2/2010-16.pdf

“Gaisi Takeuti’s Finitist Standpoint” and Its Mathematical Embodiment

Mariko Yasugi

Abstract Gaisi Takeuti’s mathematical achievements in consistency proofs of subsystems of second order arithmetic were always lined by his thoughts on the finitist standpoint. Takeuti’s insistence on the finitist standpoint originates in the requirement from the consistency proof per se. In its essence, it is to show the termination of a decreasing sequence from an order structure used in a consistency proof as clearly as possible. Takeuti has left many writings on the finitist standpoint. Some of them will be introduced and closely studied. Then a way of giving a mathematical form to Takeuti’s finitist standpoint will be proposed. This is done by means of a semi-formal theory of generalized functionals. One can explicitly present a functional which, given any decreasing sequence from the order structure in concern, evaluates the terminating point of the sequence.

Keywords The finitist standpoint · Consistency proof through reduction methods · Accessibility proof of order structure · Modulus of finiteness functional · Universe of mechanisms

1 Introduction

Takeuti’s consistency proof project is based on Hilbert’s program in its spirit and on Gentzen’s consistency proofs in its practice.[1] Both Gentzen and Takeuti respected Hilbert’s spirit of the finitist standpoint.

Gentzen’s and Takeuti’s consistency proofs are constructive except for the accessibility proofs of the order structures used in their consistency proofs. So, in order to justify their consistency proofs, it was important that the methods of these accessibility proofs be close to what was required by Hilbert’s program, that is, they be within

[1]Most of Takeuti’s proof theory is covered in his book “Proof Theory” [20]. See also [11] for its origin.

M. Yasugi (✉)
Kyoto Sangyo University Emerita, Kyoto, Japan
e-mail: yasugi@cc.kyoto-su.ac.jp

T. Arai et al. (eds.), *Advances in Mathematical Logic*, Springer Proceedings in Mathematics & Statistics 369, https://doi.org/10.1007/978-981-16-4173-2_10

the "finitist standpoint" (in some sense). A desirable accessibility proof should be therefore a *concrete* demonstration of the termination of decreasing sequences from the respective order structure.

At this point, I would like to address the reader by way of caution. In this article, expressions like'constructive','concrete' as well as '(strict) finitist (standpoint)' are seen in various contexts. As the very theme of "Takeuti's finitist standpoint" indicates that the word "finitist" does not represent a particular concept that can be rigorously defined in terms of a formula. After all, it is not a mathematical or logical concept. The word "finitist" or "concrete" occurs here and there in Takeuti's writings as a kind of parameter, and we can only interpret it from the content that surrounds each occurrence. We must do the interpreting works with good care, and that is one mission of this article. (Another is represented in the second part of the title.) "Strict finitist standpoint" is here meant to be Hilbert's requirement for his program. As for the word "constructive", very often it signifies a primitive recursive process, some times a recursive one, but it may be also intended to express some notion that does not exactly fit either of them. In each field of mathematics or the foundations of mathematics, there is often tacit understanding, if not exact, of this word. Here it can be understood primitive recursive or recursive.

Takeuti maintained a strong position of the "finitist standpoint" in his proof theory. He has left many verbal descriptions on the finitist standpoint regarding consistency proofs. The purpose of the present article is to speculate on those words, in order to make their intended meanings clear as much as possible, and then to outline a mathematical formulation of "Takeuti's finitist standpoint".

My stance is that Takeuti's metamathematical thoughts deserve deep considerations, since he was a leading figure in proof theory and contributed to a breakthrough in the consistency proofs of some formal systems of second order arithmetic along those thoughts. It is also important and desirable to speculate on Takeuti's thoughts to support his mathematical achievements from philosophical viewpoints.[2]

The basic knowledge of consistency proofs through reduction methods is assumed, and hence technical terms in this area of proof theory will be used without explanation. We refer the reader to [20].

This article contains the following, in its order: Hilbert's program as Takeuti sees it; some of Takeuti's writings on the finitist standpoint; the author's views; reflections on Takeuti's views; overviews of the framework of consistency proofs through reduction methods; an accessibility proof of ordinal diagrams; an interpretation of Takeuti's finitist standpoint and a principle HP that realizes it; some explanations; final words.[3]

[2] In [37], a collection of memorial articles dedicated to Takeuti is carried, including those on his proof theory and his standpoint.

[3] The author is grateful to the referee, according to whose advices Abstract has been extended to provide more information of the article, and attempts have been made to explain usages of some terminologies as well as to improve many expressions.

2 "Takeuti's" Hilbert's Program

Takeuti's aim in his proof theory was to execute Hilbert's program as much as possible.[4] Hilbert's program "as Takeuti interpreted it" is stated, for example, in [21], and in its English translation in Sect. 6 of [22], as quoted below ((1)∼(4)).[5]

(1) Formalize mathematics and treat it as a formal system.
For this reason, Hilbert's standpoint is called formalism.
(2) In this formalized system, a proposition is expressed with a string which is composed of some special symbols which stand for basic propositions, variables and logical notions respectively. Mathematical inferences and proofs are all described as concrete rules which apply to these strings, which are clusters of symbols as above.
(3) Only the arguments which are applied to the formalized proofs that are concretely considered as above and which are simple and reliable can be admitted. Namely, the strings which are relevant here are concrete strings consisting of finitely many symbols. This standpoint is called the finite standpoint.
(4) Complete the foundational underpinnings of the formalized mathematical system by proving that no contradiction can be derived in this formal system. That is, demonstrate that the system is consistent, within the finite standpoint.

The so-called consistency proofs of some systems of (second order) arithmetic through reduction methods have yielded many fruitful results. A consistency proof of this kind consists of the following seven processes:

1 a formal system $\mathcal{S}$ to be studied;
2 a constructive, linearly (and well-) ordered structure $\mathcal{J} = (J, <)$;
3 assignment of an element of J to each formal proof P in $\mathcal{S}$, say j_P;
4 the reduction of formal proofs in $\mathcal{S}$;
5 proof of decrease of j_P according to the reduction of P;
6 the accessibility proof of $\mathcal{J}$;
7 non-derivability of a contradiction in $\mathcal{S}$.

The accessibility statement stands: every decreasing sequence from J halts at a finite stage. An accessibility proof means a proof of the accessibility statement which can be accepted by the finitist standpoint. This can be understood to present a concrete method which, applied to any decreasing sequence from J, produces a halting number, that is, after that stage, the sequence becomes void (or constant). We will call such a number a modulus of accessibility, and a method to obtain it a modulus of finiteness functional.[6]

[4] He engaged in the consistency proofs of some subsystems of second order arithmetic, but he was also concerned with still higher order systems (cf. [11]).

[5] In this article, Hilbert's program and its execution are all "as Takeuti saw it", and every observation about Takeuti's thoughts is "as this author sees it". See [8] for Hilbert's program.

[6] Here the 'method' is not specified. It is a notion that can be specified only in each environment in concern.

Every process except **6** satisfies the strict finitist standpoint.[7] If one wishes to pursue the so-called consistency proof along Hilbert's program, one must demonstrate the accessibility of $\mathcal{J}$ in a manner explained above. The problem is, on what ground we can grant reasonableness to an attempted accessibility proof. The order structure in concern is designed to be well ordered, whose fact is shown set theoretically, but such a proof does not qualify the entire process above as a "consistency proof".

The original finitist standpoint accepts only a strictly concrete procedure in proving that a concretely defined property holds for concretely given objects, while even the consistency of first order arithmetic requires more than that. It is therefore exceedingly important that we analyze and speculate on Takeuti's finitist standpoint.

3 Takeuti's Finitist Standpoint

In admitting a consistency proof of a system as such, it is the "accessibility proof" of the order structure used there that is critical. Takeuti sought ways of establishing the well-orderedness of the order structures used in his works from the finitist standpoint. The treatment in [25] offers an approach very close to his taste, but even this did not convince him.

Takeuti has left many writings concerning his viewpoints on the (extended) finitist standpoint. Although there are some technical arguments (cf. [20], for example), we regard conceptual writings quite essential in understanding Takeuti's thoughts, and hence some of them will be introduced below ((i)∼(xiii)).[8] They will be analyzed in Sect. 5.

(i) When one is concerned with consistency proofs, their philosophical interpretation is always a paramount problem.
There is no doubt that Hilbert's "finitist standpoint", which considers only a finite number of symbols concretely given and arguments concretely given about finite sequences of these symbols (called expressions), is an ideal standpoint in proving consistency.
Therefore in a consistency proof it is interesting to see what is used that goes beyond Hilbert's finitist standpoint, and on what basis it can be justified ([20], Sect. 11).

(ii) Indeed, G. Gentzen in 1936 accomplished a consistency proof of number theory. But here the interpretation of the finite standpoint was certainly extended. That is what Hilbert meant by 'in a sharper manner' ([22], Sect. 7).

(iii) In formalism, one assumes the finitist standpoint as the purest and the most ideal standpoint ([15], p. 3).

[7] 'Strict finitist standpoint' is meant here to be Hilbert's standpoint, or what is understood so. (cf. (i) of Sect. 3).

[8] Citations from Takeuti's Japanese writings are not necessarily strict translations. They may be paraphrased and/or shortened at the author's responsibility.

(iv) The finitist standpoint is what flows out when one examines the inductively defined objects most purely and idealy ([15], p. 3).

(v) To admit only unsophisticated and most reliable arguments on the formalized and concretely thought proofs $\cdots$.
This is called the finitist standpoint ([21], [22], Sect. 6).

(vi) Our standpoint avoids abstract notions as much as possible, except those which are eventually reduced to concrete operations or Gedankenexperimente on concretely given objects ([20], Sect. 11).
Our standpoint, which has been discussed above, is like Hilbert's in the sense that both standpoints involve "Gedankenexperimente" only on clearly defined operations applied to some concretely given figures and on some clearly defined inferences concerning these operators ([20], Sect. 11).

(vii) The finitist standpoint allows the thought experiment on concretely given operations applied to concretely given expressions [18].

(viii) $\cdots$ we can regard "induction" simply as a general statement of a concrete method of proof applicable for any given expression x, and not as an axiom that is accepted a priori ([20], Sect. 11).

(ix) Modern mathematics assumes the infinite mind, but the human being is of the finite mind.
We need to justify the world of the infinite mind within the world of the finite mind [18].

(x) We believe that our standpoint is a natural extension of Hilbert's finitist standpoint, similar to that introduced by Gentzen, so we call it the Hilbert-Gentzen finitist standpoint ([20], Sect. 11).

(xi) Gentzen's consistency proof of arithmetic was an important result for Hilbert's Program or at least his finite standpoint. Let me therefore discuss Hilbert's phrase 'in a sharper manner' a bit more. In Gentzen's proof, an equivalent of transfinite induction up to ε_0 is applied. Otherwise every step is finite and elementary, which causes no problem. Thus, I will focus on transfinite induction ([22], Sect. 7).

(xii) Though nobody denies that the above way of thinking is contained in Hilbert's standpoint, there are many opinions about where to set the boundary of this standpoint: for example, assuming that transfinite induction up to each of ω, $\omega \cdot 2$, $\omega \cdot 3, \ldots$ is accepted, whether transfinite induction up to ω^2 should also be accepted ([20], Sect. 11).

(xiii) If I were to work now, I would like to consider more intrinsic matters such as what the essence of the finitist standpoint be and what it means to construct ordinals within the finitist standpoint. What are natural ordinals?
What are the finitist standpoint and the consistency that reside within me? When I define a system of ordinals along them, it should be a natural well ordering ([19]).

4 My View

Prior to speculations on Takeuti's thoughts, I state here some of my general views, which will underlie subsequent discussions.

- First of all, it is a reality that the finitist standpoint is not a clear-cut notion. Nevertheless, I will attempt to make it clear as much as possible.
- The consistency of a mathematical system means that a contradiction is *not* formally provable in it; that connotes that Hilbert's program requires a demonstration of this *negative* fact, and it should be demonstrated as clearly and easily understandably as possible. The finitist standpoint may be thought of as *this clarity and easy understandability*. Here resides the difficulty.
- Takeuti's "unsophisticated and most reliable arguments" and "what flows out when one examines the inductively defined objects most purely and idealy" may lead us, if we attentively observe those words, to a clear and easily understandable demonstration of a negative fact as above.
- Easy understandability does not necessarily mean logical simplicity, elementary notions or mechanical processes.[9] Borrowing the expression of [7], we might say it is the compactness of thoughts and expressions.

  Also, the finite mind and the elementary or mechanical process are different. The human mind has the ability to understand a clearly stated statement, to comprehend its content even if it contains infinite processes, and to turn one's imagination to it. Elementary or mechanical images are not necessarily connected to human intelligence of clear understanding.
- A statement which can be easily accepted by the human brain should be admitted as "somewhat finite", even if it contains some notion of infinity. This may mean that the finite mind can describe and understand the infinite mind. "The standpoint that allows the thought experiment" may be interpreted this way.
- Activities of human brain may be finite, but the human intelligence has the ability to understand the infinity. A thought experiment does not soley follow mechanical steps, but it can also involve infinitary notions.
- The finitist standpoint should admit various different views. It is a common practice to grasp one phenomenon from different viewpoints, and that may be why Takeuti was compelled to go on writing on the finitist standpoint, avoiding a rigid formulation.

[9] An interesting discussion on this theme is seen in Chap. 4 of [7] with unix programming as a material.

5 Some Reflections on Takeuti's Thoughts

Parallel to Takeuti's verbal expressions of the finitist standpoint, he and I formalized the well-foundedness proof of the system of ordinal diagrams. This was done in terms of the inductive definition as a principle [25], which did not satisfy the standard that the finitist standpoint requires, since the inductive definition was applied right as an abstract principle. So, I later proposed in [27] a more concrete principle, construction principle (CP), to justify Gentzen's ordinal notations, and went on to propose a principle named hyper-principle (HP: [30]) for ordinal diagrams. Such a device can be conceived only for each system in concern.

Now I realize that those principles are in a way suitable to embody Takeuti's finitist standpoint.

In order to give mathematical forms to Takeuti's views on the finitist standpoint, we need to take thought for the following concepts (p), (a), (b) and (c). The cited views of Takeuti in Sect. 3 are categorized in those four concepts.

(p) (Preliminaries: Inner necessity) The starting point of Takeuti's consistency proofs and his thoughts on the finitist standpoint is how to extend Hilbert's idea of the strictly finitist standpoint in consistency proofs, and hence extended notions of the finitist standpoint are necessitated by the consistency problems themselves: (i), (ii).[10]

(a) (Abatement of external objects) According to the (extended) finitist standpoint, it is desirable not to bring in objects or notions from outside the original concern except those which are necessary to discourse on it: (iii), (iv), (v).[11]

(b) (Clarity and concreteness) Present a clear principle which can concretely create required objects such as a modulus of finiteness functional of decreasing sequences from an order structure and according to which one can draw a clear image of some relevant processes: (vi), (vii), (viii), (ix).

(c) (Natural extension) An accessibility proof must stand on a natural extension of the strictly finitist standpoint: (x), (xi), (xii), (xiii).

Let us speculate on Takeuti's viewpoints a little more.

From (p) ((i) and (ii)), we can corroborate Takeuti's deep concern about the philosophical underpinning of consistency proofs. His theory of eliminators in [20] was an attempt to show the accessibility of Gentzen's order structure of order type ε_0. The definition of eliminators and demonstrations of their behaviors are constructive. Although it is not clear how the termination of the application of eliminators is guaranteed, I was stimulated by this endeavor and took up the accessibility proof for the theory of ordinal diagrams. The theory of ordinal diagrams was the driving vehicle of Takeuti's consistency proofs ([17, 23] for example; cf. also [16, 20]).

For my purpose, I will subsequently explain a theory HP, which defines a universe of mechanisms to which a modulus of finiteness functional of ordinal diagrams

[10] "Inner necessity" is one of the Dedekind's ideas on the development of mathematics seen in [1].

[11] "Abatement of external objects" is also at the heart of Dedekind's works.

belongs [30]. Then I will propose HP as a mathematical embodiment of Takeuti's finitist standpoint which goes along (a) and (b). (c) will be treated elsewhere (cf. [33]).

As for the last sentence of (i), the clause "that goes beyond Hilbert's finite standpoint" refers to the accessibility proof of a given order structure, and, for its confirmation, a uniform method or mechanism which produces a modulus of finiteness functional for any concretely given decreasing sequence from the order structure in concern is required, so that one can imagine how the accessibility can be attained. For that purpose, one must make clear of the nature of the uniform method.

(iii) suggests that one should take care so as not to bring in foreign objects or concepts, except perhaps some of those that can be constructed from objects and concepts of the original concern or that are strictly necessary for the purpose.

(iv) tells us that, working on inductively defined objects, one may accept only the facts that can be deduced from them or what they suggest, without resorting to abstract principles. The view in (v) is similar to that of (iv).

"Gedankenexperiment" in (vi) can be interpreted as follows. When one must show that, given some clearly defined and concretely given objects and an operation on them, the results of the application of the operation have a certain property, one should present a method according to which one can make a clear image of the process resulting from it. (vii) reinforces it, and (viii) explains it with an example.

The "finite" in (ix) probably means that one can grasp only some formal and finitary expressions. Even then a finite expression may contain some concept of infinity. One can comprehend such a concept when reading a finite expression if it is stated clearly. That is, given a clearly stated expression which is easy to understand, one can comprehend infinite processes intrinsic in it. It also leads one to thought experiments on those processes.

According to (xiii), we see Takeuti's belief that, if he constructs ordinals according to his sense of the finitist standpoint, then those ordinals will be the natural ones.

In the following, we will see that a theory, which I developed years ago and in which a modulus of finiteness functional of ordinal diagrams is concretely constructed, realizes the spirit of (a) and (b), or that of Takeuti's finitist standpoint.

6 Framework of Consistency Proofs

In this section, a very short account of the consistency proofs through reductions is given in order to make clear where, Takeuti thought, the problem lies, mainly referring to [31].

6.1 Framework of Gentzen-Takeuti Consistency Proofs

The machinery of the so-called consistency proofs through the reduction method consists of three constituents and two operations as well as the accessibility proof. (See also **1** $\sim$ **7** in Sect. 2.)

Constituents
C1 A formal system $\mathcal{S}$ to be studied.
C2 A constructive, linearly (and well-) ordered structure $\mathcal{J} = (J, <)$.
C3 The universe of mechanisms (methods) $\mathcal{U}$.

Operations
O1 The reduction of formal proofs in $\mathcal{S}$.
O2 The reduction of decreasing sequences from J.

A The accessibility proof for the order structure $\mathcal{J} = (J, <)$.

The consistency proof is completed with **A**.

It is important to be aware that everything can be carried out constructively except for **A**. An accessibility proof of a linearly ordered structure means a proof of its well-orderedness as constructive as possible, and the extent to which it exceeds strict constructiveness is the concern of the finitist standpoint.

$\mathcal{U}$ in **C3** is a semi-formal theory which yields a modulus of finiteness functional that can be applied to decreasing sequences from J and which realizes **A**.

6.2 Proof-Theoretic Order Structure

We deal with Takeuti's theory of ordinal diagrams as a proof theoretic order structure.[12] It contains Gentzen's order structure as a substructure.[13]

An elementary tool with which the theory starts is $\mathcal{I} = (I, <)$, a concretely given accessible structure,[14] and a mechanism $\mathcal{O}$ operating on $\mathcal{I}$ and producing a new order structure

$$\mathcal{O}(\mathcal{I}) = (O, <_i, i \in I \cup \{\infty\}),$$

which is called the structure of ordinal diagrams. O is the set of elements newly constructed and contains I as a subset. The elements of I are called indicators, and those of O are called ordinal diagrams. $<_i$ is a linear order, which is in fact a well-order. Both O and $<_i$ are constructive.

Now, the goal of Takeuti's consistency proofs was condensed to the accessibility proof of $<_i$ within the finitist standpoint.

---

[12] Reference [16]; explained in [20].

[13] Reference [2–4]; see also [5].

[14] $I = \{0\}, \omega, \omega^2, \omega^{\omega^{\omega}}, \cdots$ etc.

7 Accessibility Proofs

In this section, some results regarding the accessibility of some order structures are reviewed.

7.1 *Takeuti's Attempt: Eliminators*

Takeuti attempted to describe an accessibility proof of the order structure $E \equiv \mathcal{O}(\{0\})$, whose order type is ε_0, as a model case ([20], Sect. 11). It takes the following steps.

(1) Define a uniform method, called eliminator, which operates on a decreasing sequence from E and produces another decreasing sequence with a smaller initial element.

(2) From the accessibility of a substructure of E, one can deduce the accessibility of a larger substructure.

(3) Every decreasing sequence from E terminates at a finite stage.

(1) and (2) can be concretely demonstrated. To complete (3) along the finitist standpoint, the number of iterated application of eliminators must be concretely evaluated. For that purpose, a mechanism, a modulus of finiteness functional, must be concretely presented. However, it is not clear from the argument in [20] what Takeuti had in mind as to how to obtain such a mechanism.

7.2 *Accessibility Proof of Ordinal Diagrams*

I have attempted to define a modulus of finiteness functional for the accessibility of every $<_i$ of $\mathcal{O}(\mathcal{I})$ based on a theory called HP in [30].[15]

What are involved are the following 1* ~7*.[16]

1* The theory of approximations of ordinal diagrams.

2* Construction of fundamental sequences of ordinal diagrams.

3* Various properties of ordinal diagrams.

4* A formal proof of the accessibility statement for $O(\mathcal{I})$ in an intuitionistic system ASOD.[17]

5* An interpretation suitable for ASOD (an extended modified realizability).

---

[15] This section as well as Sects. 8 and 9 are devoted to a very brief outline of the work in [30], the theory of HP, for the reader's convenience. Only now it has been combined with Takeuti's concept of the finitist standpoint.

[16] 1* ~3* are treated in [24, 26, 28, 29]. 4* ~7* are treated in [30]. Some simplified descriptions of the content of [30] are seen also in [10, 31, 34–36].

[17] The name is derived from the intrinsic nature of the system: an analytic system especially designed for the theory of ordinal diagrams (cited from Definition 1.3 of [30]).

6* A semi-formal theory (HP) of generalized formal expressions, such as term-forms and formula-forms.
7* Modulus of finiteness functional for $\mathcal{O}(\mathcal{I})$.

With preparatory works 1* ∼3*, one executes 4*. ASOD is induced from the accessibility proof which is intended to be as economical as possible. Observing 4*, one is naturally led to 5*, from which to the theory HP in 6*. Then finally one can construct the functional in 7*.

The semantics of the generalized functionals in HP yields the universe $\mathcal{U}$ of mechanisms, and the modulus of finiteness functional in 7* is explicitly constructed in $\mathcal{U}$.

7.3 System ASOD

A rough description of the system ASOD is here supplied. (Section 1, Part I of [30])

Definition 71 (1) The language L of the system ASOD is that of HA (Heyting arithmetic) augmented by function variables, symbols for formalization of elementary properties of $\mathcal{O}(\mathcal{I})$ and some predicate constants for transfinite induction along $\mathcal{I}$.

(2) L-terms are the terms of the language L.

(3) L-formulas are defined from the equalities of type 0 and atomic formulas formed from predicate constants as follows.

(3.1) ∃-free L-formulas are defined as usual.

(3.2) A quantifier-free formula without the predicate constants of transfinite induction is called L-recursive.

(3.3) An L-formula will be called L-u-a (universally arithmetic) when it is constructed from L-recursive formulas by applications of $\wedge$, $\Rightarrow$ and $\forall$ of type 0.

(3.4) $\exists x B(x)$ (of type 0) is allowed only when $B(x)$ is L-u-a. ∃ on function variables is not allowed.

(3.5) The results of applications of $\wedge, \Rightarrow, \forall$ to the predicate constants and the formulas defined in (3.1)∼(3.4) are L-formulas.

Definition 72 The system ASOD is HA applied to our L-formulas and augmented by some new axioms. In particular, Bar induction BI(R; A) at the lowest type and definition by (transfinite) induction along $I \cup \{\infty\}$, $TI(I, G, U)$, are the important axioms. In BI(R; A), R is L-recursive and A is arbitrary, and a bar tree with an ordinal diagram at each node is in use. The transfinite induction is of the form

$$G(i, \alpha) \Leftrightarrow U(i, \alpha, G[i]),$$

where $i \in I, \alpha \in O, U(i, \alpha, Y)$ is an L-formula with parameter Y, and G is a predicate constant introduced for U.

Elementary properties of $\mathcal{O}(\mathcal{I})$ can be expressed in terms of L-formulas and provable in ASOD. $G(i, \alpha)$ is intended to mean that α is $<_i$-accessible in a subsystem $F(i, \alpha)$ of $\mathcal{O}(\mathcal{I})$. The bar induction is used in proving that, if $G(i, \alpha)$, then $\forall j < i G(j, \alpha)$.

Definition 73 The accessibility statement for $\mathcal{O}(\mathcal{I})$, acc($\mathcal{O}(\mathcal{I})$), is a formula which states the following.

$$\forall i \in I \cup \{\infty\} \forall f$$

$$(f \text{ is } <_i\text{-decreasing sequence from } \mathcal{O}(\mathcal{I}) \Rightarrow \exists n \forall m \geq n(f(m) = empty))$$

$acc(\mathcal{O}(\mathcal{I}))$ is a legitimate L-formula, and is a positive statement in the sense that the existential quantifier occurs only at a positive position.

A mechanism to obtain an n as above for f with respect to $<_i$ is called a modulus of finiteness functional for $\mathcal{O}(\mathcal{I})$.

Theorem 1 *4* (Conclusion in Sect. 2, Part I of [30]):* acc($\mathcal{O}(\mathcal{I})$) *is ASOD-provable.*

A formal proof, say P_0, claimed in the theorem is a faithful formalization of the (informal) positive accessibility proof, using $1^* \sim 3^*$ in Sect. 7.2.

8 Functional Interpretation

Our goal is to construct concretely a modulus of finiteness functional from P_0 (cf. 7*) via a functional interpretation of the accessibility statement.

The interpretation that suits ASOD turned out to be a generalized modified realizability (mr-) translation (Sects. 1~3 in Part III of [30]). To carry it out, a language for the translation is prepared (Part II of [30]): generalized variables $X, Y, \ldots$ and a new predicate constant Δ corresponding to G, for example. The translation of an L-formula is of the form $\exists X_1 \exists X_2 \cdots \exists X_m A$, where A is an L-u-a formula containing possibly the new constant Δ and generalized variables $X_1, X_2, \ldots, X_m$.

Let us list some crucial cases. ($B \rightarrow C$ means that B is translated into C.)

$$\exists x A(x, a) \rightarrow \exists X A(X(a), a);$$

$$\exists X A(X) \Rightarrow \exists Y B(Y) \rightarrow \exists Z \forall X (A(X) \Rightarrow B(Z(X))).$$

$\forall$ which is not followed by $\exists$ is left untouched.

As for the $G(i, \alpha)$ (corresponding to $U(i, \alpha, Y)$),

$$G(i, \alpha) \rightarrow \exists X \Delta(i, X, \alpha).$$

X is intended to represent a mechanism which guarantees the $<_i$-accessibility of a certain subset of ordinal diagrams.

The accessibility statement is translated as follows.

$$acc(\mathcal{O}(\mathcal{I})) \to \exists X \forall i, f R(i, f, X(i, f)),$$

where $R(i, f, Z)$ is an L-u-a formula with parameter Z.

Assuming the theory HP and the universe of mechanisms $\mathcal{U}$, which will be explained in Sect. 9, we obtain the following.

Theorem 2 (Sect. 4 in Part III of [30]) *Suppose* $\exists X_1 \cdots \exists X_m A(X_1, \ldots, X_m)$ *is the mr-translation of a theorem of ASOD. Then, there are mechanisms* $\Phi_1, \ldots, \Phi_m$ *in* $\mathcal{U}$ *such that* $A(\Phi_1, \ldots, \Phi_m)$ *is valid in HP, and, for each A,* $\Phi_1, \ldots, \Phi_m$ *can be concretely constructed.*

Corollary 1 There is a mechanism Φ_0 in $\mathcal{U}$ such that $R(i, f, X(i, f))$ is satisfied with $X \equiv \Phi_0$ for every $i \in I$ and every $<_i$-decreasing sequence f of ordinal diagrams. Φ_0 is concretely constructed and $\Phi_0(i, f)$ serves as a modulus of finiteness for f with respect to $<_i$.

Here and in the next section, concreteness of an object is guaranteed in [30]. Actual expressions are too lengthy to quote here.

9 Universe of Mechanisms

To complete our exposition, I make a quick tour of HP for 6* (Part II of [30]). HP is a theory of the extended formal expressions called type-forms, term-forms and formula-forms. HP is interpreted in the universe of mechanisms $\mathcal{U}$, which is just adequate for constructing a modulus of finiteness functional.

Type-forms are extended types with parameters with respect to L-recursive formulas. What are characteristic of them are the following.

(1) $\Lambda x\mathbf{T}$ (abstraction)

(2) $\mathbf{C}[A_1, \ldots, A_m; t_1, \ldots, t_m, t_{m+1}]$ (case definition)

(3) $T\langle i\rangle$ (transfinite enumeration of types along I)

(4) $\mathbf{R}[\mathbf{S}, \mathbf{T}, \nu, x; l]$ (recursion)

The rule of reductions of type-forms are defined, and the strong normalizability property holds under a certain strategy.

Term-forms are extended terms with type-forms, and they include the following.

(0) variable-forms for each type-form;

(1) bar constant $\mathcal{B}$ *at the lowest level*;

(2) case definition;

(3) transfinite construction;

(4) functional application *app*;

(5) λ-abstraction.

In (1), $\mathcal{B}$ forms a formula $\mathcal{B}[\mathbf{b}; \phi, \psi, m, f](\equiv \mathbf{B})$, where f is a number sequence and $\mathbf{b}$ evaluates the bar of a tree.

B is reduced to $app(\phi : f|m)$ if $\mathbf{b}(f) \leq m$ holds, while if $\mathbf{b}(f) > m$ holds, then to $app(app(\psi; f|m); \lambda k.\mathbf{B}[m/(m+1), f/(f|m * k)])$.

(3) is the major point of concern. We define a term-form $L(i, X)$, which represents a mechanism at stage i with parameter X. Putting $M(i) \equiv \{N(j) : j < i\}$, a meta expression of a sequence of mechanisms,

$$N(i) \equiv L(i, M(i))$$

is a mechanism constructed at stage i from the already constructed $N(j)$'s. L is designed so that N is the result of a very *tight construction*. $T\langle i\rangle$ is the type-form of $N(i)$.

Reductions of term-forms are defined in a natural way, and the strong normalizability holds under a certain strategy. Semantics of term-forms is defined along the meanings of the symbols.

Of formula-forms, atomic formulas are the equalities of type 0 and $\Delta(i, X, \alpha)$, where i and α are of type 0 and the type form of X depends on i and α. Compound formulas are obtained by applications of $\wedge, \Rightarrow, \forall X$ (X is a variable-form). $\Delta(i, X_0, \alpha)$ corresponds to $G(i, \alpha)$ (Sect. 8) with X_0 realizing $\exists X \Delta(i, X, \alpha)$.

For semantics of term-forms and formula-forms, we set a group of "axioms" and assume some principles on basic term-forms. Under those principles, the "axioms" become "valid".

The theories of type-forms, term-forms and formula forms, the principles and the validity of the "axioms" constitute HP. Although the objects involved here are strictly formal and concretely presented, HP is a semi-formal theory. The "axioms" are stated formally in its appearance, but they are abbreviated expressions of intended contents.

In HP, a modulus of finiteness functional is concretely constructed so as to satisfy the accessibility statement (its mr-translation). In Sect. 4, Part III of [30], it is concluded that the functional structure of $\mathcal{O}(\mathcal{I})$ is represented by HP. Details must be referred to Parts II and III of [30].

10 Some Explanations

Let us here analyze the above practice along (a) and (b) of Sect. 5.

The standpoint of HP in Sect. 9 requires thought experiments with regard to transfinite induction and bar recursion, but only on clearly defined operations applied to some concretely given expressions.

1* ~3* in Sect. 7.2 are carried out constructively. A completely positive proof of the accessibility statement naturally determines the system ASOD in Sect. 7.3 for 4* of 7.2. ASOD is a concise or wasteless or economical system carefully designed for the purpose. This corresponds to (a).

The axioms of HP (Sect. 9) are clearly stated and they can be interpreted in a natural way. The modulus of finiteness functional Φ_0 in Sect. 8 can be given a formal expression. This goes with (b).

The special care on $\exists$ in Sect. 7.3 defining the system ASOD is to avoid abstract existential assertions, thus securing (b).

The formal ASOD-proof P_0 of the accessibility statement in Sect. 7.3 contains only the objects and notions which are strictly related to the theory of $\mathcal{O}(\mathcal{I})$. This abides with the spirit of (a).

One reason why mr-translation in Sect. 8 suits our purpose is the following. If one compares $\forall X A(X) \Rightarrow B$ with $\exists X(A(X) \Rightarrow B)$, the former does not claim any existence (for X) in implying B, while the latter claims a concretely constructed object X. Hence from the finitist standpoint, the two formulas are of different qualities.

HP was the outcome of my endeavor to seek a desired mechanism, taking care so that objects and notions coming from outside do not sneak in (cf. (a)).

Corollary 1 in Sect. 8 secures my purpose to propose a mathematical embodiment of Takeuti's finitist standpoint, as closely to his thoughts in (b) as possible with respect to ordinal diagrams as an example.

As has been explained by Takeuti, the finitist standpoint admits the mathematical induction as a way of description, but only if, applied to concrete objects and properties, it enables one to execute thought experiments. I believe that the computational rules of Gödel's primitive recursive functionals [6] provides such a thought experiment. In Gentzen's case, the transfinite induction on the order structure E has a similar role to the mathematical induction when applied to concrete objects and properties, and an induced thought experiment is provided by "eliminators" [20] as well as by "construction principle" [27], for example. In the present case, the transfinite induction and the bar induction of the system ASOD has a similar role, and the induced thought experiment is provided by the mechanisms of HP.

Δ in Sect. 9 can be thought of as what one reaches when one considers G (Sect. 7.3) in a concrete situation.

The bar induction (Sect. 7.3) is used in proving that, given any ordinal diagram, any decreasing sequence starting from it is finite in a certain subset of ordinal diagrams. The bar recursion which realizes the mr-translation of the bar induction can be regarded as a clear principle.

The L in Sect. 9 is the essence of transfinite construction of the modulus of finiteness functional Φ_0. That is, the transfinite construction of N in Sect. 9 is expressed in the language L.

11 In Retrospect

I believe that one role of meta mathematics lies in articulating human intellectual activities concerning mathematics, but there is no definite path or short cut for refinement of the finitist standpoint. HP is only one such attempt.

The mathematical role of HP is an elucidation of the structure of ordinal diagrams, but it may also shed light on the structures of some comprehension axioms.

I might here refer the reader to [9], in which interesting considerations on the views beyond *absolute* finitist standpoint, named "*relative constructivist* proof theoretic reductions", are developed. Those considerations seem to share much with Takeuti's thoughts.

One final comment:
From his writings on the standpoint of formalism in [12–14] (cf. also [32]), I suspect that Takeuti's hard concern was in fact "what mathematics is". He must have worked in proof theory perhaps en route to his goal far away.

References

1. Dedekind, J.W.R.: Über die Einfuhrung neuer Funktionen in der Mathematik, delivered as a Habilitationsvorlesung in Göttingen on 30 June 1854; first published in "Dedekind Gesammelte mathematische Werke", vol. III, 428–438 (1932)
2. Gentzen, G.: Die Widersrpuchsfreiheit der reinen Zahlentheorie. Math. Ann. **112**, 493–565 (1936)
3. Gentzen, G.: Neue Fassung des Widerspruchsfreiheitsbeweise für tie reine Zahlentheorie, Forschung zur Logik und zur Grundlegung der exakten Wissenschaften, Heft 4(1938), Leipzig: Hirzel, 19–44
4. Gentzen, G.: Beweisbarkeit und Unbeweisbarkeit von Anfangsfällen der transfiniten Induktion in der reinen Zahlentheorie. Math. Ann. **119**, 140–161 (1943). (Habilitationsschrift)
5. Gentzen, G.: (M.E. Szabo, translator), "The collected papers of Gerhard Gentzen". North-Holland Pub. Co. (1969)
6. Gödel, K.: Über eine bisher noch nicht benüzte Erweiterung der finiten Standpunktes, Dialectica, 280–287 (1958)
7. Raymond, E.S.: The Art of Unix Programming. Pearson Education Inc (2003). http://www.faqs.org/docs/artu/index.html
8. Sieg, W.: Hilbert's Programs and Beyond. Oxford University Press, Oxford (2013)
9. Sieg, W.: Methodological Frames - Paul Bernays, Mathematical Structuralism, and Proof Theory, Manuscript
10. Takaki, O., Yasugi, M.: Validity Theorem for an MR -Complete Extension of TRDB, Manuscript, 25 pages
11. Takeuti, G.: On a generalized logic calculus. Jap. J. Math. **23**, 39–96 (1953)
12. Takeuti, G.: From the standpoint of formalism I. J. Jpn. Assoc. Philos. Sci. **2**(2), 25–29 (1956). (In Japanese)
13. Takeuti, G.: From the standpoint of formalism II. J. Jpn. Assoc. Philos. Sci. **2**(3), 15–19 (1956). (In Japanese)
14. Takeuti, G.: From the standpoint of formalism III. J. Jpn. Assoc. Philos. Sci. **2**(4), 24–27 (1956). (In Japanese)
15. Takeuti, G.: The Foundations of Mathematics. Kyoritsu Publication Co (1956); Reissue of the enlarged edition with changed title: "Introduction to Proof Theory" (with M. Yasugi) (2010). (In Japanese)
16. Takeuti, G.: Ordinal diagrams. J. Math. Soc. Jpn. **9**, 386–394 (1957)
17. Takeuti, G.: Consistency proofs of subsystems of classical analysis. Ann. Math. **2**(86), 299–348 (1967)
18. Takeuti, G.: On mathematics. J. Jpn. Assoc. Philos. Sci. **17**(4), 43–45 (1972). (In Japanese)

19. Takeuti, G.: In "Discussion Meeting: The Foundations of Mathematics, Yesterday, Today and Tomorrow", Sugaku Seminar (1979). (In Japanese)
20. Takeuti, G.: Proof Theory. North-Holland Publishing Co (1987)
21. Takeuti, G.: Hilbert and Gödel , Sugaku Seminar, 22–27 (1992, March Edition). (In Japanese)
22. Takeuti, G.: "Memoirs of a Proof Theorist: Gödel and Other LogiciansâŁž, translated by M. Yasugi and N. Passell. World Scientific, Singapore (2003)
23. Takeuti, G., Yasugi, M.: The ordinals of the systems of second order arithmetic with the provably Δ^1_2-comprehension axoiom and with the Δ^1_2-comprehension axiom respectively. Jpn. J. Math. **41**, 1–67 (1973)
24. Takeuti, G., Yasugi, M.: Fundamental sequences of ordinal diagrams. Commentarii Mathematici Universitatis St. Pauli **25**, 1–80 (1976)
25. Takeuti, G., Yasugi, M.: An accessibility proof of ordinal diagrams. J. Math. Soc. Jpn. **33**, 1–21 (1981)
26. Yasugi, M.: Some properties of ordinal diagrams. Nagoya Math. J. **70**, 143–155 (1977)
27. Yasugi, M.: Construction principle and transfinite induction up to ε_0. J. Austral. Math. Soc. **31**, 24–47 (1982)
28. Yasugi, M.: Groundedness property and accessibility of ordinal diagrams. J. Math. Soc. Jpn. **37**, 1–16 (1985)
29. Yasugi, M.: Projection and elevation of ordinal diagrams. Proc. RIMS **516**, 1–9 (1984)
30. Yasugi, M.: Hyper-principle and the functional structure of ordinal diagrams. Comment. Math. Univ. St. Pauli **34**(2), 227–263 (1985). The opening part; **35**(1), 1–38 (1986), the concluding part
31. Yasugi, M.: The machinery of consistency proofs. Ann. Pure Appl. Logic **44**, 139–152 (1989)
32. Yasugi, M.: Formalistic views' seen in "Journal of the Japan Association for Philosophy of Science". J. Jpn. Assoc. Philos. Sci. **43**(1 & 2), 17–30 (2016). (In Japanese)
33. Yasugi, M.: Extending metamathematical framework. In: Abstract, The 11th Annual Meeting of JACAP (2019), pp. 71–73. https://jacap.org/meeting/
34. Yasugi, M., Hayashi, S.: A functional system with transfninitely defined types. In: Jones, N., Sato, M., Hagiya, M., (eds.), Logic, Languages and Computation. Lecture Notes in Computer Science, vol. 792, pp. 31–60. Springer, Berlin (1994)
35. Yasugi, M., Hayashi, S.: Interpretations of transfinite recursion and parametric abstraction in types. In: Ito, M., Jurgensen, H., (eds.), Words, Languages and Combinatorics II, pp. 452–464. World Scientific Publ. Co., Singapore (1994)
36. Yasugi, M., Takaki, O.: Modified realizability for TRDB. In: Nehaniv, C.N., Ito, M., (eds.), Algebraic Engineering, pp. 186–200. World Scientific, Singapore (1999)
37. "Special Feature: Gaisi Takeuti and the Foundations of Mathematics", Sugaku Seminar (2017, February Edition). (In Japanese)

Properness Under Closed Forcing

Yasuo Yoshinobu

Abstract For every uncountable regular κ, we give two examples of proper posets which turn improper in some κ-closed forcing extension.

Keywords Proper forcing · Set mapping reflection

1 Introduction

In this paper we prove the following theorem:

Theorem 1 *Let κ be an uncountable regular cardinal. Then there exists a κ-closed poset $\mathbb{Q}$ and a proper poset $\mathbb{P}$ such that*

$$\Vdash_{\mathbb{Q}} \text{“}\check{\mathbb{P}} \textit{ is not proper.”} \tag{1}$$

This theorem negatively answers the question raised by Karagila [4], which asks if there exists a sufficiently large cardinal κ such that any proper poset remains proper after any κ-closed forcing. The motivation of this question is discussed in a recent article by Asperó and Karagila [1].

In fact, we give two examples witnessing the conclusion of Theorem 1. The first example is a rather simple mimic of the known example of a pair of proper posets whose product is improper, given by Shelah [7, XVII Observation 2.12, p. 826]. This example is briefly mentioned in [1]. The second example involves the class of posets introduced by Moore [6] in connection with the Mapping Reflection Principle (MRP) he introduced in the same paper. While the second example requires a longer argument, it has the advantage that $\mathbb{P}$ is taken to be totally proper.

Y. Yoshinobu (✉)
Graduate School of Informatics, Nagoya University, Furo-cho, Chikusa-ku, Nagoya 464-8601, Japan
e-mail: yosinobu@i.nagoya-u.ac.jp

T. Arai et al. (eds.), *Advances in Mathematical Logic*, Springer Proceedings in Mathematics & Statistics 369, https://doi.org/10.1007/978-981-16-4173-2_11

2 The First Example

Let T be the complete binary tree of height κ. Let $\mathbb{Q} = T$ with the reversed ordering. Clearly $\mathbb{Q}$ is κ-closed. Let $\mathbb{P}_0 = \mathrm{Add}(\omega)$ and $\dot{\mathbb{P}}_1 = \mathrm{Col}(\omega_1, 2^\kappa)$ in $V^{\mathbb{P}_0}$. $\dot{\mathbb{P}}_1$ is σ-closed in $V^{\mathbb{P}_0}$. Note that, no new cofinal branches are added to T by forcing with $\mathbb{P}_0 * \dot{\mathbb{P}}_1$, by a well-known argument first proposed by Mitchell [5]. Let $\dot{C}$ be a $(\mathbb{P}_0 * \dot{\mathbb{P}}_1)$-name for a cofinal subset of κ of order type ω_1, and let $T \restriction \dot{C}$ denote the subset of T consisting of all nodes of level in $\dot{C}$ (defined in $V^{\mathbb{P}_0 * \dot{\mathbb{P}}_1}$). Note that, in $V^{\mathbb{P}_0 * \dot{\mathbb{P}}_1}$, $T \restriction \dot{C}$ forms a tree of size and height ω_1, and since every cofinal branch through $T \restriction \dot{C}$ generates one through T, which is in V, the number of cofinal branches through $T \restriction \dot{C}$ is also ω_1. Now let $\dot{\mathbb{P}}_2$ be a $(\mathbb{P}_0 * \dot{\mathbb{P}}_1)$-name for the c.c.c. poset specializing $T \restriction \dot{C}$, as described in Baumgartner [2, Sect. 7], and set $\mathbb{P} := \mathbb{P}_0 * \dot{\mathbb{P}}_1 * \dot{\mathbb{P}}_2$. $\mathbb{P}$ is a three step iteration of c.c.c., σ-closed and c.c.c. posets, and thus is proper. Note that, since $T \restriction \dot{C}$ is specialized in $V^{\mathbb{P}}$, ω_1 must be collapsed in any further extension where new cofinal branches through $T \restriction \dot{C}$ are added. Since forcing with $\mathbb{Q}$ over V adds a new cofinal branch through T, forcing with $\mathbb{Q} \times \mathbb{P}$ adds a cofinal branch through $T \restriction \dot{C}$ which is not in $V^{\mathbb{P}}$. Therefore $\mathbb{Q} \times \mathbb{P}$ collapses ω_1, and thus is improper. This shows that $\mathbb{P}$ is improper in $V^{\mathbb{Q}}$.

3 The Second Example

First let us review some relevant definitions and facts we will use in construction of our second example. Let us start with the notion of total properness.

Definition 2 Let $\mathbb{P}$ be a poset, and N a countable $\in$-model of a suitable fragment of ZFC which contains $\mathbb{P}$.

(1) $p \in \mathbb{P}$ is said to be $(N, \mathbb{P})$-*generic* if for every dense subset $D \in N$ of $\mathbb{P}$, $D \cap N$ is pretense below p.
(2) $p \in \mathbb{P}$ is said to be *totally* $(N, \mathbb{P})$-*generic* if for every dense subset $D \in N$ of $\mathbb{P}$, p extends some element of $D \cap N$.

Note that $\mathbb{P}$ is proper iff for every sufficiently large regular θ, every countable $N \prec H(\theta)$ containing $\mathbb{P}$ and every $p \in N \cap \mathbb{P}$, there exists an $(N, \mathbb{P})$-generic condition of $\mathbb{P}$ which extends p.

Definition 3 A poset $\mathbb{P}$ is said to be *totally proper* if for every sufficiently large regular θ, every countable $N \prec H(\theta)$ containing $\mathbb{P}$ and every $p \in N \cap \mathbb{P}$, there exists a totally $(N, \mathbb{P})$-generic condition of $\mathbb{P}$ which extends p.

The notion of total properness has the following simple characterization.

Theorem 4 (Eisworth–Roitman [3]) *A poset* $\mathbb{P}$ *is totally proper iff* $\mathbb{P}$ *is proper and* σ-*Baire.*[1] □

[1] σ-Baire posets are sometimes referred to as σ-distributive posets.

Theorem 4 can be proved using the following lemma, which we will use later.

Lemma 5 *Suppose $\mathbb{P}$ is σ-Baire and N is countable. Then every $(N, \mathbb{P})$-generic condition can be extended to a totally $(N, \mathbb{P})$-generic condition.* □

Next let us review the class of posets first introduced by Moore [6]. They were originally designed to prove the Mapping Reflection Principle (MRP) from the Proper Forcing Axiom (PFA). In our second example, $\mathbb{P}$ will be taken from this class.

Definition 6 (*Moore* [6]) Let θ be a regular uncountable cardinal, and $X \in H(\theta)$. For a club subset $\mathcal{E}$ of $\{M \in [H(\theta)]^\omega \mid M \prec H(\theta)\}$, a function $\Sigma : \mathcal{E} \to \mathcal{P}([X]^\omega)$ is said to be an *open stationary set mapping* if for every $M \in \mathcal{E}$,

(1) $\Sigma(M)$ is M-stationary, that is, $\Sigma(M) \cap C \cap M \neq \emptyset$ for every club subset $C \in M$ of $[X]^\omega$, and
(2) $\Sigma(M)$ is open in the Ellentuck topology, that is, for every $x \in \Sigma(M)$ there exists a finite $a \subseteq x$ such that

$$[a, x] := \{y \in [x]^\omega \mid a \subseteq y\} \subseteq \Sigma(M).$$

For an open stationary set mapping $\Sigma : \mathcal{E} \to \mathcal{P}([X]^\omega)$, the poset $\mathcal{P}_\Sigma$ is defined as follows: $\mathcal{P}_\Sigma$ consists of the functions of the form $q : \alpha + 1 \to \mathcal{E}$ for some $\alpha < \omega_1$ such that

(1) $q(\gamma) \in q(\gamma + 1)$ for every $\gamma < \alpha$,
(2) $q(\gamma) = \bigcup_{\xi<\gamma} q(\xi)$ for every limit $\gamma \leq \alpha$, and
(3) for every limit $\gamma \leq \alpha$, there exists $\nu < \gamma$ such that $q(\xi) \cap X \in \Sigma(q(\gamma))$ for all ξ satisfying $\nu < \xi < \gamma$.

$\mathcal{P}_\Sigma$ is ordered by initial segment.

The following lemma is the heart of Moore's proof of MRP from PFA.

Lemma 7 (Moore [6]) *$\mathcal{P}_\Sigma$ is totally proper.* □

The following are density lemmata about $\mathcal{P}_\Sigma$, which we will use later. For $q \in \mathcal{P}_\Sigma$, we write $q(\mathrm{dom}(q) - 1)$ as $\mathrm{last}(q)$.

Lemma 8 (Moore [6]) *Let $\Sigma : \mathcal{E} \to \mathcal{P}([X]^\omega)$ be an open stationary set mapping, where $\mathcal{E}$ is a club subset of $\{M \in [H(\theta)]^\omega \mid M \prec H(\theta)\}$ for an uncountable regular cardinal θ.*

(1) For every $\alpha < \omega_1$, $D_\alpha := \{q \in \mathcal{P}_\Sigma \mid \alpha \in \mathrm{dom}(q)\}$ is dense in $\mathcal{P}_\Sigma$.
(2) For every $x \in H(\theta)$, $E_x := \{q \in \mathcal{P}_\Sigma \mid x \in \mathrm{last}(q)\}$ is dense in $\mathcal{P}_\Sigma$. □

We also use the following lemma in our construction.

Lemma 9 *Let κ be an uncountable regular cardinal. For every club subset C of $[\kappa]^\omega$, there exists a club subset $\tilde{C}$ of C such that*

(a) $\tilde{C}$ *is closed in the Ellentuck topology.*
(b) $C \setminus \tilde{C}$ *is weakly unbounded, that is, for every* $a \in [\kappa]^{<\omega}$ *there exists* $x \in C \setminus \tilde{C}$ *such that* $a \subseteq x$.

Proof Let $f : {}^{<\omega}\kappa \to \kappa$ be such that

$$C(f) = \{x \in [\kappa]^\omega \mid x \text{ is closed under } f\} \subseteq C.$$

Let

$$\tilde{C} := C(f) \setminus \{\text{cl}_f(a) \mid a \in [\kappa]^{<\omega}\},$$

where $\text{cl}_f(a)$ denotes the closure of a by f. Since every union of a strictly $\subseteq$-increasing ω-sequence of elements of $C(f)$ is in $\tilde{C}$, it easily follows that $\tilde{C}$ is a club subset of C. If $x \in [\kappa]^\omega$ is an accumulation point of $\tilde{C}$ in Ellentuck topology, x must be closed under f and also cannot be finitely generated by f, and thus is in $\tilde{C}$. This shows (9). Since $a \subseteq \text{cl}_f(a) \in C \setminus \tilde{C}$ holds for every $a \in [\kappa]^{<\omega}$, we have (9). □

Now we start to describe our construction of the second example. Suppose a regular uncountable cardinal κ is given. Let $\mathcal{C}$ be the set of club subsets of $[\kappa]^\omega$, and let $\mathbb{Q} = Col(\kappa, |\mathcal{C}|)$. Clearly $\mathbb{Q}$ is κ-closed, and $|\mathcal{C}| = \kappa$ holds in $V^{\mathbb{Q}}$. Let $\dot{C}$ be the $\mathbb{Q}$-name for a diagonal intersection of the members of $\mathcal{C}$. $\dot{C}$ is a club subset of $[\kappa]^\omega$ in $V^{\mathbb{Q}}$.

Let θ be a sufficiently large regular cardinal, and let

$$\mathcal{E} = \{M \in [H(\theta)]^\omega \mid \mathbb{Q}, \dot{C} \in M \prec H(\theta)\}.$$

Fix an arbitrary $M \in \mathcal{E}$. Pick a totally $(M, \mathbb{Q})$-generic condition $p_M \in \mathbb{Q}$. Let $\{C_n^M \mid n < \omega\}$ enumerate $M \cap \mathcal{C}$. Note that, for each $n < \omega$, we may assume that $\tilde{C}_n^M \in M \cap \mathcal{C}$, where $\tilde{C}_n^M$ is made from C_n^M as in Lemma 9. By the definition of $\dot{C}$, for each $n < \omega$ there exists a $\mathbb{Q}$-name $\dot{\alpha}_n^M \in M$ (for an ordinal below κ) such that

$$\Vdash_{\mathbb{Q}} \text{“}\{x \in \dot{C} \mid \dot{\alpha}_n^M \in x\} \subseteq (\tilde{C}_n^M)\check{}.\text{”} \tag{2}$$

By (a) and (b) of Lemma 9, there exists a $\mathbb{Q}$-name $\dot{y}_n^M$ and $\dot{a}_n^M$ both in M such that

$$\Vdash_{\mathbb{Q}} \text{“}\dot{\alpha}_n^M \in \dot{a}_n^M \in [\dot{y}_n^M]^{<\omega} \wedge \dot{y}_n^M \in (C_n^M \setminus \tilde{C}_n^M)\check{} \wedge [\dot{a}_n^M, \dot{y}_n^M] \cap (\tilde{C}_n^M)\check{} = \emptyset,\text{”} \tag{3}$$

By (2) and (3) we have

$$\Vdash_{\mathbb{Q}} \text{“}[\dot{a}_n^M, \dot{y}_n^M] \cap \dot{C} = \emptyset.\text{”}$$

By the total genericity of p_M, for each $n < \omega$ there exists $y_n^M \in C_n^M \cap M$ and $a_n^M \in [y_n^M]^{<\omega}$ such that

$$p_M \Vdash_{\mathbb{Q}} \text{“}[(a_n^M)\check{}, (y_n^M)\check{}] \cap \dot{C} = \emptyset.\text{”} \tag{4}$$

Now let us set $\Sigma(M) := \bigcup_{n<\omega}[a_n^M, y_n^M]$. It is easy to check that Σ is an open stationary set mapping ($\Sigma(M)$ is clearly open, and is M-stationary since $y_n^M \in \Sigma(M) \cap C_n^M \cap M$ for every $n < \omega$). By (4) we have

$$p_M \Vdash_{\mathbb{Q}} \text{“}(\Sigma(M))\check{} \cap \dot{C} = \emptyset.\text{”} \tag{5}$$

Now let $\mathbb{P} := \mathcal{P}_\Sigma$. $\mathbb{P}$ is totally proper by Lemma 7.

Lemma 10 *There exists $p \in \mathbb{Q}$ such that*

$$p \Vdash_{\mathbb{Q}} \text{“}\{M \in \check{\mathcal{E}} \mid p_M \in \dot{G}\} \textit{ is stationary in } [(H(\theta)^V)\check{}]^\omega,\text{”}$$

where $\dot{G}$ is the canonical $\mathbb{Q}$-name for $\mathbb{Q}$-generic filter.

Proof Suppose otherwise. Then there exists a $\mathbb{Q}$-name $\dot{D}$ such that

$$\Vdash_{\mathbb{Q}} \text{“}\dot{D} \text{ is a club subset of } \check{\mathcal{E}} \text{ and } \forall M \in \dot{D}[p_M \notin \dot{G}].\text{”}$$

Let λ be a sufficiently large regular cardinal, and let N be a countable elementary submodel of $H(\lambda)$ such that $\dot{D} \in N$ and $M := N \cap H(\theta) \in \mathcal{E}$. Now let G be a $\mathbb{Q}$-generic filter over V such that $p_M \in G$. Then by genericity of p_M it holds that $N[G] \cap (H(\theta))^V = N \cap (H(\theta))^V = M$. On the other hand, since $\dot{D} \in N$ and $\dot{D}$ is forced to be a club, it holds that $N[G] \cap (H(\theta))^V \in \dot{D}_G$. Thus it follows that $M \in \dot{D}_G$, and $p_M \notin G$ should be the case. This is a contradiction. □

Now let p be as in Lemma 10. We will show that

$$p \Vdash_{\mathbb{Q}} \text{“}\check{\mathbb{P}} \text{ is not proper.”} \tag{6}$$

Note that it is enough for our goal: we may rename $\mathbb{Q}$ below p as $\mathbb{Q}$, or use the weak homogeneity of $\mathbb{Q}$.

To this end, let G be a $\mathbb{Q}$-generic filter over V such that $p \in G$. Work in $V[G]$. Let

$$\mathcal{F} := \{M \in \mathcal{E} \mid M \cap \kappa \in \dot{C}_G\}.$$

Since $\dot{C}_G$ is a club subset of $[\kappa]^\omega$, $\mathcal{F}$ is a club subset of $\mathcal{E} \subseteq [(H(\theta))^V]^\omega$.

Now for each $\alpha < \omega_1$, let

$$\mathcal{D}_\alpha := \{q \in \mathbb{P} \mid \exists \xi \in \mathrm{dom}(q)[\xi \geq \alpha \wedge q(\xi) \in \mathcal{F}]\}.$$

It is clear that $\mathcal{D}_\alpha$ is a dense subset of $\mathbb{P}$ (by Lemma 8(1), and the unboundedness of $\mathcal{F}$; note that any $q \in \mathbb{P}$ can be extended by adding any member of $\mathcal{E}$ which contains $\mathrm{last}(q)$ as an element).

Now let $S := \{M \in \mathcal{E} \mid p_M \in G\}$. By our choice of p, S is a stationary subset of $[(H(\theta))^V]^\omega$.

Now let λ be a sufficiently large regular caridnal, and set

$$\mathcal{S} := \{N \in [H(\lambda)]^\omega \mid \mathbb{P}, (\mathcal{D}_\alpha)_{\alpha<\omega_1} \in N \prec H(\lambda) \wedge N \cap (H(\theta))^V \in S\}.$$

Then $\mathcal{S}$ is stationary in $[H(\lambda)]^\omega$. Pick any $N \in \mathcal{S}$. Set $M := N \cap (H(\theta))^V \in S$ and $\delta := N \cap \omega_1 = M \cap \omega_1$. The following claim is enough to show (6).

Claim *There are no* $(N, \mathbb{P})$*-generic conditions.*

Proof Suppose there is an $(N, \mathbb{P})$-generic condition. Since $\mathbb{P}$ remains σ-Baire in $V[G]$, by Lemma 5 such a condition can be extended to a totally $(N, \mathbb{P})$-generic condition q. By the total genericity of q, Lemma 8(1)(2), the facts that the domain of every condition of $\mathbb{P} \cap N$ is less than δ, and that $\mathcal{D}_\alpha \in N$ for every $\alpha < \delta$, we may assume that

(a) $\mathrm{dom}(q) = \delta + 1$,
(b) $\mathrm{last}(q) = q(\delta) = M$ and
(c) $\{\xi < \delta \mid q(\xi) \in \mathcal{F}\}$ is unbounded in δ.

By (3) we have

$$\{\xi < \delta \mid q(\xi) \cap \kappa \in \dot{C}_G\} \text{ is unbounded in } \delta.$$

On the other hand, by (3) there exists $\nu < \delta$ such that

$$q(\xi) \cap \kappa \in \Sigma(M) \text{ holds for all } \xi \text{ satisfying } \nu < \xi < \delta.$$

But since $M \in S$, $p_M \in G$ holds, and thus by (5) we have $\Sigma(M) \cap \dot{C}_G = \emptyset$. This is a contradiction. □

This finishes our construction.

Note that our second example is more delicate than the first one in the following sense: although the properness of $\mathbb{P}$ is destroyed by forcing over $\mathbb{Q}$, the product $\mathbb{Q} \times \mathbb{P} \simeq \mathbb{P} \times \mathbb{Q}$ remains proper unlike the first example, because $\mathbb{P}$ is σ-Baire and therefore $\mathbb{Q}$ remains σ-closed by forcing over $\mathbb{P}$.

Acknowledgements This work was supported by Grant-in-Aid for Scientific Research (C) (No. 18K03394) from JSPS.

References

1. Asperó, D., Karagila, A.: Dependent choice, properness and generic absoluteness (2018). preprint, https://arxiv.org/abs/1806.04077
2. Baumgartner, J.E. : Applications of the Proper Forcing Axiom. In: Kunen, K., Vaughan, J.E., (eds.), Handbook of Set-theoretic Topology, pp. 913–959. North-Holland (1984)
3. Eisworth, T., Roitman, J.: CH with no Ostaszewski spaces. Trans. Am. Math. Soc. **351**(7), 2675–2693 (1999)
4. Karagila, A.: Open problems. http://karagila.org/problems.html
5. Mitchell, W.: Aronszajn trees and the independence of the transfer property. Ann. Math. Logic **5**, 21–46 (1972)

6. Moore, J.T.: Set mapping reflection. J. Math. Logic **5**, 87–97 (2005)
7. Shelah, S.: Proper and improper forcing. In: Perspectives in Mathematical Logic. Springer, Berlin (1998)